Wastewater Engineering

Wastewater Engineering: Issues, Trends, and Solutions explains current treatment scenarios of wastewater in different countries across the globe, the characteristics of wastewater, and rules and regulations associated with the treatment and disposal/ reuse of wastewater. It covers the design and theory involving laying of sewerage network and different conventional and advanced treatment technologies employed to treat domestic wastewater. It overviews different types of emerging contaminants and their properties, ecological impacts, detection/quantification, treatment technologies, and circular economy.

Features:

- Gives an overview of current wastewater treatment scenarios across the world.
- Provides insights into emerging contaminants sources, procedure to sample, available methods for analyses, and possible treatments.
- Reviews existing rules and regulations on wastewater engineering and standards for wastewater disposal or reuse.
- Includes how to use wastewater as a resource in the context of circular economy.
- Describes fundamentals of wastewater conveyance and treatment.

The book is aimed at graduate students and researchers in wastewater treatment, water, and environmental engineering.

Wastewater Engineering

Issues, Trends, and Solutions

Ashok Kumar Gupta, Venkatesh Uddameri,
Abhradeep Majumder, and Shripad K. Nimbhorkar

CRC Press
Taylor & Francis Group
Boca Raton London New York

CRC Press is an imprint of the
Taylor & Francis Group, an **informa** business

Designed cover image: © Shutterstock

First edition published 2024
by CRC Press
6000 Broken Sound Parkway NW, Suite 300, Boca Raton, FL 33487-2742

and by CRC Press
4 Park Square, Milton Park, Abingdon, Oxon, OX14 4RN

CRC Press is an imprint of Taylor & Francis Group, LLC

© 2024 Ashok Kumar Gupta, Venkatesh Uddameri, Abhradeep Majumder, and Shripad K. Nimbhorkar

Library of Congress Cataloging-in-Publication Data
Names: Gupta, Ashok Kumar, author. | Uddameri, Venkatesh, author. |
Majumder, Abhradeep, author. | Nimbhorkar, Shripad K. (Shripad Khanderao), author.
Title: Wastewater engineering: issues, trends, and solutions / Ashok Kumar Gupta,
Venkatesh Uddameri, Abhradeep Majumder, Shripad K. Nimbhorkar.
Description: First edition. | Boca Raton, FL: CRC Press, 2023. |
Includes bibliographical references and index.
Identifiers: LCCN 2023002202 (print) | LCCN 2023002203 (ebook) |
ISBN 9781032399751 (hbk) | ISBN 9781032428208 (pbk) | ISBN 9781003364450 (ebk)
Subjects: LCSH: Sewerage. | Sewage–Purification. | Sewage disposal. | Water reuse.
Classification: LCC TD645 .G83 2023 (print) | LCC TD645 (ebook) |
DDC 628.3–dc23/eng/20230124
LC record available at https://lccn.loc.gov/2023002202
LC ebook record available at https://lccn.loc.gov/2023002203

ISBN: 9781032399751 (hbk)
ISBN: 9781032428208 (pbk)
ISBN: 9781003364450 (ebk)

DOI: 10.1201/9781003364450

Typeset in Times
by Newgen Publishing UK

Contents

About the Authors

Ashok Kumar Gupta earned his PhD in Environmental Science and Engineering from the Indian Institute of Technology Bombay, Mumbai, India. Currently, he is Professor (HAG) in the Environmental Engineering Division of the Civil Engineering Department and Ex-Head, School of Water Resources, Indian Institute of Technology Kharagpur, Kharagpur, India. He has over 25 years of teaching, research, and professional experience in various aspects, such as water treatment, wastewater treatment and reuse, treatment of emerging and geogenic pollutants, and others. He is credited with more than 140 publications in top-ranking international journals and his publications have appeared in more than 9900 citations with a h-index of 52. He has also published a comprehensive book titled *Fluoride in Drinking Water: Status, Issues, and Solutions* by CRC Press. Dr. Gupta is a renowned technical consultant in the arena of environmental engineering, having more than 50 completed, ongoing R&D, or consultancy projects of national and international importance to his credit. He has also carried out 30 short-term courses/workshops on water supply systems and wastewater management and reuse and others. Dr. Gupta has been awarded a Fellow of the Indian National Academy of Engineering (FNAE), a Fellow of the National Academy of Sciences (FNASc), and a Fellow of the West Bengal Academy of Science and Technology (WAST) in 2018, 2020, and 2021, respectively.

Venkatesh Uddameri earned his masters and doctoral degrees in civil engineering and environmental engineering, respectively, from the University of Maine, Orono, United States of America. Currently, he is Professor and Chair Professor in the Department of Civil and Environmental Engineering at Lamar University, Beaumont, Texas, United States of America. He has over 25 years of teaching, research, and professional experience in various aspects of water resources and environmental engineering. A significant part of his research is focused on alternative sources of water, including but not limited to the use of brackish groundwater resources and the reuse of wastewater. He is interested in the development of mathematical tools for the design and analysis of water and environmental systems. He is a fellow of the American Water Resources Association and currently serves as the Editor-in-Chief of the *Journal of the American Water Resources Association*. His research has been funded by a variety of federal, state, and local agencies, and he has received awards for both research and teaching innovations.

Abhradeep Majumder is a budding researcher in the field of water and wastewater treatment through biological and advanced oxidation processes. He earned his M.Tech in Civil Engineering with a specialization in Environmental Engineering from the National Institute of Technology Agartala, Agartala, India. Currently, he is Senior Research Scholar from the School of Environmental Science and Engineering, Indian Institute of Technology Kharagpur, Kharagpur, India. His broad field of research is wastewater treatment. His research work is dedicated to the development of photocatalytic materials for degrading emerging organic contaminants present in wastewater and using mathematical modelling to describe their degradation process. He also has experience in the biological treatment of wastewater. His scientific contribution is marked by several publications on advanced oxidation processes, emerging contaminants, biological treatment, adsorption, and others in top-ranking international journals. He has 20 publications in top-ranking international journals and is credited with more than 800 citations with a h-index of 13.

Shripad K. Nimbhorkar earned his M.Tech in Environmental Science and Engineering from the Indian Institute of Technology Bombay, Mumbai, India. He has more than 40 years of experience in the field of industrial water and wastewater treatment. He has worked with renowned industrial wastewater treatment solution companies. He has been the Associate Vice President at Kirloskar Consultants, Bangalore, India, and Reva Enviro Systems Private Ltd., Nagpur, India, between 1988 and 1995. He was also the General Manager of Biosystems India Ltd., Ahmedabad, India. Furthermore, Mr. Nimbhorkar had also held the position of Lecturer at Maulana Azad College, Aurangabad, India, and been an Environmental Management System Auditor. His field of interest is wastewater management, wastewater reuse, and the use of renewable energy in wastewater treatment.

Preface

The field of wastewater engineering has seen transformative changes in recent years. There is a growing recognition that "wastewater" is actually a valuable resource that should not be wasted! Increased frequency of droughts in many parts of the world, growing populations, and competition for water among various users have placed enormous stresses on rapidly dwindling freshwater sources. Wastewater provides a readily available and reliable source of water in many urban areas. Reuse of wastewater is increasingly becoming necessary to meet growing water demands.

Advances in analytical chemistry has also helped in understanding wastewater composition. Many potentially harmful constituents such as viruses, pharmaceutical and personal care products, and other emerging contaminants have been found in wastewater streams. Proper characterization of these streams and developing new treatment technologies therefore become imperative.

Wastewater is not just a source of water but can contain many other mineral resources that are of economic value. Profitable extraction of these resources helps reduce the global environmental footprint. These ideas form the fundamental core of the circular economy and how we treat and management wastewater directly impacts our transition to circular economy.

This book is designed keeping in mind the current requirements of undergraduate and postgraduate level students, researchers, and working engineers. The book not only covers all the elements required for a course on wastewater engineering but also dives deep into emerging treatment technologies that are being developed recently for addressing a wide range of contaminants. Most of the existing books focus on the engineering aspects of wastewater design and cover the basics of wastewater characterization, design of existing treatment technologies, conveyance of sewage, and disposal of treated sewage. This book takes a more holistic approach and addresses the need for the wastewater management and provides insights into the current global situation with regards to wastewater management, with a special emphasis on India.

The significance of different components of wastewater, particularly emerging contaminants, has been thoroughly described and their analytical methods have been described with a help of user-friendly illustrations. These illustrations are aimed at not only enhancing the fundamental knowledge of the readers but also helping them to precisely analyse these components for characterization and design purposes. The book throws light into the various existing legislations present globally for the discharge and reuse of wastewater and provides the latest permissible limits of various pollutants in different countries. Therefore, the book serves as a useful compendium of treatment standards that engineers need to know in an increasingly globalized world.

The book covers the basics and design aspects of different existing wastewater treatment technologies with user-friendly illustrations and solved problems. The design of sewage network and sludge handling has also been discussed from the grassroot level for the better and complete understanding of the readers. In addition, the book discusses the presence of different emerging pollutants in wastewater, their analysis, and their removal technologies.

Furthermore, the fundamentals and applicability of upcoming treatment technologies, such as MBR, constructed wetlands, advanced oxidation processes, etc. have been discussed, which will inspire researchers to work in the field of advanced wastewater treatment. Since water scarcity has become a global issue, reuse of wastewater has become essential. Different case studies of reuse of wastewater across the world has been compiled and provided, which may help organizations to implement such technologies for reuse of wastewater based on their requirement. This book includes numerous illustrations and worked out problems for easy comprehension of concepts and processes.

We hope the effort is beneficial to practicing engineers and students alike as they embark on designing, retrofitting, and rehabilitating wastewater treatment plants with conjunctive goals of reducing human health and environmental risks, increasing water footprint, and creating alternative, environmentally friendly resource recovery pathways that foster a global circular economy.

Ashok Kumar Gupta, Venkatesh Uddameri,
Abhradeep Majumder, and Shripad K. Nimbhorkar

Acknowledgments

We could not be where we are in life without the support of our loved ones, both from our past and from our present. The members of our family put in a lot of effort to ensure that that we were high on motivation, and we are grateful to have friends and students who share our passion for education.

1 Overview of Wastewater Management

CHAPTER OBJECTIVES

The chapter seeks to provide a broad overview of wastewater management and discuss emerging concepts like water reuse and circular economy. Challenges and opportunities for wastewater treatment in both developing and developed countries are reviewed. A holistic overview of the wastewater treatment system is also provided.

1.1 BACKGROUND

Water is an essential requirement for the sustenance of life on Earth. However, the over-exploitation of water resources and their improper usage has resulted in severe water scarcity in several places globally. Water scarcity has increased due to a rapid increase in population and accelerated industrial growth over the past few decades. In particular, the population in the urban areas has grown at an astounding pace due to migration from rural areas, resulting in a huge gap between water supply and demand. Additionally, the increased wastewater generation in urban areas has exposed the vulnerability of many cities and towns and highlighted the need for building efficient infrastructure for managing wastewater resources.

Wastewater is the water discharged by a community after it has been used for various purposes and contains contaminants that make it unfit for most reuses unless it is treated. Liquid wastewater resulting from anthropogenic activities in residential buildings, academic institutions, office buildings, shopping malls, restaurants, and other institutions is termed domestic wastewater or sewage (Mackenzie, 2010).

Around 80% of the water that is supplied for domestic purposes is discharged as wastewater (CPCB, 2021; UNESCO, 2017). The amount of wastewater generated is directly proportional to the per capita water requirement and is higher in developed countries. However, developed nations are equipped with advanced technologies to treat and reuse wastewater, On the other hand, sewage in developing countries is often partially treated, and therefore a significant portion of undesirable organic and inorganic substances do not get removed (Huang et al., 2010). These substances increase the chemical oxygen demand (COD), total organic carbon (TOC), biochemical oxygen demand (BOD), total nitrogen, total phosphorous, and other biochemical and physical parameters in the receiving water bodies. Furthermore, the wastewater may contain metabolites of pharmaceutical and personal care products that may pose risks to humans and ecological receptors.

DOI: 10.1201/9781003364450-1

The decrease in the availability of clean water has shifted the focus from treating wastewater as a nuisance to a valuable resource. This shift has led to the advancement of wastewater treatment technologies and harnessing the reuse potential of wastewater. In this context, this chapter highlights the necessity of wastewater treatment, along with associated major challenges. The current global scenario of wastewater management is outlined. The chapter also discusses the need for sustainable wastewater management and ways to harness energy from wastewater and reuse wastewater for potable and non-potable purposes. The declining availability of freshwater resources and the negative impacts of wastewater on the environment, along with some other factors, have necessitated wastewater management. The following section discusses the various factors that demand wastewater management implementation in this context.

1.2   NEED FOR WASTEWATER MANAGEMENT

1.2.1   WATER DEMAND AND AVAILABILITY

Nearly 75% of the Earth is covered by water, but most of it is unfit for human consumption. Accessible freshwater is only about 2% of all water available to us. The available freshwater is not uniformly available across the world and is limited in areas with high populations. The finite nature of freshwater resources and the disparity between available supplies and demands have created a scarcity of water (UNESCO, 2019).

The water demand and availability primarily depend on the following factors:

- Available water resources
- Population
- Agriculture
- Urbanization, population growth, and industrialization
- Environmental and ecological considerations

The most important factor determining water availability is freshwater resources, such as groundwater, lakes, or rivers in the area. Various parts of Africa, North America, South America, Asia, and Australia lack adequate freshwater resources. These countries face a severe water shortage. The shortage of water restricts overall development. Urbanization and industrialization often lead to increased water demand resulting from population migration that generates increasing demand for water.

As per the UNESCO report, the consumption of water has been increasing by 1% since 1980. The trend may go on till 2050, indicating a likely increase of 20–30% in water consumption compared to the current demand. The rise mainly arises from the increased demand in developing countries and emerging economies (UNESCO, 2019). However, the average water consumption of developed countries is still significantly higher than that of developing countries (UNESCO, 2019). The amount of water used in high-income countries and upper-middle-income countries (as per United Nations' country classifications) was higher than in lower-middle-income countries and least-developed countries (Figure 1.1a) ("AQUASTAT database," 2021; United Nations, 2021). It can also be noted that the median population of the

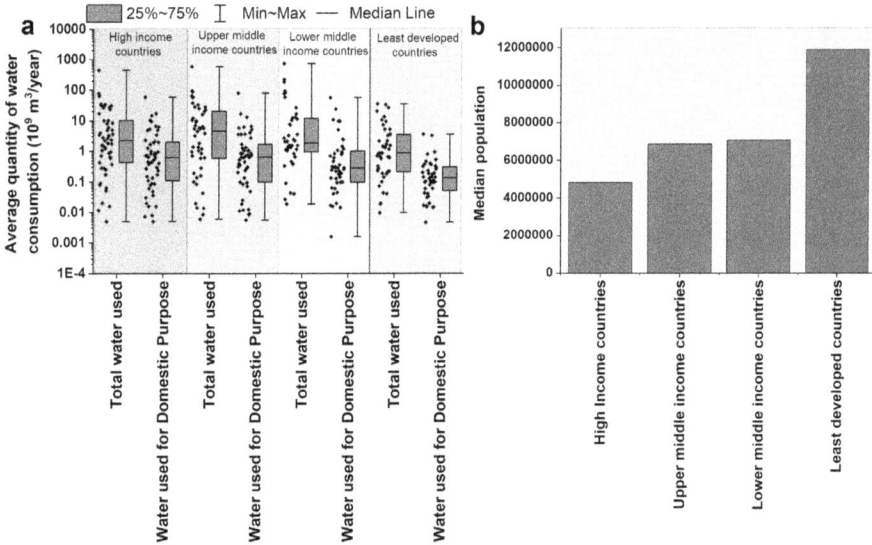

FIGURE 1.1   (a) Box plot of the average amount of water used and the average amount of water used for domestic purposes in countries of different income statuses and (b) their respective median population.

Source: AQUASTAT database for the year 2018.

countries with higher income was lower than that of the countries with lower income (Figure 1.1b). Hence, the per capita consumption of water by people in high-income countries is significantly higher as compared to low-income countries. The reason for higher water use for domestic purposes in countries with a high income is due to the higher standards of living.

Water stress provides a convenient measure to estimate the relative impacts of water supply and demand as it computes the proportion of the water withdrawn to the total amount of available renewable water (WWAP, 2019). Water stress can be calculated by taking the ratio of total freshwater withdrawn annually by all major sectors, including environmental water requirements, to the total amount of renewable freshwater resources (UNESCO, 2019). It is denoted by Eq.1.1 (Vanham et al., 2018).

$$\textbf{Water stress}(\%) = \frac{\textbf{Total freshwater withdrawn}}{\textbf{Total renewable water resources} - \textbf{Environmental flow requirements}} \times 100$$

(1.1)

The total volume of freshwater used to produce goods and services consumed by individuals or communities, or produced by businesses, is referred to as the water footprint (Mekonnen and Hoekstra, 2016).

Blue water scarcity is the "ratio of the blue water (fresh surface water and groundwater) footprint to the total blue water availability" (Mekonnen and Hoekstra, 2016).

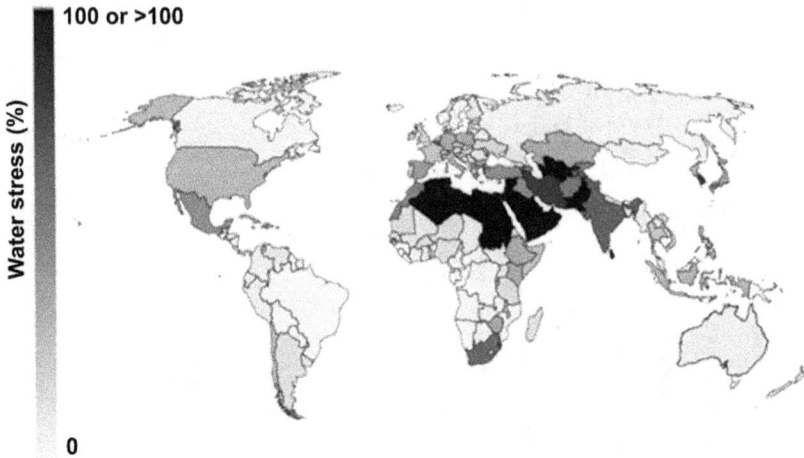

FIGURE 1.2   The global water stress scenario in 2018.

Source: AQUASTAT database.

As per Mekonnen and Hoekstra (2016), 4 billion people, almost two-thirds of the world population, experience water stress for at least 1 month in a given year. Many countries in North Africa, Middle East, a few countries in South America, and parts of Mexico suffer from severe blue water (fresh surface water and groundwater) scarcity throughout the year. Figure 1.2 depicts the global overview of water stress. Countries in the Middle East such as Kuwait, United Arab Emirates, Saudi Arabia, Qatar, Yemen, and some African countries, such as Libya, Egypt, Algeria, and Tunisia, were found to experience high water stress of above 100% ("AQUASTAT database," 2021).

Various countries like India, Pakistan, Afghanistan, Bangladesh, China and Mongolia also suffer from severe blue water scarcity during April, May, and June (summer season) (Mekonnen and Hoekstra, 2016). Many parts of Southern Asia, Northern Africa, Central Africa, South America, Mexico, and China experience more than four months of blue water scarcity (Mekonnen and Hoekstra, 2016). The current global water scarcity scenario highlights that it is imperative to reuse wastewater for potable or non-potable purposes to the maximum extent possible. However, implementing an efficient wastewater management system is necessary to achieve this goal.

1.2.2 SANITATION SYSTEMS

When untreated sewage is accumulated, the organic matter starts to decompose, leading to the generation of foul-smelling gases, which may be toxic. Untreated sewage hosts pathogens that can directly affect human health. Sewage contains nutrients that can adversely impact the growth of aquatic plants and may contain mutagenic or carcinogenic compounds that are harmful to both humans and other living beings. Proper sanitation is required to maintain appropriate conditions and processes to prevent harmful impacts on human health.

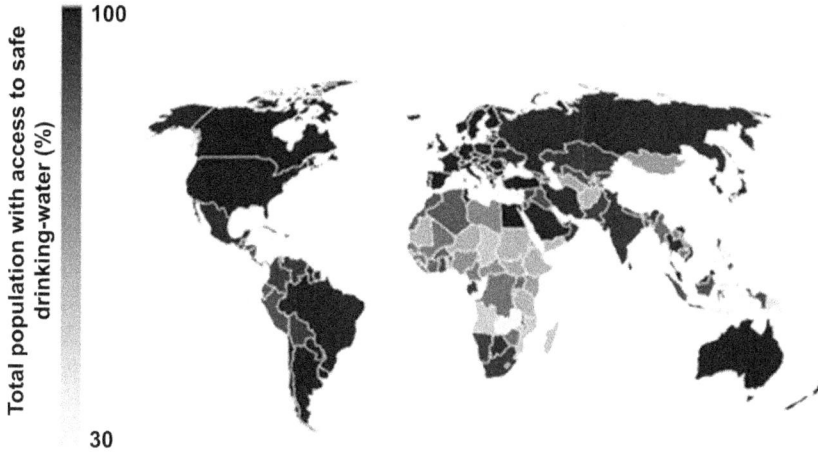

FIGURE 1.3 The global scenario on access to safe drinking water in 2018.

Source: AQUASTAT database.

A large percentage of the population of Sub-Saharan African countries, land-locked developed countries, and less developed countries do not have access to basic drinking water facilities (UNESCO, 2019). The global scenario of safe drinking water availability is presented in Figure 1.3. The population of these landlocked and small island countries suffer considerably more. Around 50% of the total population of many African countries have access to safe drinking water. Somalia was found to be the most affected in terms of safe drinking water availability, where around 30% of the total population has access to safe drinking water. In the Arab region, around 9% of the total population lacked a basic drinking water service in 2015, and 73% of the people were from rural areas (UNESCO, 2019).

One interesting observation is that the people of wealthy countries with high water stress conditions, such as Kuwait, Qatar, United Arab Emirates, and Singapore, have around 100% access to safe drinking water (Figures 1.2. and 1.3). These countries spend a lot of money and energy resources to obtain fresh water from different sources, and a significant fraction of the population of countries, such as Singapore, Qatar, United Arab Emirates, Kuwait, relies on the desalination of seawater for drinking purposes (Eke et al., 2020).

More than 50% of the population of Sub-Saharan African countries and Central and Southern Asian countries do not have basic sanitation coverage, and a considerable fraction of the population practices open defecation. Around 38% of healthcare facilities in 54 countries (mostly from Africa and Latin America) lack access to basic water sources, and 20% lack access to primary sanitary facilities (UNESCO, 2019).

Many countries lacking in sanitation also tend to exhibit heterogeneity with regards to sanitation facilities. Richer neighborhoods generally tend to have better sanitation, while poorer areas within the region lack basic amenities. This heterogeneity can have widespread consequences, including the spread of diseases, economic losses due to an unproductive workforce, and social injustice. Highly populated urban areas often tend to wield greater political power, which often leads to implementing sanitation facilities in these areas. This heterogeneous wastewater development has led to large-scale migration from rural (lower sanitation) areas to urban centers and, in the process, overloading the existing facilities. Hence, wastewater management must adopt a holistic outlook and not be carried out in a piecemeal manner.

1.2.3 IMPACT OF INADEQUATE WASTEWATER MANAGEMENT

Untreated or partially treated wastewater has the potential to disrupt the entire ecosystem. Water is an integral part of any ecosystem, and its disruption can have a long-lasting and irreversible impact on the environment. Feces and urine form a major fraction of the waste load in sewage (Taylor et al., 1996). This fraction hosts a wide range of disease-causing microorganisms. Unmetabolized part of food supplements or pharmaceuticals find their way into the wastewater through the feces and urine (Agunbiade and Moodley, 2016; Majumder et al., 2021, 2019; Yang et al., 2017). Additionally, sewage may contain other harmful organic compounds, such as personal care products, detergents, surfactants, and heavy metals, which are not easily removed even after conventional treatment processes (Al Enezi et al., 2004; Hargreaves et al., 2018; Majumder et al., 2019; Vareda et al., 2019). These substances can cause a variety of environmental and health problems.

A large part of the wastewater discharged with or without treatment finds its way to surface water, such as rivers, lakes, or ponds. A fraction of wastewater also reaches the groundwater through leakage and seepage during the conveyance of wastewater. When proper wastewater treatment is not provided, untreated wastewater may contaminate the source of drinking water, and the population faces significant health risks. The various diseases that may be caused due to improper wastewater management include typhoid, cholera, hepatitis A, polio, and diarrhea, to name a few (Taylor et al., 1996).

When biodegradable organic matter content in the water increases, the dissolved oxygen (DO) present in the water decreases. This is due to the consumption of the available oxygen by the bacteria to stabilize the organic matter. The decrease in DO may be associated with different reasons, such as the death of aquatic flora/fauna, negative effects on the predation rate of zooplankton–fish larvae–larval predator food web, and survival and asexual reproduction of microorganisms (Breitburg et al., 1997; Condon et al., 2001; Nebeker et al., 1992).

Another important aspect of improper wastewater management is eutrophication. When wastewater having high nutrient content is discharged into stagnant, or slowly flowing water bodies, the enrichment of nutrients in the water results in algal blooms (Schindler et al., 2016; Sinha et al., 2017; Wurtsbaugh et al., 2019). Several lakes and coastal marine ecosystems have reached the tipping point of the eutrophication process. Eutrophication or algal bloom can have a direct impact on the ecological health of a waterbody by lowering water quality and thus affecting aquatic biodiversity. When algae in a eutrophic lake do not receive enough sunlight, they stop producing oxygen and consume the remaining oxygen. Simultaneously, the microorganism uses the remaining oxygen to decompose the algal biomass, resulting in DO depletion (Jarvie et al., 2018). Other than nutrients, wastewater holds a wide range of other pollutants. These pollutants have the potential to seriously disrupt the ecosystem and harm humans and other living organisms exposed to the contamination for an extended period of time. The scarcity of readily available water and the adverse impacts of wastewater on the environment has necessitated the implementation of wastewater management systems across the globe. However, different countries have varying problems, such as climate, economy, social aspects, and others, which pose significant wastewater management challenges. In this context, the following section discusses the various obstacles to wastewater management.

1.3 CHALLENGES IN WASTEWATER MANAGEMENT

The challenges of wastewater management are often location-specific. Different regions have different geographic and demographic conditions. The availability of adequate financial resources is one major factor. The inexperience and lack of initiative from regulatory authorities is another limiting factor. Assessing the performance of the treatment facility requires regular monitoring of the influent and effluent parameters. The task requires skilled manpower, which may not be readily available in many parts of the world. A wide range of organic compounds is present in trace quantities (typically in the range of μg/L to ng/L). The detection of such contaminants requires sophisticated instruments that are expensive to acquire and maintain and require advanced training for operators. Subsequently, it is challenging to set up standards for such contaminants (Majumder et al., 2021, 2019). Furthermore, the lack of good governance and appropriate infrastructure add to the water quality problems (United Nations, 2015). The major challenges to wastewater management are discussed in the following sections.

1.3.1 FINANCIAL ASPECTS

Setting up and running a sewage treatment plant (STP) requires substantial costs. As a result, often countries with poor economies cannot treat a major portion of the wastewater generated. The problem is further exacerbated due to frequent power outages and a lack of skilled labor for regular operation and maintenance (Nikiema et al., 2011). Developing an extensive sewage network in cities is cost-intensive. Underdeveloped countries are therefore finding it difficult to lay proper conveyance networks (Nikiema et al., 2011). These constraints significantly bring down the performance of the treatment process. The governments in underdeveloped countries are often reluctant to spend on sanitation and wastewater treatment because this sector is capital-intensive and does not generate significant revenue (United Nations, 2015). It has been estimated that there is an annual global shortfall in funds of $56 billion between 2002 and 2025 for sewage treatment construction and maintenance (United Nations, 2015).

1.3.2 SOCIAL ASPECTS

The stakeholders and people of the community are integral partners for the proper management of an STP. However, only a handful of people are interested in wastewater management activities (Bosso and Guber, 2005; Capodaglio et al., 2017; Daniels et al., 2011). Inadequate participation from the common public and social acceptance is one of the major reasons behind improper waste management (Gupta et al., 2022). The concept of reuse of wastewater is not acceptable to a major fraction of the population (Mizyed, 2013; Shen et al., 2015; Villarín and Merel, 2020). When the laws regarding the disposal of waste or polluting water resources are not properly implemented, industries, hospitals, and households directly discharge untreated wastewater into drainage systems or nearby water bodies (Gupta et al., 2022). Such practice leads to the growth of pathogens, which impact the surrounding community and increase the pollutant load of the wastewater discharged (Boavida et al., 2016; Massoud et al., 2009). Hence, the development of stringent laws regarding waste and wastewater management is critical to overcoming the social challenges involved in wastewater treatment (Gupta et al., 2022).

1.3.3 EMERGING ASPECTS

Other than financial and social challenges, many other problems have also become more prominent in recent times. These include the rising cost associated with the operation of the treatment plants. The decline in performance and reliability of treatment plants with aging infrastructure is another concern. The performance of these treatment plants gets affected by variations in the quantity and quality of wastewater. Intractable toxic substances, such as emerging contaminants and heavy metals, are increasingly being detected in sewage. Parida et al. (2021) and Majumder et al. (2019) reported different concentrations of emerging contaminants in raw sewage. The observed concentration of the emerging contaminants was usually in the range of ng/L to μg/L (Majumder et al., 2019; Parida et al., 2021). The heavy metals, such as Pd, Cr, Hg, Ni, Zn, and others, were detected both in sewage and its sludge. The concentration of these heavy metals in wastewater and sludge is usually very low (Al Enezi et al., 2004; Du et al., 2020; Hargreaves et al., 2018; Liang et al., 2021). Quantifying these compounds requires expensive instruments. Further, removing these compounds requires additional treatment, thereby increasing the overall cost.

The different challenges to wastewater management have significantly impacted the current scenario of wastewater management across the globe. Different countries face varying challenges during wastewater collection, implementation, and running of STPs. In this context, the following section discusses the problems faced by different countries and the efficiency of their wastewater management systems.

1.4   CURRENT WASTEWATER MANAGEMENT SCENARIO

The amount of sewage treated in the higher-income countries was found to be higher as compared to low-income countries (Figure 1.4). Around 70% of the wastewater is treated in high-income countries. However, the amount of wastewater that is treated in lower middle-income countries and low-income countries is around 28% and 8%, respectively. This value is below 5% for the least developed countries (UNESCO, 2017).

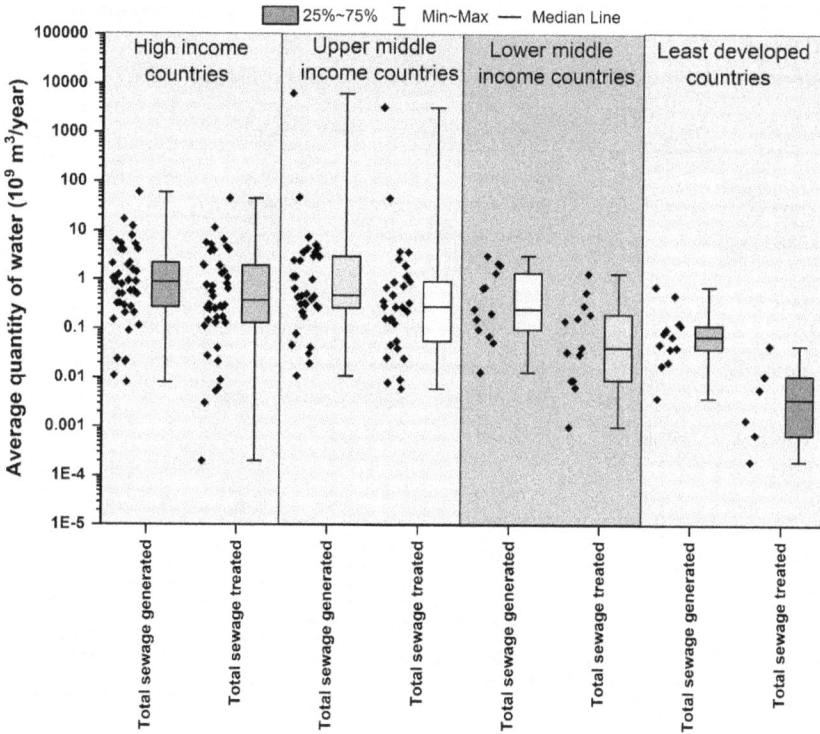

FIGURE 1.4 Box plot of the average amount of sewage generated and the average amount of sewage treated in countries of different income statuses.

Source: AQUASTAT database for the year 2018.

Countries in Asia and the Pacific region usually treat 10–20% of the wastewater generated. In 2013, it was found that countries such as Pakistan, Thailand, Vietnam, Bangladesh, Papua New Guinea, Tajikistan, and Nepal were treating around 10–20% of wastewater, while countries like Myanmar, Bhutan, Cambodia, and Samoa were reported to treat less than 10% of the wastewater generated (UNESCO, 2017). The massive difference in the availability of water and wastewater treatment facilities between high-income countries and low-income countries is primarily because of the lack of financial resources, proper awareness, and differences in prioritization of governmental expenditure (UNESCO, 2019). The treatment and management of generated wastewater offer ample opportunities for improvement and is also important to achieve sustainable development in many parts of the world.

The wastewater management scenarios are not uniform across different regions of the world. Differences in climate, population, water availability, economic conditions, and other social factors significantly affect the wastewater treatment scenario of a place. The region-specific problems in different locations are depicted in Table 1.1. In African countries, the overall lack of infrastructure for collecting and treating

TABLE 1.1
Wastewater Treatment Scenario in Different Geographical Regions

Region	Wastewater Management Scenario
Africa	Lack of infrastructure Insufficient coverage of the sewage network Financial constraints Lack of skilled labor
Arabian region	Fluctuation in wastewater flow Effluent standards do not cater to discharge standards
Asia and Pacific	A small percentage of wastewater is treated Lack of proper sanitation and sewer facilities
Caribbean and South America	Economic constraints Insufficient sewage network and wastewater collection facilities, non-stringent legislation Lack of support from the government
North America	Advanced wastewater treatment is available Certain areas have a severe shortage of water; hence reuse is a common practice
Europe	Advanced wastewater treatment is available with a primary focus on the treatment of emerging contaminants

wastewater is a major limitation. In addition, there is variation in high organic and hydraulic loading, high energy cost, power outage, and others (Nikiema et al., 2011). In the Arabian region, the sudden increase in population due to migration has rendered the existing wastewater treatment plants inadequate, thereby making them incompetent in terms of meeting discharge standards (UNESCO, 2017). High population and lack of finance are major problems that countries in Asia and the Pacific regions face. In North America, there are areas where there is a severe shortage of water. To overcome the deficit, the reuse of wastewater is a common practice. European countries and North America can afford to implement advanced treatment systems since their economy is better than many Asian and African countries (UNESCO, 2017). Recently, the detection and removal of emerging contaminants have attracted considerable attention due to the persistent and toxic nature of the emerging contaminants at low concentrations (Majumder et al., 2021, 2019).

The proper operation and maintenance of STPs is a challenging task. Inadequate planning, lack of finances, and lack of skilled labor are reasons that render a number of STPs non-operational or incapable of meeting desired standards. The major problems associated with the proper functioning of STPs are as follows:

- Inadequate fund provision for proper operation and maintenance of a STP
- Inadequate power supply
- Outdated infrastructure and lack of knowledge

- Often, there are fluctuations in the flow rate of raw wastewater. The quantity and quality of the wastewater are difficult to predict due to the irregular water supply, variation in lifestyle, storm-runoff, and unaccounted water.

Apart from these constraints, there are other factors that bring down the overall efficiency of the wastewater management system:

- Wastewater management of many urban utilities is devoid of a properly designed sewage network or inadequate sewage treatment plant. The existing sewage network is often age-old and defunct or has inadequate capacity to carry the wastewater to STP.
- The people or government have a negative mindset toward the necessity of wastewater management.

There are several hindrances to proper wastewater management. However, the treatment of wastewater is achievable. It is important to assess the quantity of wastewater generated and the characteristics of the wastewater. Based on the desired degree of treatment, a suitable treatment system can be designed. Regular monitoring of the influent and effluent parameters, operation, and maintenance enhances the performance and life of the treatment system. Proper planning and implementation of the wastewater management system can efficiently eliminate many of the health and sanitation problems. However, it should be noted that wastewater management should be carried out sustainably, such that there is no secondary pollution and no negative implications for the environment. It is also desirable to recover as much energy as possible during wastewater treatment. Hence, the following section discusses the different approaches that can be undertaken to attain sustainable wastewater management.

1.5 APPROACHES TO SUSTAINABLE WASTEWATER MANAGEMENT

In the 21st century, energy and water are becoming increasingly expensive. Hence, it is inevitable to look for sustainable solutions. The focus has shifted to the use of renewable energy sources for the operation of the treatment processes. Technological development has shown a preference toward using treatment processes with low carbon footprints. A life cycle assessment (LCA) for sustainability of the STPs has been conducted to evaluate the water footprint and carbon footprint (Gómez-Llanos et al., 2020; Zawartka et al., 2020). Resource recovery from wastewater is another aspect that has recently attracted attention (Gao et al., 2014; Guerra-Rodríguez et al., 2020; Kehrein et al., 2020). This significantly reduces the overall treatment cost by providing additional revenue from products derived from wastewater. Further, regions suffering from severe water stress have started to reuse the treated wastewater for non-potable and potable uses. The specific approach to sustainable wastewater management is discussed in detail in the following sections.

1.5.1 Circular Economy

Resource recovery is an essential component of the circular economy (Schroeder et al., 2019). A circular economy is a management process that reduces the burden on natural resources by recovering resources from waste, thereby generating social, economic, and environmental benefits (Lieder and Rashid, 2016). The concept of circular economy is based on the foundation of reuse, reduce, recover, reclaim, restore and recycle (6Rs) (Winans et al., 2017; Govindan and Hasanagic, 2018). The circular economy is nothing but rethinking a new way to recycle waste material to another valuable product. In this way, the waste generated today can be treated as resources for the future (Kakwani and Kalbar, 2020). In the circular economy model, wastewater is treated as a lucrative resource of nutrients, energy, and heavy metals (Vaneeckhaute et al., 2018; Guerra-Rodríguez et al., 2020).

Resource recovery can be achieved in several ways, that is, by recovering nutrients, metals, biochemicals, or energy. Nitrogen and phosphorous are the most commonly available nutrients in wastewater. Nitrogen primarily exists in the form of nitrite, nitrate, and ammonia, and phosphorous primarily exist in the form of phosphate. Different technologies, including constructed wetlands, electrodialysis, adsorption, ion exchange, membrane distillation, and others, have been implemented to recover nitrogen and phosphorous (Kehrein et al., 2020). Similarly, metals and biochemicals can be recovered using different technologies, such as bio-electrochemical systems, microbial electrosynthesis, and biological immobilization can be used to recover energy in the form of biofuels, heat, and electricity. Microbial fuel cells have been used to treat wastewater simultaneously and generate electricity (Kehrein et al., 2020). Resources recovery can bring down the overall cost of wastewater treatment and minimize the formation of end products. Hence, practicing resource recovery during wastewater treatment can be an effective way to attain sustainable development.

1.5.2 Reuse and Reclamation of Wastewater

Reclaiming the wastewater and reusing it is practiced in regions suffering from water scarcity. Secondary wastewater treatment processes are often not able to completely remove the recalcitrant organic and inorganic fractions of wastewater. In order to remove these contaminants, the implementation of advanced tertiary treatment is necessary. A few of the tertiary treatment systems involved are filtration, advanced oxidation, and disinfection. Filtration may be done using membranes or without membranes. Among membrane filtration, processes such as nanofiltration and reverse osmosis have been used based on the wastewater characteristics and the proposed use of the treated water. Among non-membrane filtration, there is activated carbon and biochar filtration. Activated carbon and biochar have been extensively applied in recent times for the tertiary treatment of wastewater. UV disinfection and/or chlorination are mandatory treatment processes to eliminate the microbial fraction from the wastewater before reusing. Advanced oxidation processes, such as ozonation, cavitation, Fenton, and photocatalytic oxidation, are used to remove the persistent organic fraction of the wastewater. The addition of these technologies may increase the overall cost of treatment, but that can be balanced by the economic gains obtained from reusing the water (Kehrein et al., 2020).

1.5.3 SUSTAINABILITY OF THE TREATMENT TECHNOLOGIES

The depletion of renewable resources, water shortage, highly polluted wastewater, and their other environmental implications are the driving forces behind the sustainable treatment and reuse of wastewater. Hence, the sustainability of the process is slowly becoming a key criterion for wastewater treatment and a driving force for its advancement. Sustainable treatment is defined as a type of treatment that can recover raw materials and conserve natural resources under the condition that all effluents from the treatment facility are fully utilized or recycled. Furthermore, the treatment process should not lead to secondary pollution. Unfortunately, wastewater treatment processes consume a lot of energy and involve high capital and operating costs. A significant amount of greenhouse gases, such as carbon dioxide, nitrous oxide, and other volatile substances, are released during the treatment. Many treatment techniques produce toxic sludge, which, if not disposed of properly, causes pollution. Advanced oxidation processes, such as ozonolysis or UV-based light energy, consume a large amount of energy. The formation of toxic by-products or other toxic substances released during the treatment or synthesis of adsorbent or photocatalytic materials must be minimized. Leaching from electrochemical processes adds to the pollution. As a result, alternative treatment technologies should be implemented that are powered by renewable energy sources, produce fewer greenhouse gases, and release fewer toxic end products. In this context, the LCA of different technologies is carried out to find the limitations of existing techniques and provide recommendations to overcome the problems.

LCA is a technique used to quantify the impacts associated with a product, service, or process from the cradle to the grave perspective (Corominas et al., 2013). In wastewater treatment, LCA implies analyzing the environmental impacts of wastewater treatment technologies from the start to the end of its operation period (Gallego-Schmid and Tarpani, 2019). LCA helps in identifying any scope of improvement for a particular plant and helps in comparing the performance of numerous treatment technologies (Niero et al., 2014). However, inconsistencies in assumption and methods in different LCAs reduce transparency and makes it difficult to compare different technologies (Corominas et al., 2020).

The development of a complete LCA requires four main phases as per ISO standards (ISO 14040:2006, 2006; ISO 14044:2006, 2006) (Gallego-Schmid and Tarpani, 2019). The four phases are depicted in Figure 1.5 (Corominas et al., 2013; Gallego-Schmid and Tarpani, 2019; Tabesh et al., 2019). The LCA of a treatment plant may be carried out on the performance of the system, the product. that is, the quality of the treated water, the amount of wastewater treated, and others have been described in Figure 1.5. Based on that, the functional units, system boundaries, and inventories can be defined. After defining the goal and scope, the data pertaining to the system is collected. In the next step, the impact of the system on the environment is assessed. Impact assessment is followed by interpretation. Often, the goal

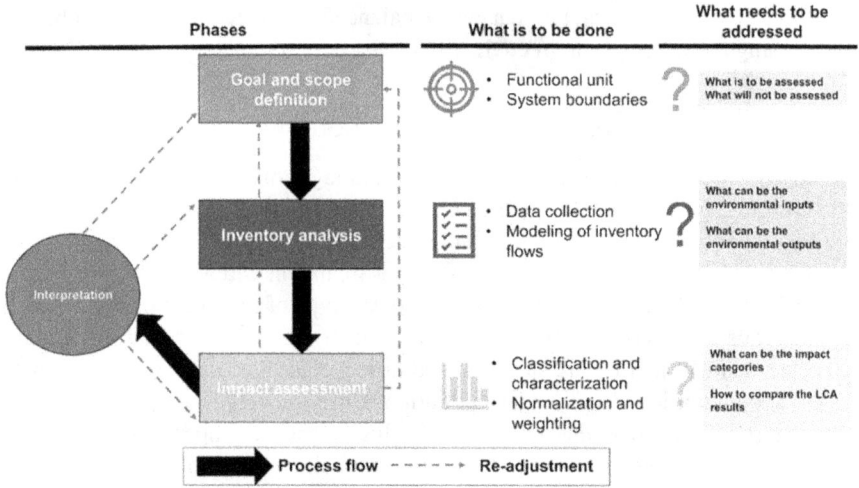

FIGURE 1.5   The general methodology for life cycle assessment.

and scope or inventory analysis need to be re-adjusted for better interpretation, and therefore the process is iterative in nature.

The most important step in implementing a sustainable domestic wastewater management system is understanding the functions of the various units in an STP. In this context, a complete overview of the different processes involved in an STP has been provided in the following section.

1.6   OVERVIEW OF PROCESSES INVOLVED IN A TYPICAL SEWAGE TREATMENT PLANT

The fundamental step in the proper management of wastewater lies in understanding the various processes involved in sewage management, which is depicted in Figure 1.6. Sewage from households is either sent to septic tanks or connected to a sewer, which conveys it to the sewage treatment plant (STP). The septic tanks are designed to hold wastewater under anaerobic conditions. The sewage from the septic tank is carried to the STP through lined drains or pipes.

The various units of a conventional STP are shown in Figure 1.6. The preliminary treatment involves screening and grit removal. Wastewater comprises rags, sticks, and heavy inorganic grit, which may cause maintenance and operational problems. They are removed in the screens and grit chamber. The screen comprises parallel bars, rods, gratings, wire mesh, or a perforated plate. The reject from the screen chamber or screenings is hauled to disposal sites or incinerated. Grit chambers are provided to remove grits (sand, gravels, cinders, and heavy solid materials with specific gravity substantially higher than organic putrescible solids in wastewater). This reduces the formation of heavy deposits in pipelines, channels, and conduits (Figure 1.6).

The primary sedimentation tank is used to remove inorganic and organic suspended solids. They help to reduce the organic load of the secondary treatment units. The

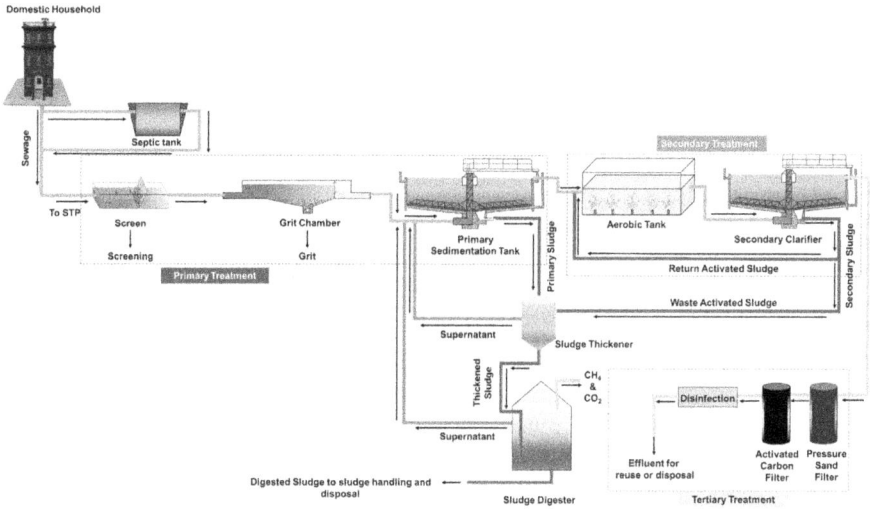

FIGURE 1.6 Overview of processes involved in sewage management.

sludge formed in the primary treatment is sent to processing units, such as sludge thickener and digester. Although the primary treatment removes a significant portion of the suspended solids, a significant portion of the dissolved organic matter remains in the wastewater. The removal of the organic matter is usually carried out in the secondary treatment or the biological treatment. Microorganisms in biological processes help to degrade the organic matter either in the presence (aerobic process) or absence (anaerobic process) of oxygen. The effluent from the biological process goes to a secondary settling tank or secondary clarifier. The secondary settling tank provides clarification to produce high-quality effluent and provides thickening of settled solids in the underflow. The concentration of suspended solids in mixed liquor in the aeration chamber is controlled by the sludge recirculation. A certain portion of the settled sludge is sent to the aeration chamber as return activated sludge. The remaining excess sludge or waste activated sludge is sent to sludge thickeners for further processing (Figure 1.6). The thickened sludge is sent to an anaerobic digester, sludge dewatering equipment, or sludge drying beds. Sludge rich in organics is anaerobically treated in a sludge digester, where organics undergo a process of acidification followed by methanization. This biochemical reaction takes place in a digester producing methane, carbon dioxide and other gases. Methane is used as biogas, that is, fuel or for electricity production. The sludge digester is provided where STP capacities are large, and gas utilization is feasible. The digester is not provided where gas utilization is not feasible. The supernatant from the sludge thickeners and sludge digester is sent to the inlet of the primary sedimentation tank, while the digested sludge is sent to the sludge handling facility for dewatering and disposal (Figure 1.6). The effluent from the secondary clarifier may be further treated using pressure sand and activated carbon filter. Disinfection is carried out to remove any remaining pathogens in the effluent. After disinfection, the effluent may be discharged or further processed by ultrafiltration, reverse osmosis or others, depending on the desired reuse purpose.

1.7   CHAPTER SUMMARY

- Around 80% of the water used is generated as wastewater.
- Countries such as Kuwait, United Arab Emirates, Saudi Arabia, Qatar, Yemen, Libya, Egypt, Algeria, and Tunisia faced a water stress above 100%.
- A large portion of the population of many African and Middle East countries do not have access to safe drinking water.
- A high percentage of the wastewater was treated by high-income countries while most of the low-income countries partially treat or do not treat their generated wastewater.
- Overpopulation, lack of infrastructure, insufficient coverage of sewage network, financial constraints, and social constraints were the major hindrances to wastewater management in many low-income countries.
- Recovery of resources, reuse and reclamation of wastewater, use of sustainable treatment technologies, and conducting LCAs are vital for sustainable wastewater management.

1.8   CONCLUDING REMARKS

Clearly, there are many components and processes associated with the treatment of wastewater. Optimizing these operations with an eventual goal to reduce the hazards posed by constituents within wastewater streams and find ways to recover and reuse water and products of economic value form the guiding principles of wastewater treatment and management. The purpose here was to provide a holistic picture of the entire operation. All of these components and processes will be discussed in great detail in the subsequent chapters.

REFERENCES

Agunbiade, F.O., Moodley, B., 2016. Occurrence and distribution pattern of acidic pharmaceuticals in surface water, wastewater, and sediment of the Msunduzi River, Kwazulu-Natal, South Africa. *Environ. Toxicol. Chem.* 35, 36–46. https://doi.org/10.1002/etc.3144

Al Enezi, G., Hamoda, M.F., Fawzi, N., 2004. Heavy metals content of municipal wastewater and sludges in Kuwait. *J Environ Sci Health A Tox Hazard Subst Environ Eng.* 39(2), 397–407. https://doi.org/10.1081/ESE-120027531

AQUASTAT database [WWW Document], 2021. www.fao.org/aquastat/statistics/query/results.html (accessed 3.24.21).

Boavida, S., Pinto, M., Salvador, T., 2016. Centralized vs. decentralized wastewater systems – potential of water reuse within a transboundary context. *New Water Policy Pract.* 2, 54–75. https://doi.org/10.18278/nwpp.2.2.6

Bosso, C.J., Guber, D.L., 2006 Maintaining Presence: Environmental Advocacy and the Permanent Campaign. *Environmental Policy.* CQ Press, pp. 78–91.

Breitburg, D.L., Loher, T., Pacey, C.A., Gerstein, A., 1997. Varying Effects of Low Dissolved Oxygen on Trophic Interactions in an Estuarine Food Web. *Ecological Monographs* 67(4), 489–507. https://doi.org/10.1890/0012-9615(1997)067[0489:VEOLDO]2.0.CO;2

Capodaglio, A.G., Callegari, A., Cecconet, D., Molognoni, D., 2017. Sustainability of decentralized wastewater treatment technologies. *Water Pract. Technol.* 12, 463–477. https://doi.org/10.2166/wpt.2017.055

Condon, R.H., Decker, M.B., Purcell, J.E., 2001. Effects of low dissolved oxygen on survival and asexual reproduction of scyphozoan polyps (Chrysaora quinquecirrha), in: *Jellyfish Blooms: Ecological and Societal Importance.* Springer, pp. 89–95. https://doi.org/ 10.1007/978-94-010-0722-1_8

Corominas, L., Byrne, D.M., Guest, J.S., Hospido, A., Roux, P., Shaw, A., Short, M.D., 2020. The application of life cycle assessment (LCA) to wastewater treatment: A best practice guide and critical review. *Water Res.* 184, 116058. https://doi.org/10.1016/j.wat res.2020.116058

Corominas, L., Foley, J., Guest, J.S., Hospido, A., Larsen, H.F., Morera, S., Shaw, A., 2013. Life cycle assessment applied to wastewater treatment: State of the art. *Water Res.* 47(15), 5480–5492. https://doi.org/10.1016/j.watres.2013.06.049

CPCB, 2021. *National Inventory of Sewage Treatment Plants.*

Daniels, D., Krosnick, J., Tompson, T., 2011. *Public Opinion on Environmental Policy in the United States.* Oxford Academic.

Du, P., Zhang, L., Ma, Y., Li, Xinyue, Wang, Z., Mao, K., Wang, N., Li, Y., He, J., Zhang, X., Hao, F., Li, Xiqing, Liu, M., Wang, X., 2020. Occurrence and fate of heavy metals in municipal wastewater in Heilongjiang province, China: A monthly reconnaissance from 2015 to 2017. *Water (Switzerland)* 12(3). https:// doi.org/10.3390/w12030728

Eke, J., Yusuf, A., Giwa, A., Sodiq, A., 2020. The global status of desalination: An assessment of current desalination technologies, plants and capacity. *Desalination.* 495, 114633. https://doi.org/10.1016/j.desal.2020.114633

Gallego-Schmid, A., Tarpani, R.R.Z., 2019. Life cycle assessment of wastewater treatment in developing countries: A review. *Water Res.* 153, 63–79. https://doi.org/10.1016/j.wat res.2019.01.010

Gao, H., Scherson, Y.D., Wells, G.F., 2014. Towards Energy Neutral Wastewater Treatment: Methodology and State of the Art. *Environmental Science: Processes & Impacts,* 16(6), 1223–1246.

Gómez-Llanos, E., Matías-Sánchez, A., Durán-Barroso, P., 2020. Wastewater treatment plant assessment by quantifying the carbon and water footprint. *Water (Switzerland)* 12(11), 1–16. https://doi.org/10.3390/w12113204

Govindan, K., Hasanagic, M., 2018. A systematic review on drivers, barriers, and practices towards circular economy: A supply chain perspective. *Int. J. Prod. Res.* 56(1–2), 278–311. https://doi.org/10.1080/00207543.2017.1402141

Guerra-Rodríguez, S., Oulego, P., Rodríguez, E., Singh, D.N., Rodríguez-Chueca, J., 2020. Towards the implementation of circular economy in the wastewater sector: Challenges and opportunities. *Water (Switzerland),* 12(5), 1431. https://doi. org/10.3390/w12051431

Gupta, A.K., Majumder, A., Ghosal, P.S., 2022. Introduction to modular wastewater treatment system and its significance. *Modular Treatment Approach for Drinking Water and Wastewater,* 81–106. https://doi.org/10.1016/B978-0-323-85421-4.00010-3

Hargreaves, A.J., Constantino, C., Dotro, G., Cartmell, E., Campo, P., 2018. Fate and removal of metals in municipal wastewater treatment: A review. *Environ. Technol. Rev.* 7(1), 1–18. https://doi.org/10.1080/21622515.2017.1423398

Huang, M.H., Li, Y.M., Gu, G.W., 2010. Chemical composition of organic matters in domestic wastewater. *Desalination* 262(1–3), 36–42. https://doi.org/10.1016/j.desal.2010.05.037

Jarvie, H.P., Smith, D.R., Norton, L.R., Edwards, F.K., Bowes, M.J., King, S.M., Scarlett, P., Davies, S., Dils, R.M., Bachiller-Jareno, N., 2018. Phosphorus and nitrogen limitation and impairment of headwater streams relative to rivers in Great Britain: A national perspective on eutrophication. *Sci. Total Environ.* 621, 849–862. https://doi.org/https://doi.org/10.1016/j.scitotenv.2017.11.128

Kakwani, N.S., Kalbar, P.P., 2020. Review of circular economy in urban water sector: Challenges and opportunities in India. *J. Environ. Manage.* 271, 111010. https://doi.org/https://doi.org/10.1016/j.jenvman.2020.111010

Kehrein, P., Van Loosdrecht, M., Osseweijer, P., Garfí, M., Dewulf, J., Posada, J., 2020. A critical review of resource recovery from municipal wastewater treatment plants-market supply potentials, technologies and bottlenecks. *Environ. Sci. Water Res. Technol.* 6(4), 877–910. https://doi.org/10.1039/c9ew00905a

Liang, Y., Xu, D., Feng, P., Hao, B., Guo, Y., Wang, S., 2021. Municipal sewage sludge incineration and its air pollution control. *J. Clean. Prod.* 295, 126456. https://doi.org/10.1016/j.jclepro.2021.126456

Lieder, M., Rashid, A., 2016. Towards circular economy implementation: a comprehensive review in context of manufacturing industry. *J. Clean. Prod.* 115, 36–51. https://doi.org/10.1016/j.jclepro.2015.12.042

Mackenzie, D.L., 2010. *Water and Wastewater Engineering: Design Principles and Practice.* McGraw Hill Education.

Majumder, A., Gupta, A.K., Ghosal, P.S., Varma, M., 2021. A review on hospital wastewater treatment: A special emphasis on occurrence and removal of pharmaceutically active compounds, resistant microorganisms, and SARS-CoV-2. *J. Environ. Chem. Eng.* 9, 104812. https://doi.org/10.1016/j.jece.2020.104812

Majumder, A., Gupta, B., Gupta, A.K., 2019. Pharmaceutically active compounds in aqueous environment: A status, toxicity and insights of remediation. *Environ. Res.* 176, 108542. https://doi.org/10.1016/j.envres.2019.108542

Massoud, M.A., Tarhini, A., Nasr, J.A., 2009. Decentralized approaches to wastewater treatment and management: Applicability in developing countries. *J. Environ. Manage.* 90, 652–659. https://doi.org/10.1016/j.jenvman.2008.07.001

Mekonnen, M.M., Hoekstra, A.Y., 2016. Sustainability: Four billion people facing severe water scarcity. *Sci. Adv.* 2, e1500323. https://doi.org/10.1126/sciadv.1500323

Mizyed, N.R., 2013. Challenges to treated wastewater reuse in arid and semi-arid areas. *Environ. Sci. Policy* 25, 186–195. https://doi.org/10.1016/j.envsci.2012.10.016

Nebeker, A.V., Dominguez, S.E., Chapman, G.A., Onjukka, S.T., Stevens, D.G., 1992. Effects of low dissolved oxygen on survival, growth and reproduction of *Daphnia, Hyalella* and *Gammarus. Environ. Toxicol. Chem.* 11, 373–379. https://doi.org/10.1002/etc.5620110311

Niero, M., Pizzol, M., Bruun, H.G., Thomsen, M., 2014. Comparative life cycle assessment of wastewater treatment in Denmark including sensitivity and uncertainty analysis. *J. Clean. Prod.* 68, 25–35. https://doi.org/10.1016/j.jclepro.2013.12.051

Nikiema, J., Figoli, A., Weissenbacher, N., Langergraber, G., Marrot, B., Moulin, P., 2011. Wastewater treatment practices in Africa – Experiences from seven countries. *Sustainable Sanitation Practice* 14, 26–34.

Parida, V.K., Saidulu, D., Majumder, A., Srivastava, A., Gupta, B., Gupta, A.K., 2021. Emerging contaminants in wastewater: A critical review on occurrence, existing legislations, risk assessment, and sustainable treatment alternatives. *J. Environ. Chem. Eng.* 9, 105966. https://doi.org/10.1016/J.JECE.2021.105966

Schindler, D.W., Carpenter, S.R., Chapra, S.C., Hecky, R.E., Orihel, D.M., 2016. Reducing phosphorus to curb lake eutrophication is a success. *Environ Sci Technol.* 50(17), 8923–8929.

Schroeder, P., Anggraeni, K., Weber, U., 2019. The relevance of circular economy practices to the sustainable development goals. *J. Ind. Ecol.* 23, 77–95. https://doi.org/10.1111/jiec.12732

Shen, Y., Linville, J.L., Urgun-Demirtas, M., Mintz, M.M., Snyder, S.W., 2015. An overview of biogas production and utilization at full-scale wastewater treatment plants (WWTPs) in the United States: Challenges and opportunities towards energy-neutral WWTPs. *Renew. Sustain. Energy Rev.* 50, 346–362. https://doi.org/10.1016/j.rser.2015.04.129

Sinha, E., Michalak, A.M., Balaji, V., 2017. Eutrophication will increase during the 21st century as a result of precipitation changes. *Science* 357, 405–408.

Tabesh, M., Feizee Masooleh, M., Roghani, B., Motevallian, S.S., 2019. Life-cycle assessment (LCA) of wastewater treatment plants: A case study of Tehran, Iran. *Int. J. Civ. Eng.* 17, 1155–1169. https://doi.org/10.1007/s40999-018-0375-z

Taylor, C., Yahner, J., Jones, D., Dunn, A., 1996. On-site wastewater disposal and public health. *Wastewater Public Heal.* 7, 1–8.

UNESCO, 2017. The United Nations world water development report, 2017: Wastewater: the untapped resource – UNESCO Digital Library.

UNESCO, 2019. The United Nations world water development report 2019: Leaving no one behind, facts and figures.

United Nations, 2015. *Wastewater Management: A UN-Water Analytical Brief.* New York.

United Nations, 2021. *World Economic Situation and Prospects 2021 | United Nations.* United Nations.

Vaneeckhaute, C., Belia, E., Meers, E., Tack, F.M.G., Vanrolleghem, P.A., 2018. Nutrient recovery from digested waste: Towards a generic roadmap for setting up an optimal treatment train. *Waste Manag.* 78, 385–392. https://doi.org/10.1016/j.wasman.2018.05.047

Vanham, D., Hoekstra, A.Y., Wada, Y., Bouraoui, F., de Roo, A., Mekonnen, M.M., van de Bund, W.J., Batelaan, O., Pavelic, P., Bastiaanssen, W.G.M., Kummu, M., Rockström, J., Liu, J., Bisselink, B., Ronco, P., Pistocchi, A., Bidoglio, G., 2018. Physical water scarcity metrics for monitoring progress towards SDG target 6.4: An evaluation of indicator 6.4.2 "Level of water stress." *Sci. Total Environ.* 613, 218–232. https://doi.org/10.1016/j.scitotenv.2017.09.056

Vareda, J.P., Valente, A.J.M., Durães, L., 2019. Assessment of heavy metal pollution from anthropogenic activities and remediation strategies: A review. *J. Environ. Manage.* 246, 101–118. https://doi.org/10.1016/j.jenvman.2019.05.126

Villarín, M.C., Merel, S., 2020. Paradigm shifts and current challenges in wastewater management. *J. Hazard. Mater.* 390, 122139. https://doi.org/10.1016/j.jhazmat.2020.122139

Winans, K., Kendall, A., Deng, H., 2017. The history and current applications of the circular economy concept. *Renew. Sustain. Energy Rev.* 68, 825–833. https://doi.org/10.1016/j.rser.2016.09.123

Wurtsbaugh, W.A., Paerl, H.W., Dodds, W.K., 2019. Nutrients, eutrophication and harmful algal blooms along the freshwater to marine continuum. *Wiley Interdiscip. Rev. Water* 6, e1373.

Yang, Y., Ok, Y.S., Kim, K.-H., Kwon, E.E., Tsang, Y.F., 2017. Occurrences and removal of pharmaceuticals and personal care products (PPCPs) in drinking water and water/sewage treatment plants: A review. *Sci. Total Environ.* 596–597, 303–320. https://doi.org/10.1016/j.scitotenv.2017.04.102

Zawartka, P., Burchart-Korol, D., Blaut, A., 2020. Model of carbon footprint assessment for the life cycle of the system of wastewater collection, transport and treatment. *Sci. Rep.* 10, 1–21. https://doi.org/10.1038/s41598-020-62798-y

2 Wastewater
Sources, Characterization, and Analysis

CHAPTER OBJECTIVES

The chapter describes the different components of sewage and its characteristics. The different physical and chemical parameters of sewage will be discussed, along with detailed methodologies for analyzing these parameters.

2.1 INTRODUCTION

Domestic wastewater or sewage is the liquid waste generated from households, commercial buildings, and residential buildings of a particular city or town. The amount of sewage generated is primarily dependent on water consumption. Approximately 80% of the water supplied for domestic use is discharged as wastewater (CPCB, 2021). The lifestyle and economic conditions of the population of the region, geographic location, and availability of water resources are some of the important factors in determining the amount of water consumed and the quantity of sewage generated (Mackenzie, 2010; Qasim and Zhu, 2017).

2.2 COMPONENTS OF SEWAGE

Sewage primarily comprises human wastes, such as feces and urine coming out of the toilet, lavatory basins, and urinals, along with the wastewater generated from other activities like laundry, cooking, washing utensils, and other domestic applications. Sewage can be classified as greywater or sullage, black water, yellow water, and brown water, as depicted in Figure 2.1. Yellow water is urine mixed with flush water, while brown water is human feces mixed with flush water. Collectively, yellow water and brown water together form black water. Greywater is the wastewater from the kitchen, laundry, shower, bathroom sink, spillage, and other household sources that do not contain human excreta (Francesco Di et al., 2020).

In addition to households, commercial establishments, such as hotels, offices, shopping complexes, railway stations, and airports, discharge a considerable amount of wastewater. The amount of wastewater generated from these establishments

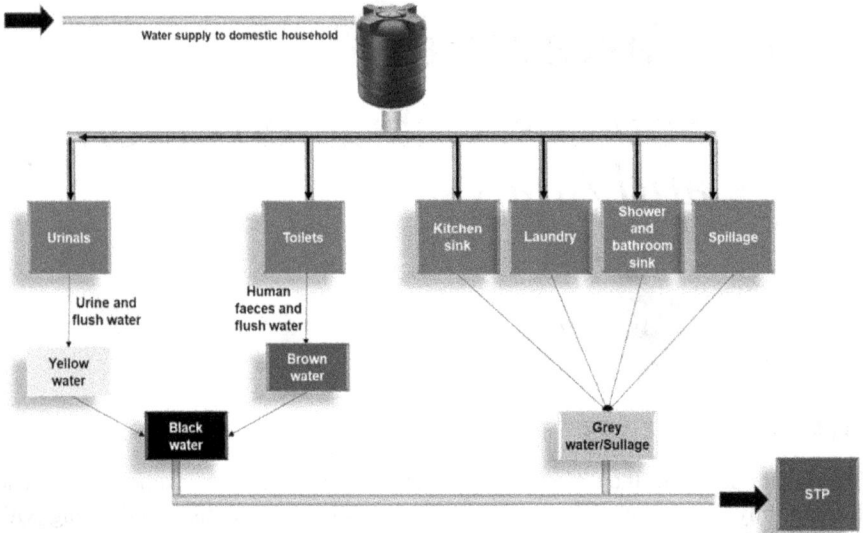

FIGURE 2.1 Water consumption and classification of sewage.

depends upon the size of the operation. Schools, colleges, city halls, and other public institutions also contribute significantly to wastewater generation within a city.

Hospitals and other healthcare facilities also discharge large quantities of wastewater. According to the World Health Organization (WHO) guidelines, roughly 40–60 L of water is required daily for every inpatient within a hospital. Operation theaters require around 100 L per intervention and around 100–400 L daily for each patient suffering from the severe acute respiratory syndrome or viral fever (WHO, 2017). It is estimated that hospitals in developed countries consume a considerably larger amount of water as compared to developing countries (Kumari et al., 2020; Majumder et al., 2021). The wastewater from healthcare facilities contains high organic matter, pathogens, and toxic compounds relative to sewage. Most organic compounds, such as pharmaceuticals, X-ray contrast media, surfactants, and personal care products, are recalcitrant in nature. The antibiotics present in the hospital effluent may lead to the formation of antibiotic-resistant genes (ARG) and antibiotic-resistant bacteria (ARB). Health problems arising due to exposure to such contaminants are quite severe. Hence, treating the wastewater generated from such institutions in situ has been mandated. However, hospital wastewater is often discharged into the sewers, where it mixes with the sewage and is sent to treatment systems for treating wastewater. These treatment plants are commonly referred to as sewage treatment plants (STPs) (Al Aukidy et al., 2018; Mackenzie, 2010).

At places where there is a combined sewerage system, during the conveyance of wastewater, the stormwater (runoff from rainfall or snowmelt) also gets mixed with wastewater via roof drains, submerged manholes, foundation drains, and others. This further adds to the total quantity of sewage generated (Mackenzie, 2010). The quantum of wastewater from all these components is largely dependent on the topography and demography of the place. The economy of the place also plays an important role in the amount of wastewater generated. The median amount of sewage generated by countries with high income was estimated to be around 0.68 km^3/year ("AQUASTAT database," 2021) in 2018. Similarly, the median sewage generation ("AQUASTAT database," 2021) from countries with upper-middle income, lower-middle income, and least-developed countries was 0.48, 0.20, and 0.07 km^3/year, respectively, in 2018.

The quantity of wastewater generated and its characteristics play a significant role in the design of an STP. The concentration of pollutants in wastewater in developing countries is much higher than in developed countries. This is because developed countries consume significantly higher quantities of water and consequently discharge an equivalent amount of wastewater. As a result, the pollutant concentration gets diluted in developed countries. Hence, it is imperative to consider the amount of wastewater generated and its quality for designing or adopting a particular treatment system.

2.3 CHARACTERISTICS OF SEWAGE

Sewage primarily comprises water (99%), while the remaining 1% is made up of suspended and dissolved organic and inorganic substances (CPHEEO, 2013; Muserere et al., 2014; Pereira et al., 2014). Carbohydrates, lignin, fat, proteins, soaps, surfactants, and other synthetic and natural organic compounds are also present in sewage. Inorganic substances, such as heavy metals, can also be seen in sewage (Pereira et al., 2014). Many chemical constituents in wastewater are in low concentrations that do not pose acute health risks to humans. However, pathogenic macro -and microorganisms present in the wastewater are of great concern to humans (CPHEEO, 2013; Muserere et al., 2014; Tchobanoglous et al., 2014). Raw sewage usually has pathogenic viruses, bacteria, protozoa, and helminths and can survive for long durations. The possible concentration of these pathogens and their survival duration have been depicted in Figure 2.2. At these levels, many of these pathogens can cause illnesses and diseases to humans and other living beings. Furthermore, these microorganisms can survive in wastewater for months.

Blackwater and greywater primarily make up the major part of the generated sewage. The organic and nutrient concentrations of the blackwater are related to the amount of flush water used. The more the amount of flush water used, the more the dilution, thereby lower the organic matter. The organic and solid content in greywater is significantly lower than in blackwater since greywater does not comprise excreta, which is a major contributor to the organic component. However, as stated earlier, the greywater quality may vary significantly based on the amount of water consumed in domestic activities. In Chapter 1, it was observed that the amount of water used is more in high-income countries than in low-income countries. As a result, COD and biological oxygen demand (BOD) of greywater concentration in low-income countries are usually higher than in high-income countries (Ghaitidak and Yadav, 2013). The wastewater can be classified into high-, medium-, and low-strength wastewater depending on the contaminant concentration. The COD content of the high-, medium-, and low-strength wastewater is typically 1,000, 500, and 250 mg/L, respectively

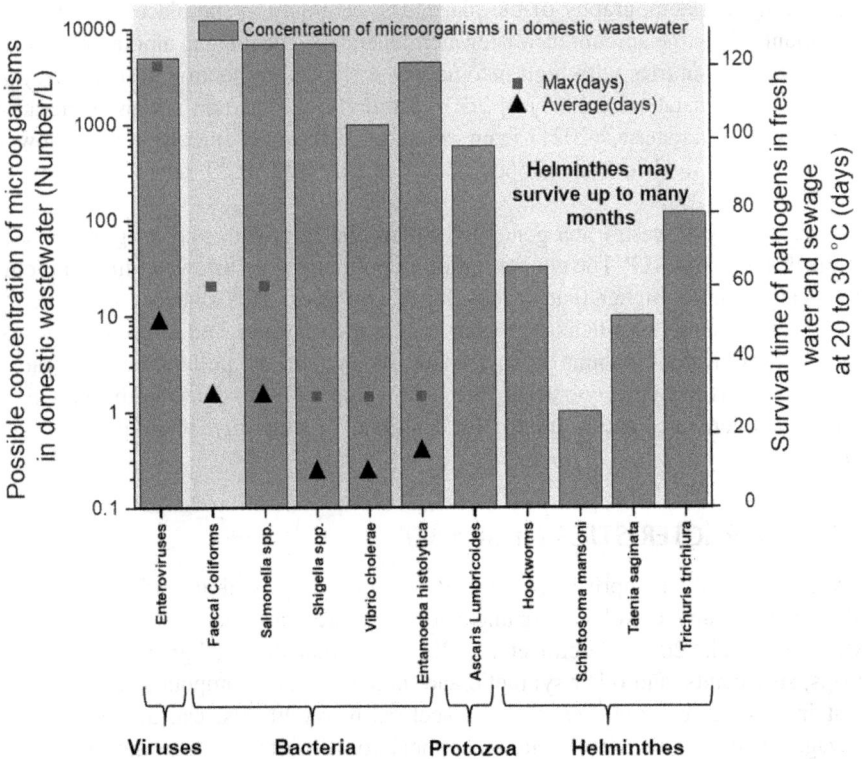

FIGURE 2.2 Possible levels of pathogens in sewage and their survival time.

Source: Feachem et al., 1983.

(Vianna et al., 2012). The COD/BOD ratio of the high-strength wastewater (3.3) is higher than that of medium-strength (2.5) and low-strength wastewater (2.5), thereby indicating the low biodegradability of the contaminants present in high-strength wastewater (Vianna et al., 2012). Similarly, the total suspended solids (TSS) content of the high-, medium-, and low-strength wastewater are typically 350, 200, and 100 mg/L, respectively (Vianna et al., 2012).

2.4 IMPORTANT PARAMETERS OF WASTEWATER AND THEIR ANALYSIS

Sewage contains several hundreds of chemicals, and it is hard to individually characterize each and every compound. Hence, it is generally characterized by a few standard parameters. The physicochemical parameters, such as pH of the wastewater, biochemical oxygen demand (BOD), chemical oxygen demand (COD), nitrogen, phosphorous, and solids are crucial to understanding the type and strength of wastewater and hence the type of treatment required. The significance of these parameters in the context of wastewater treatment is discussed in the following sections.

2.4.1 BIOCHEMICAL OXYGEN DEMAND (BOD)

The amount of dissolved oxygen (DO) required by microorganisms for the biological degradation of organic matter present in a given water sample over a period of 5 days at 20°C or 3 days at 27°C is known as BOD. It is frequently utilized to analyze or express the level of organic contamination in the sample. The total BOD primarily comprises carbonaceous BOD (CBOD) and nitrogenous BOD (NBOD). CBOD is the amount of oxygen required to oxidize the carbonaceous organic matter, while NBOD is the amount of oxygen required to oxidize nitrogenous compounds. However, conventionally the BOD tests are carried out for a period of 3 or 5 days. This BOD test value excludes the NBOD since the NBOD cycle starts after the period of 3–5 days, that is, when the CBOD cycle nears its end. It is customary to solely measure CBOD when performing the BOD_5 test since nitrogenous demand does not accurately reflect the oxygen demand from organic matter. As a result, only the CBOD has been covered in this chapter (Delzer and Mckenzie, 2003; Tchobanoglous et al., 2014).

2.4.1.1 BOD Model

In order to obtain a BOD model, a general assumption is made that at any given time, the oxygen consumption rate is directly proportional to the concentration of the biodegradable organic matter. Using the first-order reaction kinetics, BOD reaction kinetics can be established as per Equation 2.1 (APHA, 2017; Tchobanoglous et al., 2014).

$$dL_t / dt = - KL_t \qquad (2.1)$$

where L_t is the amount of organic matter remaining in wastewater at time "t" expressed as oxygen equivalents (mg/L) and K is the first-order BOD reaction rate constant (Equations 2.2 to 2.5).

$$\text{Hence, } \int_0^t dL_t = -\int_0^t KL_t \, dt \qquad (2.2)$$

$$\int_0^t \left[dL_t \right] / \left[L_t \right] = -\int_0^t K dt \qquad (2.3)$$

$$\left[\log L_t \right]_0^t = -K \cdot t \qquad (2.4)$$

$$L_t / L_0 = e^{-K \cdot t} \qquad (2.5)$$

where L_0 or BOD_u is the BOD at time $t = 0$. It is also referred to as the ultimate BOD, present initially in the sample.

At any time t, BOD remaining is given by Equation 2.6.

$$L_t = L_0 \left(e^{-K \cdot t} \right) \qquad (2.6)$$

The amount of BOD at any time "t" is given by Equation 2.7.

$$BOD_t = L_0 - L_t = L_0(1 - e^{-K \cdot t}) \tag{2.7}$$

Hence, for 5 days,

$$BOD_5 = L_0 - L_5 = L_0(1 - e^{-5K}) \tag{2.8}$$

Usually, the K values for wastewater range from 0.23 to 0.7 day^{-1} (Tchobanoglous et al., 2014).

The curves representing the amount of BOD consumed, amount of BOD remaining, and ultimate BOD have been depicted in Figure 2.3. The value of K has been assumed to be 0.5 day^{-1} and the ultimate BOD has been assumed to be 500 mg/L.

The amount of BOD present in the wastewater before oxidization occurs is defined as ultimate BOD (L_0). The amount of BOD removed plus the amount of BOD remaining at any time must be equal to L_0. Theoretically, the oxidation of L_0 will take infinite time. Hence, practically the time at which the BOD curve becomes parallel to the x-axis, it is called the "L_0" achievement time. The wastewater characteristics, such as the amount of organic matter present, its chemical composition, and incubation temperature, determine the time needed to achieve ultimate BOD. At high temperatures, the ultimate BOD is reached faster as compared to a lower temperature for the same wastewater. This is due to increased microbial activity and degassing of dissolved oxygen (APHA, 2017; Tchobanoglous et al., 2014).

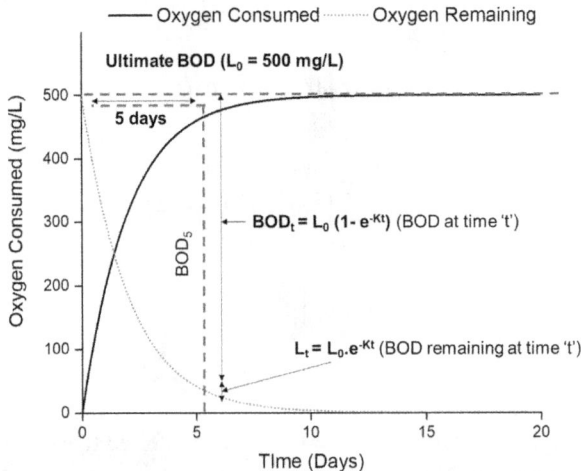

FIGURE 2.3 BOD profile variation during when value of $K = 0.5$ day^{-1} and ultimate BOD (L_0) = 500 mg/L.

2.4.1.2 BOD Test

In the BOD test, it is assumed that microorganisms utilize the dissolved oxygen to produce CO_2 and H_2O while completely oxidizing the organic matter. The BOD test includes estimating the change in DO concentration as organic matter is degraded by microorganisms in a sample stored in a BOD bottle after incubating it for 5 days at 20°C or for 3 days at 27°C. The oxygen demand of the water or wastewater sample is directly proportional to the concentration of the organic matter. The more organic matter is present, the more oxygen consumption will be, and vice versa. Henceforth, the BOD test indirectly estimates organic matter (APHA, 2017; CPCB, 2011).

Water from different sources, including freshwater from lakes and rivers, domestic and industrial wastewater, polluted water bodies, and coastal and estuarine water, can all be evaluated for BOD. The determination process described has been adapted from Standard Methods for the Examination of Water and Wastewater (APHA, 2017) and *Central Pollution Control Board Guide Manual: Water and Wastewater Analysis*. For detailed information, the mentioned manuals may be referred (APHA, 2017; CPCB, 2011).

Principle of Determination

The BOD test is an indirect estimation of organic matter. It estimates the change in the concentration of DO in the wastewater held in a BOD bottle for the duration of incubation. The change in DO concentration occurs due to the microorganisms present in the wastewater consuming the oxygen to degrade the organic matter. The 5 days incubation period at 20°C is mostly used for BOD determination. At 20°C, most of the carbonaceous wastes are oxidized during the first 5 days. As a result, the test is directed for 5 days at a temperature of 20°C (APHA, 2017; CPCB, 2011).

Requirement of Seed

If the microbial concentration in the water to be tested is low, then a seed BOD test is conducted in which a known amount of microorganisms (as a seed) are introduced to the sample. Seed can be obtained from primary settling units or purchased commercially. When the wastewater contains a high population of microorganisms, then seeding is not required. During BOD estimation, the amount of oxygen consumed by the seeded microorganisms in the sample is deducted since organic matter present in the microorganism will consume some oxygen (APHA, 2017; CPCB, 2011).

Requirement of Dilution Water and Its Preparation

- In order to prevent all the oxygen from being depleted in the BOD bottle during incubation, dilution is carried out.
- The amount of dilution is dependent on the organic matter content in the sample. As per APHA manual, for a valid BOD test, at least 1.0 mg/L of DO should remain in the bottle, and DO uptake should be at least 2 mg/L at the end of 5 days of incubation for reliability. In the case of a higher organic load, most of the oxygen will be consumed, and the final DO will be less than 1 mg/L. This will give inaccurate results. As a result, dilution should be carried out. Similarly, if the organic matter is less and the sample is diluted, then there may be a very small drop in DO concentration. This also has the chance of giving

erroneous results. Hence, selecting the amount of dilution water is a hit-and-trial method and should be carried out systematically to obtain the best BOD results.

- The dilution required is estimated by utilizing the expected BOD values for specific wastewater.
- The dilution water is prepared by aerating distilled water in a container to get DO saturation for 8 to 12 h.
- Further, a small quantity of $FeCl_3$, $MgSO_4$, $CaCl_2$, and phosphate buffer is added to 1 L of dilution water to introduce nutrients in the dilution water.

Sample Preparation and Inhibition of Nitrogenous Demand

- A 300 mL BOD bottle is taken and filled with a limited quantity of test wastewater sample. The remaining portion is filled up with oxygen-saturated dilution water and nutrients needed to aid biological development.
- Dilution of the sample is necessary and carried out by adding water with an adequate amount of pre-arranged dilution water to get desired outcomes and meet the prerequisite of oxygen and nutrients during the incubation time frame.
- The sample must be free of chlorine. If chlorine is present, the sample should be dechlorinated using Na_2SO_3. The pH of the water sample should be neutral (6.5–7.5) (APHA, 2017; CPCB, 2011).
- Allylthiourea is added to kill autotrophic bacteria. As a result, nitrification is inhibited, resulting in only the oxidation of carbonaceous organic matter (Liu et al., 2020).

2.4.1.3 Disoolved Oxygen Determination by Winkler's Method and Role of Chemicals Added

The DO of the sample is determined using Winkler's test. The wastewater is collected in a BOD bottle and is filled completely. Divalent manganese is added to the sample, followed by strong alkali iodide. The DO present in the sample quickly oxidizes an equivalent amount of the divalent manganese and converts it to manganic oxide. A strong acid is added to the solution. The iodine ions present in the solution in the presence of acid convert the manganic oxide back to a divalent state and releases iodine, equivalent to the amount of the original DO. The iodine is titrated against standard thiosulphate solution to find the amount of iodine. A starch indicator may be used to obtain the endpoint of the titration (APHA, 2017; CPCB, 2011).

The procedure of determining BOD by Winkler's method is provided as follows and illustrated in Figure 2.4.

- In the first step, two BOD bottles are filled with a test sample, and dilution water is added up to 300 mL. One BOD bottle is kept for incubation, while DO is estimated for the other sample.
- A small amount of manganese sulfate and alkali-iodide-azide are added to the sample whose DO is to be measured.

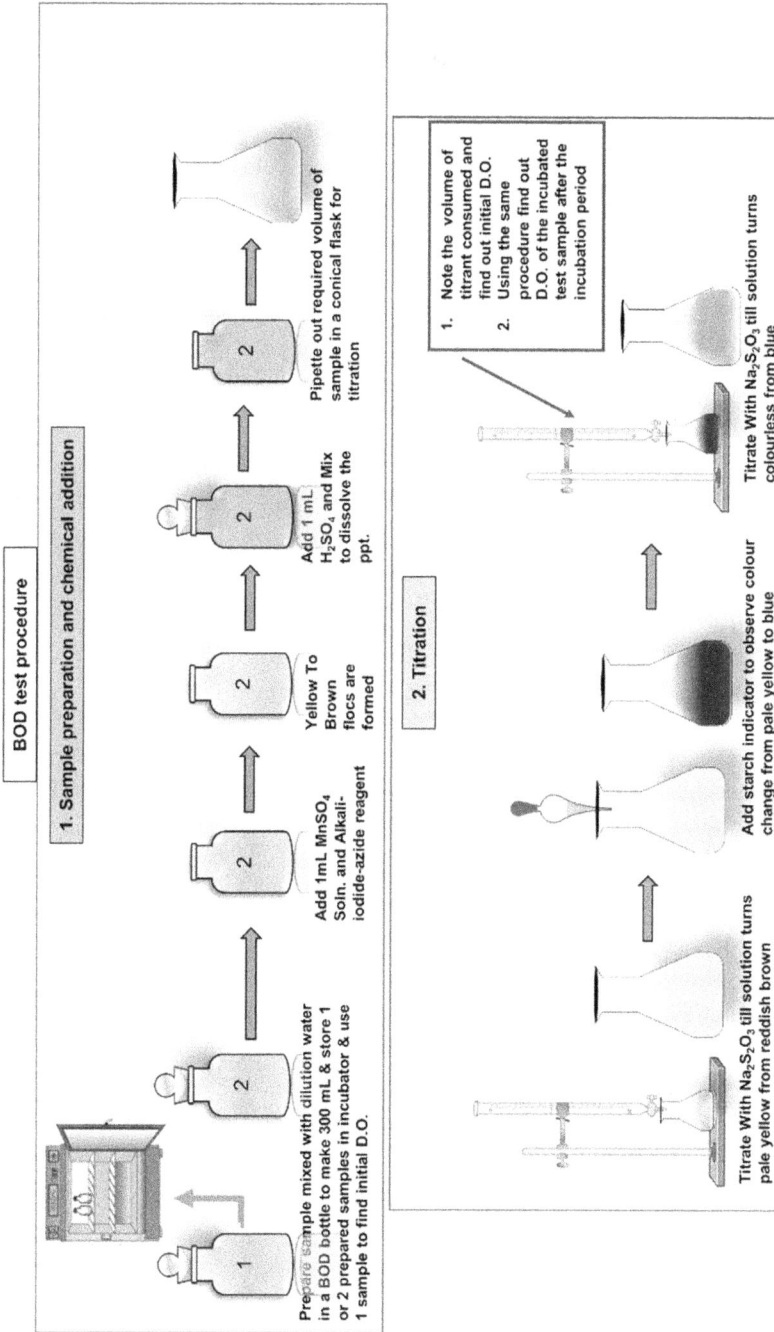

FIGURE 2.4 Depiction of standard procedure for determination of DO using Winkler's test.

Source: APHA, 2017, CPCB, 2011.

- $MnSO_4$ reacts with KOH and forms a white precipitate of $Mn(OH)_2$, thus utilizing the oxygen present in the water sample (Equation 2.9).

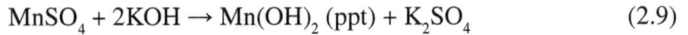

$$MnSO_4 + 2KOH \rightarrow Mn(OH)_2 \text{ (ppt)} + K_2SO_4 \qquad (2.9)$$

- Oxygen present in the sample oxidizes the Mn^{2+} to Mn^{4+} and forms brown hydrated oxide precipitates (Equation 2.10).

$$2Mn(OH)_2 \text{ (ppt)} + O_2 \rightarrow 2MnO(OH)_2 \text{ (ppt) (brown)} \qquad (2.10)$$

- The second part of Winkler's test involves the addition of an acid, that is, concentrated H_2SO_4 which is used to dissolve the precipitate back into the solution (Equation 2.11).
- Since the addition of alkali-iodide-azide introduces iodine to the solution, and under acidic conditions, MnO_2 oxidizes to produce I_2 which is insoluble and forms a complex as excess iodide (Equation 2.12).

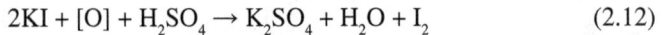

$$MnO(OH)_2 \text{ (ppt)} + H_2SO_4 \rightarrow MnSO_4 + 2H_2O + [O] \qquad (2.11)$$

$$2KI + [O] + H_2SO_4 \rightarrow K_2SO_4 + H_2O + I_2 \qquad (2.12)$$

- Due to acid expansion, the iodine in the form of KI is broken down, which gives a circuitous proportion of the measure of oxygen present in the water sample.
- The measurement of iodine is carried out using iodometric titration performed using $Na_2S_2O_3$ as a titrant.
- A certain quantity of the sample with the above-performed steps is pipetted out in a conical flask and titrated until the shade of the sample changes from brown color to pale yellow.
- The burette knob is then closed, and a few drops of the starch indicator are added at this point to the sample solution. The shade of the solution changes from light yellow to dark blue.
- The flask is again positioned under the burette and titrated until the shade of the sample changes from dark blue to colorless.
- The reading of the burette is noted down to record the amount (volume) of $Na_2S_2O_3$ consumed. The basic purpose of adding a starch indicator is to achieve a sharp endpoint.
- On the basis of the following calculations, the amount of DO present in the sample is calculated (Equation 2.13).

$$N_1V_1 = N_2V_2 \qquad (2.13)$$

where
  N_1 = Normality of $Na_2S_2O_3$
  V_1 = Volume of $Na_2S_2O_3$ consumed for titration
  N_2 = Normality of water sample
  V_2 = Volume of sample used for titration

Normality is defined as the no. of gram equivalent / volume of solution liter

No. of gram equivalent = Weight of Solute (gram) / Equivalent weight.

Hence, Normality = Weight of Solute (gram) / (Equivalent weight × volume of solution liter)

Hence, DO in mg/L = $N_1 V_1 \times 8 \times 1000 / V_2$

N_2 is the weight of solute (DO) (gram) / (Equivalent weight (O_2) × volume of solution liter) or $N_1 V_1 / V_2$

Hence, DO in g/L = $N_1 V_1 \times 8$(Equivalent weight of O_2)/V_2

Or, DO in mg/L = $N_1 V_1 \times 8 \times 1000/V_2$

Similarly, after incubation (5 days at 20°C ± 1°C or at 27°C ± 1°C for 3 days), the final DO in one incubated bottle is calculated.

2.4.1.4 Calculation of BOD

The BOD calculations are as follows (APHA, 2017; CPCB, 1986)

a. When seeding is not carried out (Equation 2.14)

$$BOD(mg/L) = \frac{\left(\begin{array}{c} \text{Initial DO of sample} - \text{Final DO of the} \\ \text{sample after incubation} \end{array} \right)}{\% \text{ dilution}} \times 100 \qquad (2.14)$$

b. When seeding is carried out (Equation 2.15)

$$BOD\ (mg/L) = \left[\frac{(\text{Initial DO} - \text{Final DO})_{Sample} - \left[(\text{Initial DO} - \text{Final DO})_{Seeded\ dilution\ water} \times P \right]}{\text{Dilution}\ (\%)} \right] \times 100 \qquad (2.15)$$

where $P = 1 -$ Dilution (%)

Example 1

Find the CBOD of wastewater (seeded) when 15 mL of wastewater is diluted to 300 mL and the drop in BOD over the incubation period was 6.9 mg/L. The drop in DO level drop for the seeded dilution water was 1.2 mg/L over the incubation period.

Solution 1

Change in DO of the seeded dilution water = 1.2 mg/L

Change in DO of the seeded wastewater solution = 6.9 mg/L

Dilution (%) = 15/300 = 0.05 × 100 = 5%

$P = 1 - 0.05 = 0.95$

Hence, from Equation 2.15

$$BOD_5O_2 mg/L = \frac{(6.9) - \left[(1.2) \times 0.95\right]}{5} \times 100$$
$$= 115.2 mg/L$$

Precision and Bias in BOD Tests

Following conditions are necessary to be met for accurate results:

- At least 2 mg/L of DO must be depleted.
- At the end of the test period, the lowest amount of DO residual should be 1 mg/L.
- The 5 mg/L of BOD value of 5 days 20°C or 3 days 27°C can be measured directly without dilution.
- Multiple tests must be carried out to reproduce accurate results.
- Dilution water must be prepared using good-quality reagents.

The lowest detection limit is 1 mg/L. The outcomes of the BOD test are affected by numerous elements, such as working pH, nutrients, buffers, microorganism-inhibiting substances, seed material, and other similar factors. The standard check with the glucose-glutamic acid should be carried out to assess the reliability of analytical methods (APHA, 2017).

2.4.1.5   Measurement of BOD Using DO Probes

The DO can also be measured using DO probes. The DO of the raw sample with dilution water and the sample after incubation may be measured separately to calculate the BOD. Dissolved oxygen probes measure how much oxygen diffuses into a probe across a permeable (or semi-permeable) membrane (sensor). Once oxygen is present inside the sensor, a chemical reduction event occurs, resulting in the generation of an electrical signal. The DO probe detects this signal, which is then shown on a meter (Wei et al., 2019).

Although the DO probe can be used to measure the DO very easily as compared to Winkler's method and do not require the use of so many chemicals, they have certain disadvantages. These disadvantages include poor stability, poor anti-interference ability, large drift, and low accuracy. Furthermore, these instruments require regular calibration (Wei et al., 2019).

2.4.2   CHEMICAL OXYGEN DEMAND (COD)

COD is defined as the amount of oxygen equivalents consumed in the chemical oxidation of organic matter by a strong oxidant. COD gives us the total amount of organic matter (biodegradable and non-biodegradable) in a water or wastewater sample.

In COD test, both organic and inorganic matter is subjected to oxidation, but the oxidation of organic component predominates. Hence, the COD of a sample usually depicts the total amount of biodegradable and non-biodegradable organic matter present in the sample. In a BOD test, only biologically reactive carbon is oxidized, while in a COD test, all organic matter is converted to carbon dioxide. The total amount of organic carbon in a sample is called total organic carbon (TOC), and the amount of oxygen consumed by all the elements in a sample when total (complete) oxidation is achieved is called total oxygen demand (TOD).

The COD determination process described has been adapted from Standard Methods for the Examination of Water and Wastewater (APHA, 2017) and *Central Pollution Control Board Guide Manual: Water and Wastewater Analysis*. Please refer to these documents for additional details (APHA, 2017).

2.4.2.1 Principle of Determination

When organic matter reacts with a mixture of potassium dichromate ($K_2Cr_2O_7$) and silver sulfate as a catalyst in the presence of concentrated H_2SO_4, the majority of it gets oxidized, yielding CO_2 and H_2O. In a sulfuric acid medium, a sample is refluxed with a known amount of $K_2Cr_2O_7$, and the excess dichromate is titrated against ferrous ammonium sulfate (FAS). The amount of dichromate consumed is proportional to the amount of oxygen consumed.

2.4.2.2 Reagents

Standard potassium dichromate solution, 0.25 N: Potassium dichromate is a powerful oxidizing agent in an acidic medium (primary digestion catalyst). As the oxidation state of an atom increases, it becomes more electronegative when elements come into contact with it in a chemical reaction. In the presence of dilute sulfuric acid, it liberates oxygen (Equation 2.16).

$$2K_2Cr_2O_7 + 8H_2SO_4 \rightarrow 2K_2SO_4 + 2Cr_2(SO_4)_3 + 8H_2O + 3O_2 \qquad (2.16)$$

During digestion, the organic matter gets oxidized by hexavalent dichromate ion ($Cr_2O_7^{2-}$) present due to the ionization of $K_2Cr_2O_7$. Oxygen (O_2) reacts with carbon to generate carbon dioxide (CO_2) (Equation 2.17) and the hexavalent ion ($Cr_2O_7^{2-}$) concentration to the trivalent ion (Cr^{3+}) (Equation 2.18).

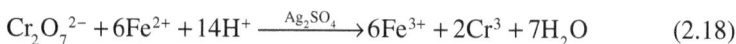

$$C_6H_{12}O_6 + 6O_2 \rightarrow 6CO_2 + 6H_2O \qquad (2.17)$$

$$Cr_2O_7^{2-} + 6Fe^{2+} + 14H^+ \xrightarrow{\ Ag_2SO_4\ } 6Fe^{3+} + 2Cr^3 + 7H_2O \qquad (2.18)$$

Sulfuric acid reagent: Ag_2SO_4 is added to concentrated H_2SO_4 and allowed to stand for one to two days for complete dissolution. Ag_2SO_4 is added to concentrated H_2SO_4 as a secondary catalyst, which accelerates the oxidation of straight-chain aliphatic and aromatic compounds such as diesel fuel and motor oil.

FAS solution, 0.1 N: Excess dichromate reacts with FAS (a reducing agent) after digestion. Excess dichromate is converted to its trivalent form when FAS is added slowly, oxidizing the organic matter present in the sample.

Ferroin indicator: It exhibits a green hue when adding the Ferroin indicator to the digested sample containing $K_2Cr_2O_7$. The color changes from a green hue to a bright blue hue and then to a brick red on reaching the endpoint during the titration.

Mercuric sulfate: Mercuric sulfate ($HgSO_4$) is added to the sample before adding other reagents to avoid interference caused by chlorides. Sulphamic acid (H_3NSO_3) is added to $K_2Cr_2O_7$ solution to eliminate interference caused by nitrite.

2.4.2.3 Procedure for COD Determination

The procedure for COD determination has been provided in Figure 2.5.
The COD calculation can be carried out using Equation 2.19.

$$COD \text{ (mg/L)} = (a - b) \times N \times 8000 / V_0 \qquad (2.19)$$

where a = Volume of FAS required for the blank solution (mL)
$\quad b$ = Volume of FAS required for the sample (mL)
$\quad N$ = Normality of FAS (Equation 2.20)

$$\frac{\left(\text{mL } K_2Cr_2O_7\right)(0.25)}{\text{ml FAS required}} \qquad (2.20)$$

V_0 = Volume of sample (ml)
8000 = Milli eq. wt. of $O_2 \times 1000$ mL/L.

2.4.3 NITROGEN

Nitrogen is one of the important nutrients that both plants and animals require. An excess of nitrogen in a waterway, on the other hand, can cause low levels of DO and negatively impact various forms of plant life and organisms. The runoff from fertilized lawns and croplands, failing septic systems, runoff from animal manure and storage areas, and industrial discharges containing corrosion inhibitors are all sources of nitrogen in wastewater. Nitrogen is commonly measured in water bodies in three forms: ammonia, nitrates, and nitrites (Canfield et al., 2010; Francis et al., 2007; Guo et al., 2014; Soler-Jofra et al., 2021).

Total nitrogen is the sum of total Kjeldahl nitrogen (TKN), nitrate, and nitrite. Ammonia nitrogen and organic nitrogen are collectively called TKN. The total nitrogen can be found out by measuring organic nitrogen compounds, ammonia nitrogen, and nitrate-nitrite individually and then adding them.

The nitrogen cycle in the atmosphere involves several processes, such as nitrogen fixation, nitrification, assimilation, denitrification, and decay (Francis et al., 2007; Kamilya et al., 2022; Soler-Jofra et al., 2021).

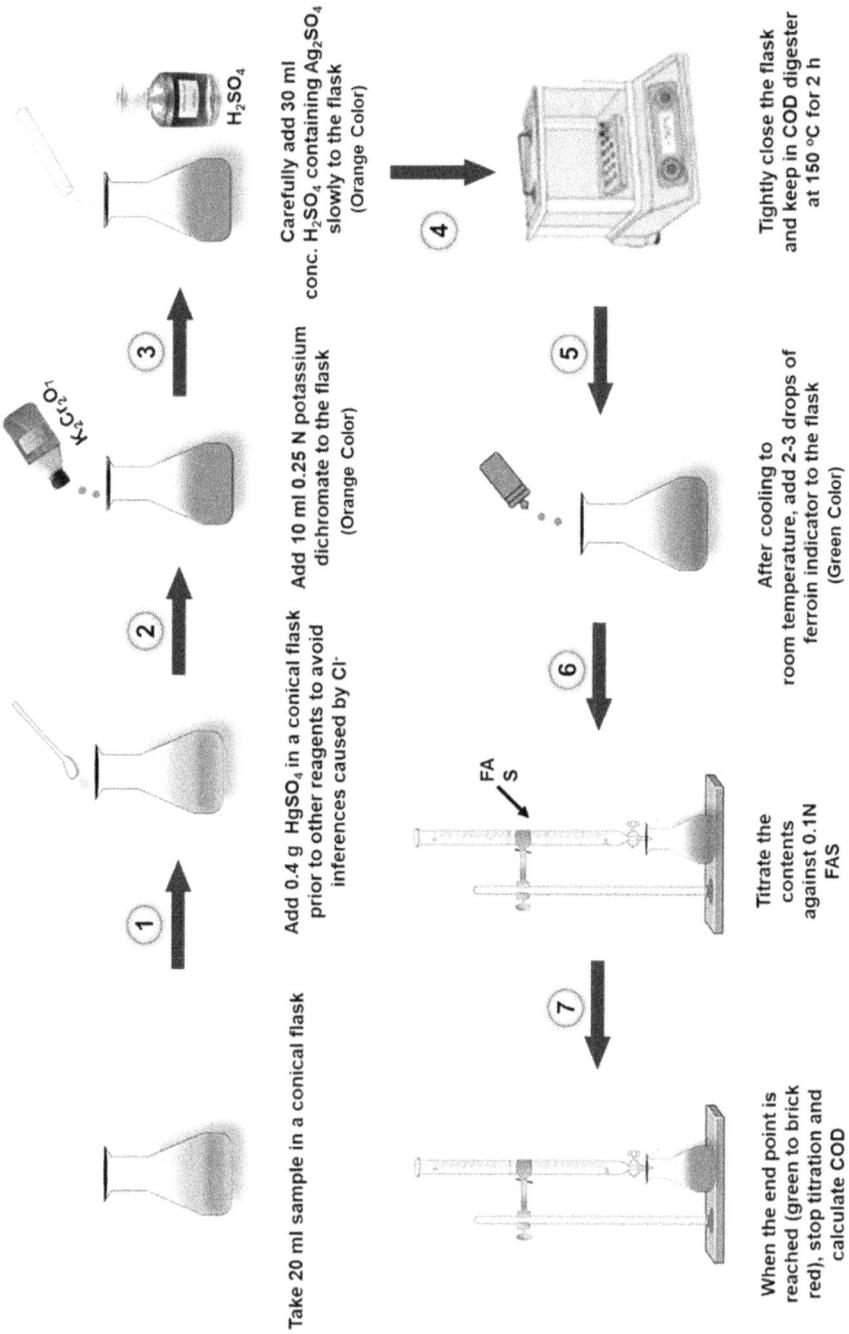

FIGURE 2.5 Depiction of standard procedure for determination of COD.

Source: APHA, 2017; CPCB, 2011.

Atmospheric fixation: A natural phenomenon where N_2 breaks to form nitrogen oxides in the presence of light, and plants can readily use this.

Biological fixation: Bacteria like rhizobium, azotobacter transform atmospheric nitrogen into usable forms like ammonia in the soil.

Nitrification: Nitrifying bacteria (Nitrosomonas and Nitrobacter) converts NH_3 into NO_3^- (Equation 2.21).

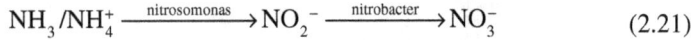

$$NH_3/NH_4^+ \xrightarrow{\text{nitrosomonas}} NO_2^- \xrightarrow{\text{nitrobacter}} NO_3^- \qquad (2.21)$$

Nitritation: In this process, ammonium oxidizing bacteria convert ammonia to nitrite in the presence of oxygen and the absence of nitrite oxidizers. The nitrite formed is then converted to molecular nitrogen without being converted to nitrate via denitrification, and this denitrification step requires less amount of carbon as well (Rodriguez-Caballero et al., 2013).

Assimilation: Plants take the nitrogen compound (ammonia, nitrate, nitrite, and ammonium ion) from the soil and utilize it for making plant and animal proteins.

Ammonification: When plants and animals die, the decomposers convert organic matter into ammonium ions or ammonia.

Denitrification: In anaerobic conditions, nitrate/nitrite is converted to nitrogen by denitrifying bacteria (clostridium, pseudomonas) to gain oxygen by which N_2 is produced as by-products.

Anammox (anaerobic ammonium oxidation): Nitrite and ammonium ions are transformed directly into diatomic nitrogen and water in the anammox process. The bacteria that perform the anammox process belong to the *Planctomycetes phylum* of bacteria (Bassin and Dezotti, 2018; Kamilya et al., 2022) (Equation 2.22).

$$NH_4^+ + NO_2^- \rightarrow N_2 + 2H_2O \qquad (2.22)$$

2.4.3.1 Determination of Ammonia Nitrogen

The determination process described has been adapted from the *Central Pollution Control Board Guide Manual: Water and Wastewater Analysis*. For additional information, the reader is directed to the following references (CPCB, 2011). The process commonly used for ammonia detection is the Nessler method.

Ammonia (present in the sample) produces a yellow-brownish-colored compound when it reacts with Nessler reagent (K_2HgI_4) under strongly alkaline conditions. The intensity of the color is directly proportional to the ammonia concentration. The sample is then measured photometrically using a spectrophotometer at 400–425 nm. The chemical reaction of the method is given in Equation 2.23.

$$2K_2HgI_4 + NH_3 + 3KOH \rightarrow Hg_2OINH_2 + 7KI + 2H_2O \qquad (2.23)$$

The proportion of ammonia nitrogen (NH_3 and NH_4^+) present in water depends on pH. Above pH 11 the NH_4^+ gets converted to NH_3 gas. Hence, the process is carried out in alkaline condition.

2.4.3.1.1 Apparatus and Equipment

Spectrophotometer (range of 300–700 nm) and Nessler tubes. A spectrophotometer is a device that detects the quantity of photons absorbed after a sample solution has passed through. This instrument may also use light intensity to determine the quantity of a known chemical compound (concentrations). Beer–Lambert law asserts that the absorption and concentration of a sample have a linear relation (Equation 2.24).

$$A = \epsilon lc \tag{2.24}$$

where A is the measure of absorbance (no units), ϵ is the molar extinction coefficient or molar absorptivity (or absorption coefficient), l is the path length, and c is the concentration. The solution is placed in a transparent container called the cuvette. The number of photons that pass through the cuvette and into the detector is determined by the length of the cuvette and the sample concentration. Once the intensity of light after it passes through the cuvette is known as transmitted intensity. The transmitted intensity can be used to calculate transmittance (T). The transmittance of a sample is the fraction of light that passes through the cuvette containing the sample (Equation 2.25). Similarly, the absorbance of a sample is the fraction of light that is not transmitted through the cuvette containing the sample (Equation 2.26).

$$\text{Transmittance}(T) = \frac{I_t}{I_0} \tag{2.25}$$

where I_0 and I_t is the intensity of light before and after it passes through the cuvette.

$$\text{Absorbance}(A) = -\log(T) = -\log\frac{I_t}{I_0} \tag{2.26}$$

where absorbance represents the number of photons absorbed. Using Beer–Lambert Law and the amount of absorbance known from the above equation, we can calculate the unknown concentration of the sample.

2.4.3.1.2 Reagents, Procedure, Calibration, and Interference

- Pre-treatment with $ZnSO_4$ helps in removing residual chlorine. Also, to get a pH of 10.5, 0.4–0.5 ml of 6 N sodium hydroxide solution is added and stirred thoroughly.
- One drop of EDTA reagent and one or two drops of Rochelle salt solution is added if the undistilled part contains enough Ca, Mg, or other ions to cause turbidity or precipitation with the Nessler's reagent.
- Nessler's reagent
- Stock ammonia solution and standard ammonia solution.

The procedure for the determination of ammonia nitrogen is illustrated in Figure 2.6. The calibration curve is prepared using suitable aliquots of standard solution using

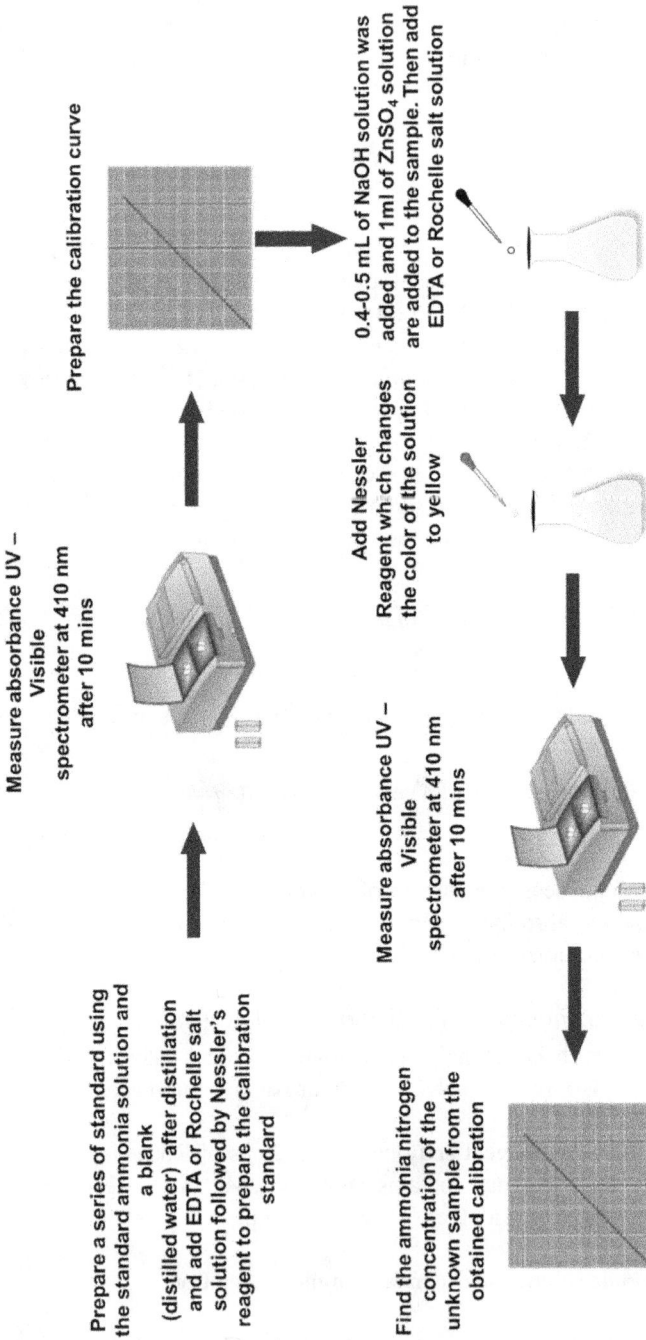

FIGURE 2.6 Depiction of standard procedure for determination of ammonia nitrogen.

Source: CPCB, 2011.

the standard ammonium solution prepared and following the procedure shown in Figure 2.6. At least three samples are essential to forming a calibration curve. However, for better accuracy, more standards should be prepared. Ideally, five to eight samples are preferred. Also, it should be noted that while making the calibration curve, the difference between the maximum and minimum concentrations should be as low as possible.

Interferences are primarily induced by color, turbidity, Ca, Mg salts, and Fe in the sample. Pre-treatment with $ZnSO_4$ helps in removing residual chlorine. Also, to get a pH of 10.5, 0.4–0.5 ml of 6 N sodium hydroxide is added to the solution and stirred thoroughly. In undistilled samples, EDTA reagent or Rochelle salt solution is added if the undistilled part contains enough Ca, Mg, or other ions to cause turbidity or precipitation with Nessler's reagent.

2.4.3.2 Determination of Nitrate (NO_3^-)

The determination process described has been adapted from the *Central Pollution Control Board Guide Manual: Water and Wastewater Analysis* (CPCB, 2011). The absorbance at 220 nm of a sample containing 1 mL of hydrochloric acid (1 N) in a 100 mL sample is used to evaluate nitrate. The first measurement is made at a wavelength of 220 nm. At 220 nm, nitrate and organic materials absorb light. At 275 nm, the second measurement is made. At 275 nm, nitrate does not absorb. The second measurement is used to account for the absorption of organic matter. The use of 1 N hydrochloric acid is intended to avoid interference from hydroxide or $CaCO_3$ concentrations of up to 1,000 mg/L.

2.4.3.2.1 Apparatus and Reagents, Procedure, and Interference

UV-visible spectrophotometer, for use at 220 nm and 275 nm with matched silica cells of 1 cm or longer light path. The wavelengths 220 and 275 nm are in the UV range. Hence, a UV-spectrophotometer supporting UV range detection should be used for this study. The reagents required are stock nitrate solution, standard nitrate solution, and aluminum hydroxide suspension. A calibration curve using suitable aliquots of standard solution using the standard nitrate solution is prepared.

- 1 mL of hydrochloric acid is taken, and a water sample is added to make 50 mL.
- The solution prepared above can be used to determine nitrate at 220 nm.
- If the presence of organic matter is suspected, absorbance can also be measured at 275 nm.

Two times the reading at 275 nm is subtracted from the reading at 220 nm to get the absorbance due to nitrate as compensation for dissolved organic materials. The nitrate values obtained from a standard calibration curve can be used to convert this absorbance value to equivalent nitrate. The calibration curve is prepared using suitable aliquots of standard solution using the standard nitrate solution prepared from stock nitrate solution and following the procedure described above. The nitrate concentration of an unknown sample can be calculated from the calibration curve.

Commonly color, chlorides, and nitrite act as interfering agents. In order to prevent NO_2^- interference, sulphamic acid was added to the sample. In order to remove chlorine, a suitable amount of Ag_2SO_4 is added to the solution to precipitate out $AgCl$. Colored samples are treated with aluminum hydroxide suspension or diluted to minimize color interference.

2.4.3.3 Determination of Nitrite (NO_2^-)

The determination process described has been adapted from *Central Pollution Control Board Guide Manual: Water and Wastewater Analysis*. For detailed information, these manuals may be referred to for additional details (CPCB, 2011). Nitrite is determined by combining diazotized sulphanilamide with *N*-(1-naphthyl) ethylenediamine dihydrochloride and forming a reddish-purple azo dye with Griess reagent at pH 2.0–2.5. (NED dihydrochloride). The chemical reactions involved in the determination of nitrite are provided in Equation 2.27 and Equation 2.28.

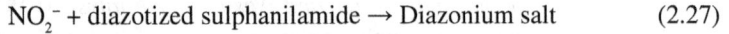

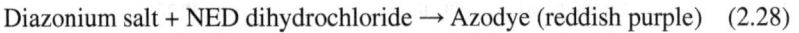

$$NO_2^- + \text{diazotized sulphanilamide} \rightarrow \text{Diazonium salt} \qquad (2.27)$$

$$\text{Diazonium salt} + \text{NED dihydrochloride} \rightarrow \text{Azodye (reddish purple)} \quad (2.28)$$

2.4.3.3.1 Apparatus, Reagents, and Procedure

Spectrophotometer or photometer with a green filter and maximal absorption near 540 nm for usage at 543 nm. The reagents required are sulphanilamide reagent, NED dihydrochloride solution, stock nitrite solution, and standard nitrite solution.

- If the sample contains suspended solids, it is to be filtered via a membrane filter with a pore width of 0.45 µm.
- 1 mL sulphanilamide solution is to be added to 50 mL clear sample neutralized to pH 7, and the reagent is allowed to react for 2–8 min.
- 1.0 mL NED dihydrochloride is added to the solution right away.
- The absorbance is measured, after 10 min but before 2 h, at 543 nm.
- The procedure is repeated for blank by replacing the sample with distilled water.

The calibration curve is prepared using suitable aliquots of standard solution using the standard nitrite solution prepared from and following the procedure described above. The nitrite value of an unknown sample can be calculated from the calibration curve.

When the reagents are added, nitrogen trichloride (NCI_3) produces a false red color. It can be reduced by first using NED dihydrochloride and then with sulphanilamide acid reagent.

2.4.3.4 Determination of Total Kjeldahl Nitrogen

The amino nitrogen of many organic materials is transformed to ammonium sulfate in the presence of sulfuric acid, potassium sulfate, and mercuric sulfate catalyst. Ammonium sulfate is made from free ammonia and ammonium nitrogen. A mercury ammonium complex is produced during sample digestion, which is then destroyed by sodium thiosulfate. Digestion in an acidic medium is carried out to convert the organic nitrogen to ammonium sulfate in the presence of a digestion reagent (mixture of mercuric sulfate/cupric sulfate and potassium sulfate). By increasing pH above 11, the ammonium sulfate is converted to ammonia; then, the distillate is collected using boric acid. The ammonia is distilled from an alkaline medium and absorbed in boric acid. The ammonia measured using this process indicated the total of the already present ammonia nitrogen and the organic nitrogen. The ammonia nitrogen can be determined using the method described in (Section 2.4.3.1). The ammonia nitrogen and the organic nitrogen are collectively called the total Kjeldahl nitrogen. In order to estimate the organic nitrogen, the total ammonia nitrogen can be determined and subtracted from the total Kjeldahl nitrogen.

2.4.4 SOLIDS

Solids comprise both organic and inorganic matter. A few examples of inorganic solids are silt, sand, gravel, and clay, while those of organic matter are plant fibers, microorganisms, and others. Solids in wastewater can be classified as suspended, settleable, colloidal, or dissolved based on their size. They are also classified as volatile or non-volatile.

Solids play a vital role in wastewater treatment processes. Mixed liquor volatile suspended solids (MLVSS) are the concentration of active biomass, while mixed liquor suspended solids (MLSS) are the concentrations of active biomass and inert solids. However, solids can have various negative impacts and are designed to be removed in STPs. Solids in wastewater can be further classified into various types (Qasim and Zhu, 2017; Tchobanoglous et al., 2014).

2.4.4.1 Determination of Solids

The determination process described has been adapted from Standard Methods for the Examination of Water and Wastewater (APHA, 2017) and *Central Pollution Control Board Guide Manual: Water and Wastewater Analysis*. Please refer to these manuals for additional details involved in the measurement of solids (APHA, 2017; CPCB, 2011). The different types of solids found in water and wastewater have been depicted in Figure 2.7.

Fixed solids: After the ignition for a specific duration at a certain temperature, the total suspended or dissolved solids remaining in the sample.

Volatile solids: After the ignition for a specific duration at a certain temperature, the total suspended or dissolved solids removed from the sample.

Total dissolved solids (TDS): Under the specific condition, the part of total solids that pass through a filter 2.0 µm (or smaller) nominal pore size.

Total solids: The material left in a sample vessel after evaporation and oven drying at a specified temperature.

Total suspended solids (TSS): The part of total solids in the water sample that is retained on the filter.

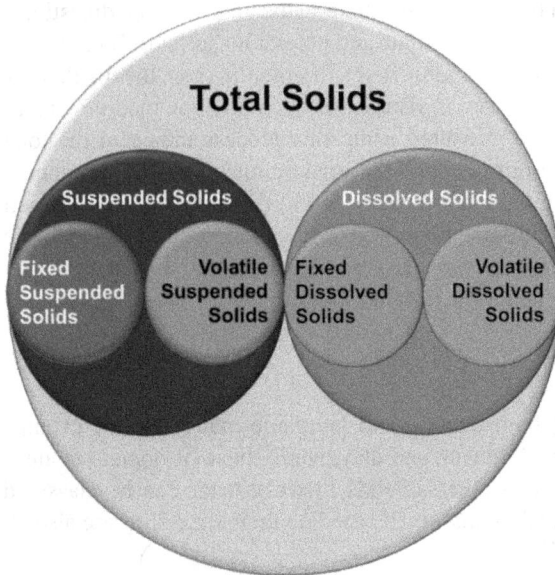

FIGURE 2.7   Classification of solids in wastewater.

Total suspended and dissolved solids combine to make up total solids, which are physically separated through filtration. Whether a solid particle is filtered into the "suspended" or "dissolved" part mainly relies on a filter's pore size, area, thickness, porosity, physical nature, solids quantity to be filtered, and their particle size.

2.4.4.2 Procedure for the Determination of Total Solids

- A dish is taken and heated to 103–105°C for more than 1 h. After cooling the dish to ambient temperature, its weight is measured (Q). The weighted dishes are placed in the desiccator or oven as required to avoid the deposition of moisture or any other foreign particles.
- The sample is stirred and mixed, and the required amount of sample is placed in the pre-weighted dish. The sample is evaporated to dryness at 103–105°C.
- The dish is cooled in a desiccator to ambient temperature and weighed (P). P is the sum of the final weight of dried residue and dish (mg), and Q is the weight of the dish (mg). The total solids can be calculated using Equation 2.29.

$$\text{Total solids}\,(\text{mg/L}) = \frac{(P-Q)\times 1000}{\text{volume of sample }(\text{mL})} \tag{2.29}$$

2.4.4.3 Procedure for the Determination of Suspended Solids

- A filtration device consisting of the filter (2.0 μm or smaller) with the wrinkled side up is inserted. In order to eliminate any particles adhering to the glass interference, a vacuum is applied, and the disc is cleaned with the help of reagent-grade water thrice.

- The wastewater or water sample is added to the filtration setup. Continuous suction is applied to eliminate all traces of water. The filter is removed from the mechanical filtration assembly and placed on an inert weighing dish.
- The filter is dried at 103–105°C in an oven for more than 1 h and subsequently cooled in a desiccator to normal temperature, and the weight is measured.
- The filters are then weighed. X is the sum of the final weight of filter and dried residue (mg), and Y is the weight of the filter (mg), before wastewater was passed through it. The total suspended solids can be calculated using Equation 2.30.

$$\text{Total suspended solids } (mg/L) = \frac{(X - Y) \times 1000}{\text{Volume of sample } (mL)} \qquad (2.30)$$

2.4.4.4 Procedure for Determination of Total Dissolved Solids

- The sample is mixed well, and a pipette or graduated cylinder is used to move a fixed volume of sample onto a glass fiber filter (2.0 μm or smaller).
- The surface of the filter should be washed with filter water, and a vacuum is applied to filter the sample.
- The total filtrate (with washings) is transferred to a pre-weighted evaporating dish and evaporated until dry in a drying oven.

The evaporated dish is further dried for 1 h in an oven at 180 ± 2°C, cooled in a desiccator to ambient temperature, and weighed. The dish now contains dissolved solids. The weight of the dish (Q), which is subtracted from the weight of the dish and dissolved solids (R), will give the total weight of dissolved solids. The total dissolved solids can be calculated using Equation 2.31. It can also be said that total dissolved solids is total suspended solids subtracted from total solids (Equation 2.32).

$$\text{Total dissolved solids } (mg/L) = \frac{(R - Q) \times 1000}{\text{Volume of sample } (mL)} \qquad (2.31)$$

$$\text{Total dissolved solids (mg/L)} = \text{Total solids (mg/L)} - \text{Total suspended solids (mg/L)} \qquad (2.32)$$

2.4.4.5 Process for the Determination of Total Volatile Solids

- The filter paper used for weighing total solids is placed in a crucible and heated in a muffle furnace at 550 ± 50°C for 15 min.
- The residue is cooled in a desiccator or oven. The weight of residue and dish or filter prior to ignition (C) and the weight of residue and dish or filter after ignition (D) is measured.

The total volatile and fixed solids can be calculated using Equation 2.33 and Equation 2.34, respectively.

$$\text{Total volatile solids}\,(\text{mg}/\text{L}) = \frac{(C-D)\times 1000}{\text{Volume of sample}\,(\text{mL})} \qquad (2.33)$$

$$\text{Total fixed solids (mg/L)} = \text{Total solids (mg/L)} - \\ \text{Total volatile solids (mg/L)} \qquad (2.34)$$

Similarly, volatile dissolved solids and volatile suspended solids can be estimated by following the same process but using the filter paper used to measure TDS and TSS, respectively.

2.4.4.6 Reason for Choosing Different Temperatures

A drying temperature above 104°C is adequate to eliminate the fluid and water adsorbed on the particle surface, whereas reaching a temperature around 180°C is important to evaporate all the water that is bound mechanically to solids. However, some water of crystallization may remain, particularly when sulfates are present. Organic matter may volatilize and be lost, but not totally eliminated. Loss of CO_2 happens when bicarbonates are converted to carbonates, and carbonates might partially disintegrate into oxides or essential salts. As a result, evaporation and drying of the water sample at $180 \pm 2°C$ yields TDS value much more accurate as compared to 103–105°C (APHA, 2017).

Volatile solids assurance requires heating at $550 \pm 50°C$ because by heating at such temperature, the organic part of deposits will be changed to water vapor, releasing different gases, including CO_2. The leftover matter is comprised of inorganic or fixed residue (APHA, 2017).

2.4.5 pH

The acidic or basic/alkaline nature of a solution can be identified by measuring the pH value. Pure water exists in a partially ionized state as depicted in Equation 2.35. pH can be represented as per Equation 2.36

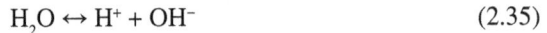

$$H_2O \leftrightarrow H^+ + OH^- \qquad (2.35)$$

$$pH = -\log_{10}[H^+] \qquad (2.36)$$

Since a substance that produces hydrogen ions is known as an acid and one that produces hydroxyl ions is known as a base, water may be classified both as an acid and a base. The presence of these ions is in the same quantity in pure water, which makes it neutral (Kohlmann, 2003; SATO, 2020; Surat, 2020).

Experiments have shown that H^+ concentration in pure water is 1×10^{-7} M. This implies that in every 10 million liters of pure water 1 molecular weight (1.008 grams) of hydrogen ions is present. The concentration of hydroxyl ions is also of an equivalent amount, that is, 1×10^{-7} M. The negative log of the concentration of hydrogen ions is defined as pH. An increase in the H^+ concentration decreases the pH. Neutral solutions have a pH of 7. Solutions with a concentration of H^+ higher than a neutral solution have a pH of less than 7 and are classified as acidic. Whereas, when the concentration of H^+ in a solution is less than that of a neutral solution, the pH is more than 7 and is classified as basic or alkaline. In water, the product of the concentration of H^+ and OH^- yields a constant value of 1×10^{-14}. Upon taking a negative logarithm, the following results can be attained (Equations 2.37 to 2.40) (Kohlmann, 2003; SATO, 2020; Surat, 2020).

$$[H^+] \times [OH^-] = 10^{-14} \tag{2.37}$$

$$-\log_{10} [[H^+] \times [OH^-]] = -\log_{10} 10^{-14} \tag{2.38}$$

$$-[\log_{10} [H^+] + \log_{10} [OH^-]] = -14 \tag{2.39}$$

$$pH + pOH = 14 \tag{2.40}$$

Electrometrical evaluation of pH is carried out using pH-sensitive glass tip electrodes. As the glass electrode is put in a solution having different pH from the solution's pH inside the electrode, an electrical potential is generated between the reference electrode and glass electrode. This potential, which is proportional to the difference in pH, is estimated, and the instrument gives an output corresponding to the pH of the sample.

Since a potentiometric scale is used to define the pH, potentiometric calibration of the instrument has to be carried out with an indicating (glass) electrode and a reference electrode using standard buffers of pre-defined pH value such that Equation 2.41 and Equation 2.42 are satisfied.

$$pH_B = -\log_{10} [H^+] \tag{2.41}$$

where pH_B = assigned pH of standard buffer.

The operational pH scale is used to measure sample pH and is defined as:

$$pH_S = pH_B + F (E_S - E_B) / 2.303 \, RT \tag{2.42}$$

where pH_S = potentiometrically measured sample pH
F = Faraday's constant (9.649×10^4 coulomb/mole)
E_S = sample electromotive force (V)
E_B = buffer electromotive force (V)
T = absolute temperature (K)
R = gas constant

The process for determination of pH has been provided in Figure 2.8.

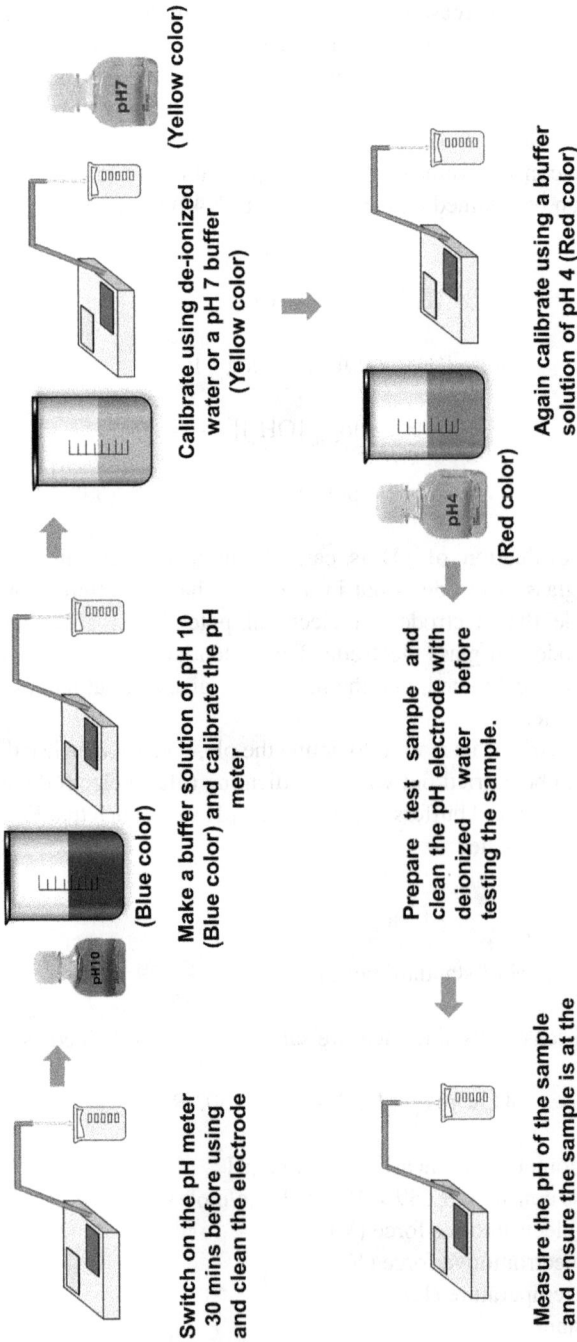

Switch on the pH meter
30 mins before using
and clean the electrode

(Blue color)

Make a buffer solution of pH 10
(Blue color) and calibrate the pH
meter

Calibrate using de-ionized
water or a pH 7 buffer
(Yellow color)

(Yellow color)

Measure the pH of the sample
and ensure the sample is at the
same temperature as the buffers
used for calibration

Prepare test sample and
clean the pH electrode with
deionized water before
testing the sample.

(Red color)

Again calibrate using a buffer
solution of pH 4 (Red color)

FIGURE 2.8   Process for determination of pH.

2.5 CHAPTER SUMMARY

- Sewage can be classified into four categories, which are yellow water, brown water, greywater, and black water.
- The concentration of organic matter and solids in wastewater generated from low-income countries or developing countries are usually higher than that of the high-income countries or developed countries.
- The amount of wastewater generated from high-income countries or developed countries are usually higher than that of the low-income countries or developing countries.
- Certain viruses and bacteria can survive in wastewater for up to 2–4 months.
- The amount of DO required by aerobic biological organisms for the biological degradation of organic matter existing in a water sample in a certain duration at a particular temperature is known as BOD. The oxygen demand of the sample is directly proportional to the organic matter present in it.
- COD is defined as the amount of oxygen equivalents consumed in the chemical oxidation of organic matter by a strong oxidant. COD gives us the total amount of organic matter (biodegradable and non-biodegradable) present in a water or wastewater sample. In a BOD test, only biologically reactive carbon is oxidized, while in a COD test, all organic matter is converted to carbon dioxide.
- Nitrogen is commonly measured in water bodies in three forms: ammonia, nitrates, and nitrites. Total nitrogen is the sum of total Kjeldahl nitrogen, nitrate, and nitrite. Ammonia nitrogen, and organic nitrogen are collectively called total Kjeldahl nitrogen.
- TDS and TSS are collectively called total solids. Both TDS and TSS have volatile fraction and non-volatile (fixed) fraction.

2.6 CONCLUDING REMARKS

The analysis of the different components of wastewater are essential to understand the quality of the wastewater entering the STP and the effluent of the STP. In this way the performance of the STP can be assessed. Moreover, strict implementation of rules and regulations regarding assessing the quality of the wastewater before being discharged has mandated the analysis of wastewater parameters. Different wastewater parameters should be analyzed to check whether they are satisfying the discharge or reuse standards. In this context the following chapter discusses about the different rules, regulations, and standards pertaining to wastewater discharge and reuse.

REFERENCES

Al Aukidy, M., Al Chalabi, S., Verlicchi, P., 2018. Hospital wastewater treatments adopted in Asia, Africa, and Australia, in: *Handbook of Environmental Chemistry*. Springer Verlag, pp. 171–188. https://doi.org/10.1007/698_2017_5

APHA, 2017. Standard Methods for the Examination of Water and Wastewater Standard Methods for the Examination of Water and Wastewater. *Public Health* 16(10). https://doi.org/10.2105/AJPH.51.6.940-a

AQUASTAT database [WWW Document], 2021. www.fao.org/aquastat/statistics/query/resu lts.html (accessed 3.24.21).

Bassin, J.P., Dezotti, M., 2018. *Moving Bed Biofilm Reactor*. Springer, Cham. https://doi.org/ 10.1007/978-3-319-58835-3

Canfield, D.E., Glazer, A.N., Falkowski, P.G., 2010. The evolution and future of Earth's nitrogen cycle. *Science* 330, 192–196.

CPCB, 1986. *General Standards for Discharge of Environmental Pollutants Part-A: Effluents, The Environment (Protection) Rules, 1986*. CPCB, Govt of India.

CPCB, 2011. *Guide Manual: Water and Wastewater Analysis*.

CPCB, 2021. *National Inventory of Sewage Treatment Plants*.

CPHEEO, 2013. *Manual on Sewerage and Sewage Treatment Systems: Part A Enigneering*, Ministry of Urban Development, Government of India.

Delzer, G.C., Mckenzie, S.W., 2003. *Chapter A7. Section 7.0. Five-Day Biochemical Oxygen Demand*, Techniques of Water-Resources Investigations. U.S. Geological Survey. https://doi.org/10.3133/twri09A7.0

Feachem, R.G., Bradley, D.J., Garelick, H., Mara, D.D., 1983. *Sanitation and Disease: Health Aspects of Excreta and Wastewater Management*. John Wiley and Sons.

Francesco, Di, M., Ghosh, S.K., Das Saha, P. (Eds.), 2020. *Recent Trends in Waste Water Treatment and Water Resource Management*. Springer Nature Singapore.

Francis, C.A., Beman, J.M., Kuypers, M.M.M., 2007. New processes and players in the nitrogen cycle: The microbial ecology of anaerobic and archaeal ammonia oxidation. *ISME J.* 1(1), 19–27. https://doi.org/10.1038/ismej.2007.8

Ghaitidak, D.M., Yadav, K.D., 2013. Characteristics and treatment of greywater – a review. *Environ. Sci. Pollut. Res.* 20, 2795–2809.

Guo, W., Fu, Y., Ruan, B., Ge, H., Zhao, N., 2014. Agricultural non-point source pollution in the Yongding River Basin. *Ecol. Indic.* 36, 254–261. https://doi.org/10.1016/j.ecol ind.2013.07.012

Kamilya, T., Majumder, A., Yadav, M.K., Ayoob, S., Tripathy, S., Gupta, A.K., 2022. Nutrient pollution and its remediation using constructed wetlands: Insights into removal and recovery mechanisms, modifications and sustainable aspects. *J. Environ. Chem. Eng.* 10, 107444. https://doi.org/10.1016/J.JECE.2022.107444

Kohlmann, F., 2003. What is pH, and how is it measured. *A Tech. Handb. Ind.* 86(2), 94–99 .

Kumari, A., Maurya, N.S., Tiwari, B., 2020. Hospital wastewater treatment scenario around the globe, in: *Current Developments in Biotechnology and Bioengineering*. Elsevier, pp. 549–570. https://doi.org/10.1016/b978-0-12-819722-6.00015-8

Liu, C., Morrison, C., Jia, J., Xu, X., Liu, M., Zhang, S., 2020. Does the Nitrification-Suppressed BOD5 Test Make Sense? *Environ. Sci. Technol.* 54(9), 5323–5324. https:// doi.org/10.1021/acs.est.0c00997

Mackenzie, D.L., 2010. *Water and Wastewater Engineering: Design Principles and Practice*. McGraw-Hill Education.

Majumder, A., Gupta, A.K., Ghosal, P.S., Varma, M., 2021. A review on hospital wastewater treatment: A special emphasis on occurrence and removal of pharmaceutically active compounds, resistant microorganisms, and SARS-CoV-2. *J. Environ. Chem. Eng.* 9, 104812. https://doi.org/10.1016/j.jece.2020.104812

Muserere, S.T., Hoko, Z., Nhapi, I., 2014. Characterisation of raw sewage and performance assessment of primary settling tanks at Firle Sewage Treatment Works, Harare, Zimbabwe. *Phys. Chem. Earth* 67–69, 226–235. https://doi.org/10.1016/j.pce.2013.10.004

Pereira, L.S., Duarte, E., Fragoso, R., 2014. Water use: Recycling and desalination for agriculture, in: *Encyclopedia of Agriculture and Food Systems*. Elsevier, pp. 407–424. https:// doi.org/10.1016/B978-0-444-52512-3.00084-X

Qasim, S.R., Zhu, G., 2017. *Wastewater Treatment and Reuse, Theory and Design Examples, Volume 1 ... – Syed R. Qasim, Guang Zhu – Google Books*. CRC Press.

Rodriguez-Caballero, A., Ribera, A., Balcázar, J.L., Pijuan, M., 2013. Nitritation versus full nitrification of ammonium-rich wastewater: Comparison in terms of nitrous and nitric oxides emissions. *Bioresour. Technol.* 139. https://doi.org/10.1016/j.biort ech.2013.04.021

SATO, G., 2020. What is pH? [WWW Document]. *Chem. Educ.* 139, 195–202.

Soler-Jofra, A., Pérez, J., van Loosdrecht, M.C.M., 2021. Hydroxylamine and the nitrogen cycle: A review. *Water Res.* 190, 116723. https://doi.org/10.1016/j.watres.2020.116723

Surat, P., 2020. What is a pH Meter and How does it work? | AZoLifeSciences. AZO life Sci. www. azolifesciences.com/article/What-is-a-pH-Meter-and-How-Does-it-Work.aspx#:~:text= The%20pH%20meter%20operates%20like,with%20the%20loss%20of%20H%2B.

Tchobanoglous, G., Burton, F.L., Stensel, H.D., 2014. *Wastewater Engineering: Treatment and Resource Recovery, Metcalf & Eddy, Inc.* McGraw-Hill Education.

Vianna, M.R., de Melo, G.C.B., Neto, M.R.V., 2012. Wastewater treatment in trickling filters using Luffa cyllindrica as biofilm supporting medium. *J. Urban Environ. Eng.* 6, 57–66. https://doi.org/10.4090/juee.2012.v6n2.057066

Wei, Y., Jiao, Y., An, D., Li, D., Li, W., Wei, Q., 2019. Review of dissolved oxygen detection technology: From laboratory analysis to online intelligent detection. *Sensors (Switzerland)* 19(18), 3995. https://doi.org/10.3390/s19183995

WHO, 2017. *Water and Sanitation for Health Facility Improvement Tool*. World Health Organization.

3 Legal Aspects of Municipal Wastewater Management and Standards

CHAPTER OBJECTIVES

The chapter describes the different existing legislature mandates for wastewater treatment in different countries. The wastewater discharge standards and reuse standards in different countries have also been highlighted in this chapter. Furthermore, the disposal regulations of sludge generated from sewage treatment plants and acceptable heavy metal concentration in treated sludge for land application have also been discussed.

3.1 INTRODUCTION

The disposal of municipal wastewater has been a challenge throughout the history of humankind. Many early civilizations were developed around major rivers; therefore, raw sewage was disposed into these water bodies. As the populations were low, the discharged wastewater was well within the assimilative capacity of the streams. Therefore, rivers provided the necessary treatment. Evidence of dug wells from the Neolithic period suggests that early humans tried to separate the water used for consumption and places of disposal.

Land application of wastewater, especially on agricultural farms, became an established practice from 15th century onward and perhaps was practices informally even before this period (IWA, 2014). This practice continues to prevail in many underdeveloped and developing countries even today. While wastewater offers nutrient-rich waters for plant growth, the practice creates exposures to pathogens both via direct (ingestion, inhalation, and dermal contact) and indirect (food consumption) pathways.

The industrial revolution led to the creation of large cities in many parts of America and Europe in the 18th and 19th centuries. Conveyance systems that carried wastewater from homes were constructed in these cities. Many times, these systems were open to the atmosphere and created significant odor problems. In addition, wastewater from conveyance systems was discharged into rivers, and these loadings often exceeded the assimilative capacities of many rivers. Comingling of wastewater with drinking water was also a widespread problem. All these factors led to the widespread prevalence of water-borne illnesses such as cholera (Abellan, 2017).

DOI: 10.1201/9781003364450-3

The growing concerns of pollution and associated loss of economic productivity caused by water-borne illnesses led to research on wastewater treatment. Early efforts focused on physical methods such as settling and sedimentation to separate water from solids (which were collected and disposed on land). The activated sludge process was formulated in 1912 at the University of Manchester (Benidickson, 2011). Over the last 100 years, these processes have been refined to improve efficiencies to treat municipal wastewater.

While the technologies for treating municipal wastewater were available since early 1900s, they were not adopted widely. Many smaller cities and towns relied on antiquated methods such as primary treatment using cesspools and lagoons. In addition, treatment standards varied widely, and many cities discharged inadequately treated wastewater into receiving water bodies, which led to widespread pollution. This situation was exacerbated over much of the 1950s and 1960s in the United States, as cities continued to grow, and other industrial wastes were also dumped into surface water bodies alongside domestic wastewater.

The uncontrolled discharge of wastewater into surface water bodies and the associated pollution led to the realization that a set of regulations and standards are necessary to properly dispose of wastewater. These effluent discharge standards would specify the concentrations of constituents that would be legally permissible for discharge. Treatment systems could be designed and sized appropriately to meet the discharge standards. The enactment of laws also meant legal action could be pursued against those municipalities that did not comply with the prescribed standards.

Today, countries worldwide have passed laws and regulations requiring the proper treatment of wastewater before discharge. Acceptable concentration thresholds have been developed to enforce these regulations. These regulations will continue to evolve as new chemicals continue to be discovered in wastewater streams. In what follows, the relevant regulations and effluent discharge standards in various countries across the world are presented. The chapter not only shows how different countries are tackling the issue of wastewater treatment but also provides a handy reference for the design of wastewater treatment systems.

3.2   EXISTING LEGISLATIVE MANDATES FOR WASTEWATER MANAGEMENT ACROSS THE WORLD

3.2.1   THE UNITED STATES OF AMERICA

The Clean Water Act (1972) was enacted to protect the surface waters of the country. The Act establishes the framework for regulating pollutant discharges into the water bodies of the United States of America. The United States Environment Protection Agency (US EPA) has developed pollution control programs under the Clean Water Act, such as defining effluent regulations for industries. Furthermore, the US EPA set up the national water quality criteria for various contaminants in surface waters ("National Pretreatment Program I US EPA," 2021; UN-HABITAT, 2009).

The Clean Water Act controls the discharges from sewage treatment plants (STP). The discharge of salty brines (or concentrate) from membrane treatment operations (e.g., reverse osmosis) to freshwater lakes and streams is mostly regulated under the Clean Water Act. National Pollutant Discharge Elimination System (NPDES) governs the discharge of any pollutants into navigable waters from a point source. NPDES permits are mandatory when discharges are made directly to surface waters from industrial, municipal, and other facilities. However, permits are not required by individual homes that have a septic system, are linked to a municipal system, or do not have a surface discharge ("National Pretreatment Program I US EPA," 2021).

The NPDES program includes the national pretreatment program. It is a collaborative effort established by government environmental regulatory agencies to protect water quality. The NPDES permits the regulatory agencies to carry out managerial and law implementation tasks involving discharges to surface waters and publicly owned treatment works (POTWs) of the municipalities. The national pretreatment program aims to protect POTW infrastructure while also lowering the levels of conventional and toxic pollutants discharged into the sewer systems and the environment ("National Pretreatment Program I US EPA," 2021).

3.2.2 EUROPE

The major legislative mechanism regulating the quality of treated municipal wastewater discharged into recipient aquatic bodies in the European Union is the Urban Waste Water Treatment Directive (UWWTD) (EU, 2019). In order to preserve water bodies, this directive restricts the degree of treatment of treated wastewater discharged into receiving bodies in the member states. The degree of treatment requirements was established depending on the population equivalent (PE), the type of receiving wastewater, and the eutrophication sensitivity of the receiving wastewater. Following the emergence of the European Union, the member states were dedicated to implementing EU regulations. Furthermore, the existing quality standards were reviewed, and aquatic environment protection legislation was changed (EU, 2019).

3.2.3 INDIA

The Ministry of Environment Forest and Climate Change (MoEF&CC), the Ministry of Housing and Urban Affairs (MoHUA), and the Jal Shakti Ministry are jointly responsible for the control and prevention of pollution activities in India. MoEF&CC is the leading agency, and the ministries mentioned are the nodal organizations for establishing policies, acts, and related standards in conjunction with the Central Pollution Control Board (CPCB). The different laws and regulations pertaining to the protection of the environment are provided in the following section.

The Water Prevention and Control of Pollution Act, 1974, amended 1988: The Water Prevention and Control of Pollution Act was implemented in order to control water pollution, as well as to maintain or recover the sanctity of water. This act also provided for the establishment of water pollution control boards. Furthermore, it gave the authority to form as many committees as necessary to carry out specific functions. It explicitly prohibits the discharge of any polluting matter into any well or stream. Approval from State Board is required to discharge any new type of contaminant into a new well or stream. Standards have been stated separately for small-scale industries. Penalties for non-compliance or causing pollution in any way include fines and/ or imprisonment. It may also include any inspection, sampling, and analysis charges incurred by the Board. The CPCB plays an important role in promoting the cleanliness of wells and streams across India. It also provides consultation to the Central government regarding matters related to safeguarding water resources ("CPCB | Central Pollution Control Board," 2019).

The State Pollution Control Boards (SPCB) are required to prepare and execute comprehensive plans for the prevention, control, or abatement of pollution of streams and wells in the states. Similar to the CPCB, the SPCBs also advise the state governments on water pollution-related matters. Moreover, it also collects and provides information related to water pollution and carries out research and pollution abatement work.

The Environmental Protection Act, 1986 ("CPCB | Central Pollution Control Board," 2019): It governs hazardous wastes and chemicals, harmful microbes, and the transportation of toxic compounds. It provides power to the Central government to take action for the control and abatement of pollution. Standards of air, water, or soil quality for various locations, maximum permitted limits of concentration of different environmental contaminants, procedures, and guidelines for the management of hazardous substances are also included in the law. The act gives the government the authority to set standards for environmental quality in all of its components, including maximum permitted concentrations of various contaminants in various regions. Schedule VI of the Environment (Protection) Rules, 1986 contains the General Standards for Discharge of Environmental Pollutants. The act also establishes a cess on water consumed by certain industries and local authorities in order to supplement the funds available to the Central and State Boards for the Prevention and Control of Water Pollution, which was established under the Water Prevention and Control of Pollution Act, 1974 ("CPCB | Central Pollution Control Board," 2019).

The National Environment Tribunal Act, 1995 ("CPCB | Central Pollution Control Board," 2019): In 1995, through the National Environment Tribunal Act 1995, the Central government established the National Environment Tribunal providing it with strict responsibility regarding taking care of cases where damage has been caused by accidents involving mishandling of hazardous substances ("CPCB | Central Pollution Control Board," 2019).

National Urban Sanitation Policy, 2008: The vision of this policy was that all the cities and towns of India should become totally sanitized and maintain healthy and livable conditions. With a special emphasis on urban poor and women, all the citizens should be ensured good public health, environmental conditions, and affordable sanitation facilities (Ministry of Urban Development Government of India, 2008).

National Water Policy, 2012: This policy suggests the reuse and recycling of water to meet the increasing water demand. It involves optimizing the use of water resources by implementing a system of laws and institutions (Central Water Commission, 2012).

3.2.4 CHINA

In 1973, the "discharge standard of the three industrial wastes" (GBJ 4–73) was published. The GBJ 4–73 regulates the wastewater treatment effluent discharge for water pollutants in China (Xu et al., 2020; Zhou et al., 2018). Currently, 62 water pollution discharge standards exist in China, including one for STPs (GB 18918–2002) and 61 standards related to various industries (Xu et al., 2020; Zhou et al., 2018). These standards, when combined, provide China's comprehensive model for water pollution control and form the basis for managing water pollution discharges. The standards for wastewater treatment improved significantly from 1973 to 2002, with more strict conditions on organic compounds and suspended solids, which were also incorporated in GBJ 4–73. The introduction and changes in standards for phosphates and nitrogenous compounds are done to protect the aesthetic value of water (Xu et al., 2020; Zhou et al., 2018).

3.3 WASTEWATER DISCHARGE STANDARDS

The discharge standards of conventional environmental parameters, such as pH, biochemical oxygen demand (BOD), chemical oxygen demand (COD), total suspended solids (TSS), total phosphorous (TP), total nitrogen (TN), and ammonia nitrogen set by different countries have been provided in Table 3.1. The Australian and New Zealand Environment and Conservation Council have provided the allowable values for BOD, TP, TN, and TSS after each stage of treatment in an STP (Australian and New Zealand Environment and Conservation Council, 1997). As per the Australian and New Zealand norms, the values of these parameters after advanced treatment were significantly lower as compared to other countries. The Government of United Kingdom has also implemented discharge limits for BOD, COD, TP, and TN (Gov. UK, 2019). These values were higher than that of Australia and New Zealand (Table 3.1). There are separate discharge standards for countries belonging to the Helsinki Commission (Denmark, Estonia, Finland, Germany, Latvia, Lithuania, Poland, Russia, and Sweden) and the European Union (Austria, Belgium, Bulgaria, Croatia, Republic of Cyprus, Czech Republic, Denmark, Estonia, Finland, France, Germany, Greece, Hungary, Ireland, Italy, Latvia, Lithuania, Luxembourg, Malta, Netherlands, Poland, Portugal, Romania, Slovakia, Slovenia, Spain, and Sweden) (Table 3.1). Apart from the discharge limits set by European Union and Helsinki Commission, Sweden, Denmark, Germany, and Switzerland have their own discharge standards. These values were higher than the Australian standards but significantly lower than those in Brazil, Japan, and India (Table 3.1). The values suggested that Australia and New Zealand had the strictest discharge policy followed by European countries. However, among the Asian countries, China and Singapore have also implemented strict regulatory discharge standards for pH, BOD, COD, and TSS, which were at par with the standards of many European countries.

Similarly, the discharge standards of metals, non-metals, and other substances which directly affect human health have been provided in Table 3.2. Regarding the discharge standards pertaining to substances that directly affect human health, the standards set by the Indian government are much more lenient as compared to other countries. The standards set by Japan and Brazil were comparable (Table 3.2). The discharge standards set by the National Environment Agency of Singapore were much stricter. The discharge limits of most metals and non-metals are at least 10 times lower as compared to the discharge standards of India and Brazil (NEA, 2020).

TABLE 3.1
Discharge Standards for Conventional Environmental Pollutants

Country	Type of Standard	pH	BOD (mg/L)	COD (mg/L)	TSS (mg/L)	Total Phosphorous (mg/L)	Total Nitrogen (mg/L)	Ammonical Nitrogen (mg/L)	References
United Kingdom			25	125		2 (10,000 – 100,000 PE) 1(>100,000 PE)	15 (10,000 – 100,000 PE) 10(>100,000 PE)		(Gov.UK, 2019)
Australia	Post pretreatment		140 to 350		140 to 350				(Australian and New Zealand Environment and Conservation Council, 1997)
	Post Primary treatment		120 to 250		80 to 200	6 to 14	30 to 55		
	Post Secondary treatment		20 to 30		25 to 40	6 to 12	20 to 50		
	Post Nutrient Removal		5 to 20		5 to 20	<2	10 to 20		
	Post advanced wastewater treatment		2 to 5		2 to 5	<1	<10		
European Union	10,000–100,000 PE		25	125		2	15		(The council of European communities, 1991)
	>100,000		25	125		1	10		(Trittin, 2005)
Germany	WWTP Size category 1: BOD$_5$ < 60 kg/d (< 1000 PE)		40	150					
	Size category 2: BOD$_5$ < 300 kg/d (<5000 PE)		25	110					
	Size category 3: BOD$_5$ < 1200 kg/d (< 20,000 PE)		25	90				10	
	Size category 4: BOD$_5$ < 6000 kg/d (<100,000 PE)		20	90		2	18	10	
	Size category 5: BOD$_5$ < 6000 kg/d (>100,000 PE)		15	75		1	13	10	

Country	Category	pH							Reference
Sweden	>2000 PE		15			0.5	15		(Preisner et al., 2020)
	2000–100,000 PE		15			0.5	15		
	>100,000 PE		15			0.5	10		
Denmark			10	75		0.4	8		(Preisner et al., 2020)
HELCOM signatory countries	300–2000 PE		25			2	35		(Preisner et al., 2020)
	2000–10,000 PE		15			1	30		
	10,000–100,000 PE		15			0.5	15		
	>100,000 PE		15			0.5	10		
Switzerland	200–10,000 PE		20	60		0.8	0.3***		(Swiss Federal Council Waters Protection Ordinance, 1998)
	>10,000 PE		15	45		0.8	0.2		
Brazil		5 to 9	120						(Brazilian NR, 2011)
India	Inland surface water	5.5 to 9.0	30	250	100	5	100*	20	(CPCB, 1986)
	Public sewers	5.5 to 9.0	350		600			50	
	Land of irrigation	5.5 to 9.0	100		200			50	
China	Surface water	6–9	20	60	20			15	(MEP, 2002)
Japan	Surface water	5.8 to 8.6	120	120	150	8	60	100	(MOE, 2015)
Singapore	Surface water or watercourse	6 to 9	20	60	30	2	20**		(NEA, 2020)

Note: PE: Population Equivalent, * Total nitrogen (as N), **Nitrate, ***Nitrite.

TABLE 3.2
Discharge Standards for Heavy Metals, Non-metals and Other Substances Which Directly Affect the Human Health

Pollutants	European Union (EU, 2013)	India (CPCB, 1986)	Japan (MOE, 2015)	Singapore (NEA, 2020)	Brazil (Brazilian NR, 2011)
Arsenic (mg/L)		0.2	0.1	0.01	0.5
Mercury (mg/L)	0.07*	0.01	0.005	0.001	0.01
Lead (mg/L)	1.3*	0.1		0.1	0.5
Cadmium (mg/L)	0.2*	2	0.03	0.003	0.2
Hexavalent chromium (mg/L)		0.1	0.5		0.1
Total chromium (mg/L)		2	2	0.05	1.1
Copper (mg/L)		3	3	0.1	1
Zinc (mg/L)		5	2	0.5	5
Selenium (mg/L)		0.05	0.1	0.01	0.3
Nickel (mg/L)	8.6*	3		0.1	2
Cyanide (mg/L)		0.2	1	0.1	1
Fluoride (mg/L)		2	15mg/L (coastal areas) 8 mg/L (non-coastal areas)		10
Phenolic compounds (mg/L)		1	5	Nil	
Manganese (mg/L)		2	10	0.5	1
Iron (mg/L)		3	10	1	15
Vanadium (mg/L)		0.2			
Beryllium (mg/L)				0.5	
Boron (mg/L)			230 mg/L (coastal areas) 10 mg/L (non-coastal areas)	0.5	5

Note: *Concentrations in μg/L.

However, European Union is the most stringent when it comes to the discharge of heavy metals into the environment. The values of 0.07 μg/L, 1.3 μg/L, 0.2 μg/L, and 8.6 μg/L set by the European Union for mercury, lead, cadmium, and nickel were significantly lower than that by other countries (EU, 2013).

3.4 WASTEWATER REUSE STANDARDS

The increase in urban population and concentration of population in certain areas has led to an increase in demands for freshwater supply. This has necessitated the reuse of water. Water reuse is reusing treated wastewater (reclaimed water) for various beneficial purposes. Recently, the reusability of the water has been facilitated by the development of different advanced wastewater treatment processes (Donnaz, 2020; Sun et al., 2011). These processes can now effectively remove a wide range of contaminants including biodegradable materials, nutrients, and pathogens, allowing the treated water to be used in a variety of ways. Water reuse can reduce water stress and provide alternative water resources for various purposes, including agricultural, irrigation, industrial, environmental, and others (Li et al., 2019). As an alternative water source, the reuse of treated water can provide major socio-economic and environmental advantages, all of which are important motivators for developing reuse programs. The reuse of water can have the following advantages:

- Increased availability of water
- Integrated management of water resources and their sustainable use
- Substitution of drinking water used for non-drinking purposes
- Reduction in overuse of ground and surface water
- Reduction in energy consumption in comparison to the use of deep groundwater resources, desalination, or water transportation
- Reduction in nutrient loads in receiving waters
- Increase in agricultural production and less use of fertilizers when nutrient containing wastewater is reused for agriculture
- Improved environmental protection through stream, pond, and wetlands restoration
- Improved local economy and increase in job opportunities

In the urban sector, where public access to water is limited, the reclaimed or reused water can be used for non-potable purposes in municipal settings. The reclaimed water can be used to irrigate food crops, which are meant for human consumption and non-food crops. Irrigation of food crops, which need to be processed before human consumption, can also be carried out using reclaimed water. The reclaimed water may also be used for recreational activities, where there is no body contact with the reclaimed water. Reclaimed water may also have various implications in enhancing or augmenting water bodies, such as streamflow, aquatic habitats, or wetlands. Industries may recycle the produced wastewater for power generation, cooling towers, extraction of fossil fuels, and other purposes. The recharge of aquifers, whose water is not used for potable purposes, may also be carried out using reclaimed water. Reclaimed water may be indirectly used for potable purposes by augmenting it with surface water or groundwater and applying suitable drinking water treatment. The reclaimed water may also be directly sent from the advanced wastewater treatment plant to the drinking water treatment plant and may be used for potable purposes only after it satisfies the necessary quality requirements (USEPA, 2012).

In order to reuse wastewater for different purposes, the wastewater should be able to satisfy certain criteria. The standards pertaining to turbidity, suspended solids, total dissolved solids (TDS), pH, oil and grease, minimum residual chlorine, total Kjeldahl Nitrogen, BOD, COD, dissolved phosphorus as P, nitrate (mg/L), fecal coliform for reuse of reclaimed wastewater in different activities have been provided in Table 3.3. The United States and India have different standards for various reuse applications. The United States has standards for different urban reuse, agricultural reuse, impoundments, environmental reuse, groundwater reuse, and potable reuse. Unrestricted urban reuse refers to the use of reclaimed or treated water for

TABLE 3.3
Allowable Concentration of Different Pollutants in Reclaimed Water for Their Reuse

Use of Reclaimed Water (Broad Classification)	Use of Reclaimed Water (Detailed Classification)	Country	Turbidity (NTU)	SS (mg/ L)	TDS (mg/ L)	pH
Urban reuse	Unrestricted reuse	United States	<2			6 to 9
	Restricted reuse	United States		<30		6 to 9
	Toilet flushing	India	<2	Nil	2100	6.5–8.3
	Fire protection	India	<2	Nil	2100	6.5–8.3
	Vehicle exterior washing	India	<2	Nil	2100	6.5–8.3
Impoundments	Unrestricted impoundments	United States	<2			6 to 9
	Restricted impoundments	United States		<30		
	Non-contact impoundments	India	<2	Nil	2100	6(6.5) to 8.3
Agricultural and irrigation	Food Crops	United States	<2			6 to 9
	Raw food crops	India	<2	Nil	2100	6.5–8.3
	Cooked food crops	India	AA	30	2100	6.5–8.3
	Processed Food Crops	United States		<30		6 to 9
	Non-Food Crops	United States		<30		6 to 9
	Non-Food Crops	India	AA	30	2100	6.5–8.3
	Irrigation of green areas	Italy		10		6 to 9.5
	Agricultural uses	Spain	1 to 15	5 to 35		
	Irrigation of green areas and certain agricultural uses	France		15		
	Irrigation	Portugal		60	640	6.5 to 8.4
	Cooked vegetables, parking areas, playgrounds and side of roads inside cities	Jordan	10	50	1500	6 to 9

Oil and Grease (mg/L)	Minimum Residual Chlorine (mg/L)	Total Kjeldhal Nitrogen (mg/L)	BOD (mg/L)	COD (mg/L)	Dissolved Phosphorus as P (mg/L)	Nitrate (mg/L)	Faecal Coliform/ 100 ml	References
	1		<10				Nil	(USEPA, 2012)
	1		<30				<200	USEPA, 2012
10	1	10	10	AA	1	10	Nil	(CPHEEO, 2013)
Nil	1	10	10	AA	1	10	Nil	(CPHEEO, 2013)
Nil	1	10	10	AA	1	10	Nil	(CPHEEO, 2013)
	1		<10				Nil	USEPA, 2012)
	1		<30				<200	USEPA, 2012)
Nil	0.5	10	10	AA	1		5	Nil	(CPHEEO, 2013)
	1		<10				Nil	USEPA, 2012)
Nil	Nil	10	10	AA	2	10	Nil	(CPHEEO, 2013)
Nil	Nil	10	20	30	5	10	230	(CPHEEO, 2013)
	1		<30				<200	USEPA, 2012)
	1		<30				<200	USEPA, 2012)
10	Nil	10	20	30	5	10	230	(CPHEEO, 2013)
		15*	20	100	2		10	(Alcalde Sanza and Gawlik, 2014)
							0 to 10000	(Alcalde Sanza and Gawlik, 2014)
				60			250 to 100000	(Alcalde Sanza and Gawlik, 2014)
								(Alcalde Sanza and Gawlik, 2014)
		45*	30	100		30	100	(WHO, 2006)

(continued)

TABLE 3.3 (Continued)
Allowable Concentration of Different Pollutants in Reclaimed Water for Their Reuse

Use of Reclaimed Water (Broad Classification)	Use of Reclaimed Water (Detailed Classification)	Country	Turbidity (NTU)	SS (mg/L)	TDS (mg/L)	pH
	Plenteous trees and green areas, side of roads outside cities			150	1500	6 to 9
	Field crops, Industrial crops and forestry			150	1500	6 to 9
	Raw food crops	Oman		15	1500	6 to 9
	Cooked food crops			30	2000	6 to 9
	Unrestricted irrigation	Saudi Arabia	5	10		6.5(6) to 8.5
	Restricted irrigation			40	2000	
Landscaping and horticulture	Horticulture, golf course	India	<2	Nil	2100	6.5–8.3
	Areas with public access	Oman		15	1500	6 to 9
	Areas with no public access			30	2000	6 to 9
Environmental Reuse	Sustaining stream flow, aquatic habitats, or wetlands	United States		<30		
Industrial Reuse	Cooling water	India				6.8 to 7
	Once-through Cooling	United States		<30		6 to 9
	Recirculating Cooling Towers	United States		<30		6 to 9
Indirect potable use	Groundwater Recharge by Spreading into Potable Aquifers	United States	<2			6.5 to 8.5
	Groundwater Recharge by Injection into Potable Aquifers	United States	<3			6.5 to 8.5
	Augmentation of Surface Water Supply Reservoirs	United States	<4			6.5 to 8.5
	Aquifer recharge	Spain				
	Aquifer recharge	Oman		15	1500	6 to 9

Note: *TN, AA – as arising when other parameters are satisfied.

Oil and Grease (mg/L)	Minimum Residual Chlorine (mg/L)	Total Kjeldhal Nitrogen (mg/L)	BOD (mg/L)	COD (mg/L)	Dissolved Phosphorus as P (mg/L)	Nitrate (mg/L)	Faecal Coliform/ 100 ml	References
		70*	200	500		45	1000	(WHO, 2006)
		70*	300	500		45		(WHO, 2006)
		5	15	150		50	200	(WHO, 2006)
		10	20	200		50	1000	(WHO, 2006)
			10			10		(WHO, 2006)
			40					(WHO, 2006)
10	1	10	10	AA	2	10	Nil	(CPHEEO, 2013)
		5	15	150		50	200	(WHO, 2006)
		10	20	200		50	1000	(WHO, 2006)
	1		30				<200	USEPA, 2012)
Nil			<5					CPHEEO, 2013)
	1		<30				<200	USEPA, 2012)
	1		<30				<200	USEPA, 2012)
	1						Nil	USEPA, 2012)
	1						Nil	USEPA, 2012)
	1						Nil	USEPA, 2012)
		10*			2			(Alcalde Sanza and Gawlik, 2014)
		5	15	150		50	200	(WHO, 2006)

non-potable uses in the municipality where public access is not restricted. Restricted urban reuse refers to the reuse of reclaimed water in settings with restrictions to the common public. Reuse of treated water in unrestricted impoundment refers to the reuse of water in recreational or other activities where contact with the human body is allowed. On the other hand, reuse in unrestricted impoundments does not allow the water to be contacted with the human body. As a result, the standards for unrestricted uses are much less strict than those for restricted uses. In the Indian context, there are standards for the use of treated water in urban sectors, agricultural sectors, and landscaping as well. Compared to the U.S. standards for unrestricted urban reuse and food crops, the Indian reuse standards are less strict (Table 3.3). European countries, such as Italy, Spain, France, and Portugal, and countries in the Middle East also have standards for the reuse of treated wastewater for different agricultural, irrigation, and landscaping activities. The European standards are much more stringent compared to the Middle East countries (Table 3.3). Although the reuse of treated wastewater in growing crops, urban uses, and landscaping is common in various countries, the reuse of treated wastewater for potable purposes is not very common. In the United States, there are standards for the reuse of treated water in groundwater recharge and augmentation of surface water supply reservoirs. These activities can be treated as indirect potable reuse of reclaimed water (Table 3.3). Spain and Oman also have standards pertaining to aquifer recharge (Table 3.3). Table 3.3 does not describe the acceptable concentrations of toxic substances in treated wastewater for reuse purposes. In this context, the maximum concentrations of toxic elements for reuse in different sectors have also been provided in Table 3.4. Direct exposure to contaminants like heavy metals may lead to detrimental health effects. As a result, the permissible limit set by different countries for these contaminants in treated water for their reuse application in agriculture, irrigation, livestock drinking water, and others is very low.

3.5 SLUDGE DISPOSAL REGULATIONS

Wastewater treatment leads to the generation of a considerable amount of sludge. Wastewater treatment and sludge management are global concerns with mounting challenges that must satisfy the interests of all stakeholders, including facility managers and operators, regulators, lawmakers, scientists, wastewater generators, taxpayers, and the general public. The higher the amount of treatment, the greater the volume of solids generated in the wastewater. Increased efficacy of excreta management at the household level may have the greatest health implications and be the biggest challenge for countries with limited or no access to basic sanitation.

Sludge can be disposed of in landfills, used as a source of energy, processed and used as a fertilizer, soil conditioner, or even used as a raw material to extract valuable components. When sludge is treated properly to achieve an acceptable quality, it is commonly called "biosolids." Recently, improvements in biosolids utilization have become more important in densely populated regions where the assimilation capacity of the ecosystem (lakes and rivers) is exhausting. Establishing improved systems for wastewater and solids management would reduce the risk of water-borne diseases and threats to the natural environment due to different contaminants present (Collivignarelli et al., 2019; UN-HABITAT, 2009).

TABLE 3.4
Allowable Elemental Concentration in Reclaimed Water for Its Reuse Purposes

Toxic Elements	US Livestock Drinking Water (USEPA, 2012)	Jordan Agriculture Irrigation (WHO, 2006)	Saudi Arabia Unrestricted Irrigation (WHO, 2006)	Oman Raw Food Crops, Areas with Public Access, Aquifer Recharge (WHO, 2006)	Oman Cooked Food Crops, Areas with Public Access (WHO, 2006)	India Continuous Irrigation (CPHEEO, 2013)
Al (mg/L)	5	5	5	5	5	1
As (mg/L)	0.1	0.1	0.1	0.1	0.1	0.1
Ba (mg/L)			1	1	2	
Be (mg/L)	0.1	0.1	0.1	0.1	0.3	0.1
B (mg/L)	5	1	0.5	0.5	1	0.5
Cd (mg/L)	0.05	0.01	0.01	0.01	0.01	0.01
Cr (mg/L)	1	0.1	0.1	0.05	0.05	0.1
Co (mg/L)	1	0.05	0.05	0.05	0.05	0.05
Cu (mg/L)	0.5	0.2	0.4	0.05	1	0.2
Fe (mg/L)		5	2	1	5	5
Pb (mg/L)	0.1	5	0.1	0.1	0.2	5
Li (mg/L)		2.5	0.07	0.07	0.07	2.5
Mn (mg/L)	0.05	0.2	0.2	0.1	0.5	0.2
Mo (mg/L)	0.3	0.01	0.01	0.01	0.05	0.01
Hg (mg/L)	0.01	0.002	0.001	0.001	0.001	
Ni (mg/L)		0.2	0.2	0.1	0.1	0.2
Se (mg/L)	0.05	0.05	0.02	0.02	0.02	0.005
V (mg/L)	0.1	0.1	0.1	0.1	0.1	0.1
Zn (mg/L)	24	5	2	5	5	2

Management of sludge is practiced to a limited extent in regions of sub-Saharan Africa, Asia, and South America. In parts of eastern Europe, Russia, Mexico, and Turkey, wastewater treatment has advanced, but the management of biosolids and sludge has become a vital concern, and more complex regulatory structures are being developed. Hence, scientists, engineers, water quality professionals, and agricultural experts are developing strategies to manage biosolids properly. Upon successful implementation of these strategies, biosolids may be used for varying purposes, which have been described as follows (Collivignarelli et al., 2019; UN-HABITAT, 2009):

1. *Land reclamation*: Biosolids can be used for mine-land reclamation as top layer soil for landfills. It can also be used for restoration works, bioremediation of polluted sites, and the establishment of wetlands.
2. *Landscaping and horticulture*: Biosolids can be used in parks, lawns, gardens, and sports grounds. It can be used as potting mixes, compost, and fertilizers.

3. *Resource recovery from biosolids*: Biosolids can act as a source of metals and minerals. Nutrients and metals may be recovered from the biosolids using varying technologies.

4. *Energy recovery*: Bioenergy can be obtained from the digestion of biosolids in digesters. Incineration with heat recovery and electricity generation, pyrolysis, gasification, and the development of technologies for energy recovery are the areas that are gaining the attention of researchers.

5. *Agriculture*: Biosolids application on agricultural land can be a promising strategy for increasing productivity of crops by increasing organic matter content in soil, fertility, and nutrient availability. Additionally, biosolids can improve the physical properties of soil, particularly in heavy-textured and poorly structured soils. Furthermore, applying biosolids on farmland minimizes the impact of soil organic matter loss, particularly in southern Europe, where depletion of organic matter in soil is one of the most important soil degradation processes.

Despite the proven advantages of recycling biosolids, it is not widely practiced. There is an urgent need to adopt the promising uses of biosolids and not regard the by-products as "trash" in a global context of increased concern over disease, climate change, environmental pollution, and resource shortages. Construction of treatment plants to deal with wastewater is relatively less challenging than its operation and maintenance. Hence, even before the treatment plant comes into operation, comprehensive planning for the treatment and disposal of biosolids is required. Different countries have different rules and regulations for the disposal and management of sludge. The different standards set by different countries regarding the use of treated sludge for land applications are provided in Table 3.5. The permissible limits for different metals in sludge are different. The permissible limit for Cu, Zn, Pb, Cr, and Ni are usually higher than As, Ba, Cd, Hg, and Mo (Table 3.5). This suggests that these metals are more toxic as compared to other metals.

TABLE 3.5
Heavy Metals Concentration (in mg/kg Dry Weight) in Treated Sludge for Land Application

Country	As	Ba	Cd	Co	Cr	Cu	Pb	Hg	Mo	Ni	Se	Zn	References
						(mg/kg Dry Weight of Sludge)							
USA	41		39			1500	300	17	75	420	100	2800	(UN-HABITAT, 2009)
Canada	170		34	340	2800	1700	1100	11		420		4200	(UN-HABITAT, 2009)
Spain			20–40		1500	1000–1750	750–1200	16–25		300–400		2500–4000	(Collivignarelli et al., 2019)
Finland	25		1.5		300	600	100	1		100		1500	(Collivignarelli et al., 2019)
Belgium	20–150		6–10		100–150	600–800	300–500	1–1.6		100		1500–2000	(Collivignarelli et al., 2019)
Sweden			0.75		40	300	25	1.5		25		600	(Collivignarelli et al., 2019)
Netherlands	15		1.25		75	75	100	0.75		30		300	(Collivignarelli et al., 2019)
France			20		1000	1000	800	10		200		3000	(Collivignarelli et al., 2019)
Italy	20		20			1000	750	10		300		2500	(Collivignarelli et al., 2019)
Austria	20		2–10		50–500	70–500	45–500	0.4–10		25–100		200–2000	(Collivignarelli et al., 2019)
Ireland			20			1000	750	16		300		2500	(Collivignarelli et al., 2019)
Portugal			20		1000	1000	750	16		300		2500	(Collivignarelli et al., 2019)
Germany			10		900	800	900	8		200		4000	(Collivignarelli et al., 2019)
Japan	50		5		500		100	2		300			(UN-HABITAT, 2009)
Russia	10		15		7.5	1750	250	500		200		750	(UN-HABITAT, 2009)
Brazil	41	1300	39		1000	1500	300	17	50			2800	(UN-HABITAT, 2009)
Mexico	41		39		1200	1500	300	17		420	100	2800	(UN-HABITAT, 2009)

3.6   CHAPTER SUMMARY

- As per various reuse standards, treated wastewater may be reused for various purposes, such as toilet flushing, fire protection, vehicle exterior washing, food crops, irrigation of green areas, cooling water, groundwater recharge, aquifer recharge, and others.
- The reuse of water can have various advantages, such as increased availability of water, reduction in overuse of ground and surface water, reduction in nutrient loads in receiving waters, increase in agricultural production and less use of fertilizers when nutrient-containing wastewater is reused for agriculture, and others.
- Biosolids may be used for varying purposes, such as land reclamation, landscaping, horticulture, and agriculture. Biosolids can be used as potting mixes, compost, and fertilizers.

3.7   CONCLUDING REMARKS

The rules, regulations, and standards pertaining to the discharge and reuse of wastewater in different countries are getting increasingly stringent in order to protect the environment. In some places, these standards are already strictly implemented, while in other places, the authority has started to mandate these standards. As a result, simply discharging wastewater without any treatment is no longer an option. The wastewater generated has to be treated in a decentralized or centralized manner. In this context, the different wastewater treatment systems and the components of wastewater treatment systems have been discussed in the following chapters.

REFERENCES

Abellan, J., 2017. Water supply and sanitation services in modern Europe: developments in 19th–20th centuries. 12th Int. Conf. Spanish Assoc. Econ. Hist. Univ. Salamanca.

Alcalde Sanza, L., Gawlik, B.M., 2014. *Water Reuse in Europe: Relevant Guidelines, Needs for and Barriers to Innovation. JRC Science and Policy Reports.* https://doi.org/10.2788/29234

Australian and New Zealand Environment and Conservation Council, 1997. *Australian Guidelines for Sewerage Systems – Effluent Management*, National Water Quality Management Strategy, 46.

Benidickson, J., 2011. *The Culture of Flushing – Google Books.* UBC Press.

Brazilian NR, 2011. CONAMA Resolution 430/11 – Brazilian NR. www.braziliannr.com/brazilian-environmental-legislation/conama-resolution-43011/.

Central Water Commission, 2012. Action Plan for National Water *Policy* 37, 37–48.

Collivignarelli, M.C., Abbà, A., Frattarola, A., Miino, M.C., Padovani, S., Katsoyiannis, I., Torretta, V., 2019. Legislation for the reuse of biosolids on agricultural land in Europe: Overview. *Sustain.* 11(21), 6015. https://doi.org/10.3390/su11216015

CPCB | Central Pollution Control Board [WWW Document], 2019. https://cpcb.nic.in/water-pollution/ (accessed 9.8.21).

CPCB, 1986. General standards for discharge of environmental pollutants. *Environ. Rules* 2, 545–560.

CPHEEO, 2013. Chapter 7: Recycling and reuse of sewage. *Man. Sewerage Sew. Treat. Syst.* Part A: Engineering 7.1–7.53.

Donnaz, S., 2020. Water reuse practices, solutions and trends at international, in: *Advances in Chemical Pollution, Environmental Management and Protection.* Elsevier, pp. 65–102. https://doi.org/10.1016/BS.APMP.2020.07.012

EU, 2013. Water environmental quality standards. *Off. J. Eur. Union* 16, 11–14.

EU, 2019. *Evaluation of the Urban Waste Water Treatment Directive* 186.

Gov.UK, 2019. Waste water treatment works: treatment monitoring and compliance limits [WWW Document]. www.gov.uk/government/publications/waste-water-treatment-works-treatment-monitoring-and-compliance-limits/

IWA, 2014. *Evolution of Sanitation and Wastewater Technologies through the Centuries.* IWA Publishing.

Li, L., Yan, G., Wang, H., Chu, Z., Li, Z., Ling, Y., Wu, T., 2019. Denitrification and microbial community in MBBR using A. donax as carbon source and biofilm carriers for reverse osmosis concentrate treatment. *J. Environ. Sci. (China)* 84, 133–143. https://doi.org/10.1016/j.jes.2019.04.030

MEP, 2002. *Discharge Standard of Pollutants for Municipal Wastewater Treatment Plant, GB 18918-2002.* Beijing, China.

Ministry of Urban Development Government of India, 2008. *National Urban Sanitation Policy* 44.

MOE, 2015. *National Effluent Standards.* Japan.

National Pretreatment Program | US EPA [WWW Document], 2021. www.epa.gov/npdes/national-pretreatment-program (accessed 9.8.21).

NEA, 2020. *Allowable Limits for Trade Effluent Discharge to Watercourse or Controlled Watercourse.*

Preisner, M., Neverova-Dziopak, E., Kowalewski, Z., 2020. An analytical review of different approaches to wastewater discharge standards with particular emphasis on nutrients. *Environ. Manage.* 66, 694–708. https://doi.org/10.1007/s00267-020-01344-y

Sun, F., Chen, M., Chen, J., 2011. Integrated management of source water quantity and quality for human health in a changing world, in: *Encyclopedia of Environmental Health.* https://doi.org/10.1016/B978-0-444-52272-6.00286-5

Swiss Federal Council Waters Protection Ordinance, 1998. SR 814.201 – Waters Protection Ordinance of 28 October 1998 (WPO) [WWW Document].

The council of European communities, 1991. COUNCIL DIRECTIVE of 21 May 1991 concerning urban waste water treatment (91/271/EEC).

Trittin, J., 2005. Federal ministry for the environment, nature conservation and nuclear safety, Germany promulgation of the new version of the ordinance on requirements for the discharge of waste water into waters on the basis of Article 2 of the sixth ordinance for Amen. *Environ. Nat. Conserv. Nucl. Saf.* 5, 1–121.

UN-HABITAT, 2009. Global atlas of excreta, wastewater sludge, and biosolids management: moving forward the sustainable and welcome uses of a global resource. Choice Rev. Online 47. https://doi.org/10.5860/choice.47-1767

USEPA, 2012. EPA Guidelines for Water Reuse.

WHO, 2006. *A compendium of standards for wastewater reuse in the Eastern Mediterranean Region World Health Organization Regional Office for the Eastern Mediterranean Regional Centre for Environmental Health Activities CEHA.* World Health Organization, pp. 1–19.

Xu, A., Wu, Y.-H., Chen, Z., Wu, G., Wu, Q., Ling, F., Huang, W.E., Hu, H.-Y., 2020. Towards the new era of wastewater treatment of China: Development history, current status, and future directions. *Water Cycle* 1, 80–87. https://doi.org/10.1016/j.watcyc.2020.06.004

Zhou, Y., Duan, N., Wu, X., Fang, H., 2018. COD discharge limits for urban wastewater treatment plants in China based on statistical methods. *Water (Switzerland)* 10(6), 777. https://doi.org/10.3390/w10060777

4 Wastewater Treatment System

CHAPTER OBJECTIVES

The chapter briefly explains centralized and decentralized wastewater treatment systems and their advantages and disadvantages. The chapter also discusses the steps involved in the estimation of sewage discharge, the design of the sewerage system, and different sewer appurtenances.

4.1 INTRODUCTION

The complex characteristics of wastewater and the stringent discharge and reuse standards have necessitated the development of efficient wastewater treatment systems. In order to efficiently treat the wastewater, it needs to be properly collected, transported, and brought to the treatment plant where it is to be treated. Hence, the conveyance of wastewater from the point of generation to the treatment plays a key role in wastewater management. This wastewater transport is carried out by a network of pipelines known as sewers. A sewerage system is a network of sewers, manholes, lifting stations, pumping stations, and other appurtenances involved in carrying the wastewater from the point of generation to the treatment facility.

The wastewater treatment can be carried out using centralized or decentralized treatment systems. A schematic of a centralized and decentralized treatment system is provided in Figure 4.1. Centralized treatment systems are more suitable for a large population, while decentralized treatment systems are suitable only for a small community. However, the level of treatment achieved in the decentralized treatment system is much better than in the centralized treatment system (Gupta et al., 2022). In centralized treatment systems, the sewage is collected and transported to the treatment facility via the sewerage system. However, while conveyance of the wastewater sewers may face some specific problems, including the clogging of the pipelines through sediment deposition, leaching of the wastewater through cracks and joints, corrosion of pipe material due to the presence of dissolved gases, and others. Hence, designing the wastewater conveyance system or the sewer system for the centralized system is imperative (Bradley et al., 2002; Suhak and Chimyshenko, 2021). On the other hand, decentralized treatment systems have a treatment facility

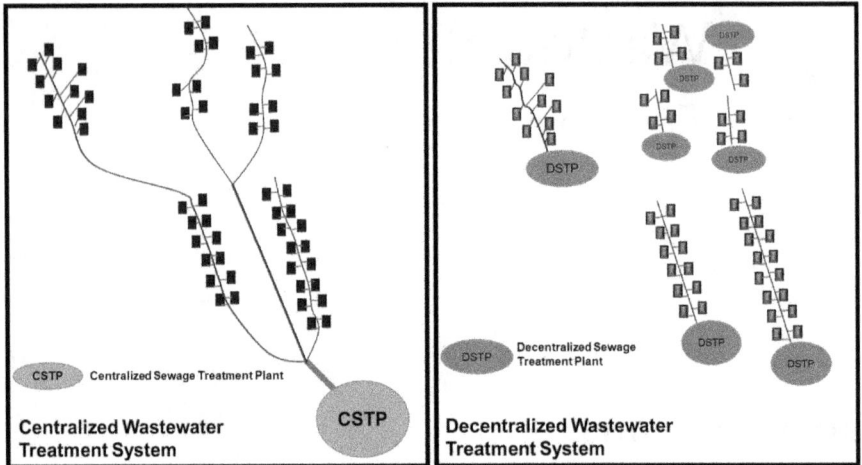

FIGURE 4.1 Schematic showing a centralized and decentralized wastewater treatment system.

TABLE 4.1
Salient Features of Centralized and Decentralized Wastewater Treatment Systems

Centralized Treatment System	Decentralized Treatment System
• Difficult to fabricate and install	• Easy to fabricate and install
• Can serve a large population	• Can serve a small community
• High land requirement	• Suitable for places where land availability is low
• Requires extensive sewer for transport of wastewater from the point of origin to the treatment plant	• An extensive sewer network is not necessary
• Treatment efficiency is not good when there is a variation in wastewater flux and quality	• They are flexible systems and can be tuned according to the wastewater flux and quality
• Very low scope for upgradation of technology	• The technology can be upgraded based on the desired level of treatment
• High operation and maintenance cost	• Low operation and maintenance cost
• Low initial cost (INR/MLD capacity)	• High initial cost (INR/MLD capacity)

for individual multi-storeyed buildings or a cluster of houses, offices, and shopping centers. In decentralized systems, the amount of piping or sewers is much less. Also, the sewage treatment plant (STP) for the decentralized treatment system is smaller since it has to deal with much lower hydraulic loads. The following section describes the salient features of the centralized and decentralized treatment systems, and Table 4.1 highlights the pros and cons of these systems.

4.2 TYPES OF TREATMENT SYSTEMS

4.2.1 CENTRALIZED WASTEWATER SYSTEMS

The public sector usually owns centralized wastewater systems. They collect the wastewater generated by a particular community and transport it to a common sewage treatment plant (Eggimann et al., 2018). In order to collect and transport the wastewater to the treatment plant, the centralized wastewater system may need several pumping stations and extensive sewer networks. As a result, the initial capital investment is high. Furthermore, the STPs and the extensive sewer networks require regular maintenance, increasing the overall expenditure (Jung et al., 2018; Kumar and Tortajada, 2020). The lack of land available that is appropriate for the installation of treatment facilities near the communities is another major issue in developing countries (Massoud et al., 2009; Wilderer and Schreff, 2000). Hence, the STPs are being forced to be constructed at a distance far away from the source of the wastewater generation. This necessitates an extensive sewer network, thereby raising the cost of the wastewater treatment project (Massoud et al., 2009). Apart from the high capital cost associated with the sewerage system in centralized wastewater treatment systems, regular sewer maintenance is also essential. Improper management of these systems may lead to wastewater overflow, endangering the health of both humans and animals in the neighborhood (Carroll et al., 2006). However, the capital cost for the STPs in the centralized system is lower as compared to the decentralized treatment units. The average capital costs for an activated sludge process, moving bed biofilm reactor, sequencing batch reactor, upflow anaerobic sludge blanket, and waste stabilization pond along with tertiary treatment are in the range of INR 6–11.5 million per million liter per day (MLD) capacity. Only membrane-based centralized STPs have a high capital cost of around INR 30 million per MLD capacity (Kumar and Tortajada, 2020).

However, the operation and maintenance costs of the STPs are significantly high. Moving bed biofilm reactors requires about INR 37.2 million/MLD, while the activated sludge process requires about INR 35.3 million/MLD (Kumar and Tortajada, 2020). In developing countries, a number of large-scale STPs have become non-functional or do not give the desired level of treatment because of the high cost incurred in operation and maintenance. (Bassan et al., 2015; Kumar and Tortajada, 2020).

4.2.2 DECENTRALIZED WASTEWATER SYSTEMS

The private sector often owns and operates decentralized STPs. The amount of wastewater handled in these systems is different from centralized systems. They often manage wastewater produced by a small community, as well as from commercial, institutional, and recreational facilities. Hence, decentralized wastewater treatment systems have a smaller sewage network as compared to centralized wastewater treatment systems (Capodaglio et al., 2017; Eggimann et al., 2018). As a result, the cost of wastewater collection, transportation, and treatment is significantly lower. Jung et al. (2018) compared centralized and decentralized wastewater systems, and it was observed that the cost of operation and maintenance of decentralized systems were far lower than those of the centralized systems. Additionally, they can be developed according to the type and volume of wastewater generated by a particular community (Capodaglio et al., 2017; Eggimann et al., 2018). As a result, the treatment efficiency of these systems is much better, and effluent reuse is a common feature of many decentralized systems (Boavida et al., 2016; Lienhoop et al., 2014). The two most commonly used decentralized primary treatment solutions are Imhoff tanks and on-site septic tanks. These systems are affordable, simple to use, and straightforward to maintain. However, these treatment systems are not very efficient and only provide preliminary treatment.

Among the decentralized systems, secondary treatment methods, such as aerobic lagoons, anaerobic lagoons, constructed wetlands, upflow anaerobic sludge blanket, and membrane bioreactors have proven to be efficient (Kumar and Tortajada, 2020; Leong et al., 2017). Decentralized STPs often entail greater capital costs (INR/MLD capacity) than centralized STPs (Jung et al., 2018; Kumar and Tortajada, 2020; Stoklosa et al., 2017). The capital cost for centralized STPs was reported by Kumar and Tortajada in 2020. Decentralized treatment units cost around INR 35 to 70 million/MLD (Kumar and Tortajada, 2020). The higher cost associated with the decentralized STPs is because of the use of advanced technology. Furthermore, they treat a smaller amount of water as compared to the centralized STPs. Hence, the cost per MLD of wastewater treated in decentralized STPs becomes higher than STPs. The high costs associated with decentralized STPs have led to a poor opinion of these systems. All these issues have severely restricted the development of decentralized STPs in developing countries (Eggimann et al., 2018). As a result, in developing countries, a centralized wastewater treatment system is preferred. In this context, the following sections discuss the different components of the sewerage system, which is the backbone of centralized wastewater treatment.

4.3   SEWERAGE SYSTEM

Sewage usually contains around 99% water. The combination of sewers used to collect and carry produced wastewater from the community for treatment and disposal is called the sewerage system. The design of the sewerage system is dependent on the quantity of wastewater it carries and the topography. As a result, the various factors affecting the discharges need to be considered (Bogusz et al., 2020; CPHEEO, 2013; Mackenzie, 2010). The various factors that affect the generation of sewage are as follows:

- *Climate conditions*: The amount of water used for domestic purposes in a place with a hot climate is significantly higher compared to places with a cold climate.
- *Standard of living*: In the previous chapter, it was observed that the amount of domestic wastewater generated by high-income countries was more than that in the countries with lower average income. Hence, a better standard of living leads to more wastewater generation.
- *Community size*: The amount of wastewater generated from a particular community is directly proportional to the population of the community.

- *Price of water*: At places where water pricing or rules and regulations regarding water utilization are implemented, the amount of water wastage is reduced. Hence, wastewater generation is also low.

The sewerage system may have different types of sewers based on the amount of wastewater generated and the kind of community being served. The different types of sewers in a typical sanitary collection system are depicted in Figure 4.2. The capacity of the sewers keeps on increasing as they go further away from the source of wastewater generation. The capacity of the house sewer is lower than that of the lateral sewer. This is because the lateral sewer is joined by multiple house sewers. Similarly, as the sewer goes further away from the source, they are joined by multiple other smaller sewers. Hence, the capacity of the sewer increases as it nears the STP (Bogusz et al., 2020; CPHEEO, 2013; Mackenzie, 2010).

House sewers: A house sewer is a collection system that discharges wastewater from a house or building to the lateral sewer.

Lateral sewers: They form the first component of a wastewater collection system. They are generally laid in the streets and are used to collect wastewater from one or more house sewers and convey it to the branch or main sewer.

Branch or sub-main sewers: They are an integral element of the domestic sewerage system, which collects wastewater from lateral sewers and conveys it to a main or trunk sewer. This separation of sewer lines facilitates repairing work in case of blockages.

Main or trunk sewers: They are large sewers used to collect and convey wastewater from two or more branch sewers to large intercepting sewers. They are normally large enough to handle the load of the community they serve.

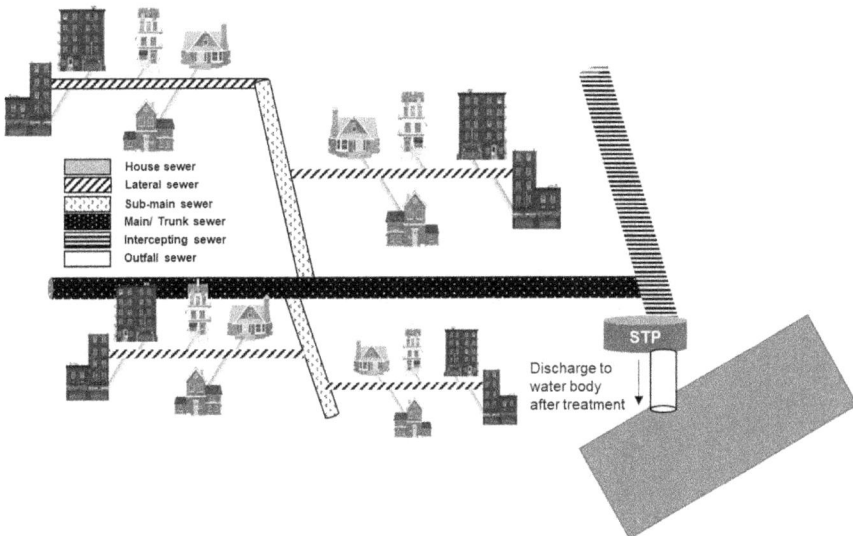

FIGURE 4.2 Various types of conduits in a typical sewage collection system.

Intercepting sewers: They are large conduits that intercept a number of main or trunk sewers and convey the wastewater to water reclamation plants or other disposal facilities.

Outfall sewers: These are the conduits used to carry the treated wastewater from the STPs to the final point of discharge or any water body.

4.3.1 TYPES OF SEWERAGE SYSTEM

4.3.1.1 Separate Sewerage System

The separate sewerage employs two separate sets of sewers. One is for the collection and conveyance of dry weather flow (sanitary sewage) to the treatment plant, and the other is for wet weather flow (stormwater) to be disposed of in various surface water matrices without any treatment. They are suitable for those locations which receive high-intensity rains (Butler et al., 2018). The schematic of the separate sewerage system has been presented in Figure 4.3.

In recent times, the separate system has been the most commonly adopted sewerage system due to its various advantages. However, there are certain limitations to this system as well. The various advantages and disadvantages of the separate sewerage system have been provided in Table 4.2.

4.3.1.2 Combined Sewerage System

The combined sewerage system employs only one sewer for the collection and conveyance of both sanitary sewage and stormwater to the treatment plant (Butler et al., 2018). When the stormwater exceeds its specified limit, the excess is diverted into natural water courses. This kind of system is suitable for locations that receive average and uniform rainfall throughout the year, as the stormwater will provide the self-cleansing velocity. Also, in congested places where there is not enough space for

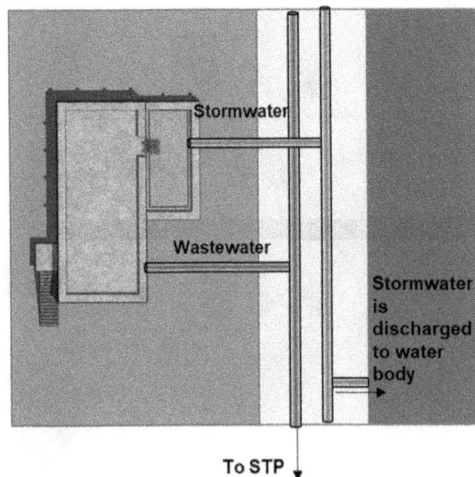

FIGURE 4.3 Separate sewerage system.

TABLE 4.2
Advantages and Disadvantages of Separate and Combined Sewerage System

Advantages	Disadvantages
Separate Sewerage System	
• Sewage runs through separate pipes. Therefore, there is less sewage to treat at sewage treatment facilities, which leads to cost-effective treatment.	• In some areas of sewers, self-cleaning velocity may not have developed, necessitating sewer flushing.
• Since only sanitary sewage is delivered through closed conduits and stormwater can be collected and moved through open drains, this system may be less expensive.	• Two sets of pipes must be laid for this system, which could be challenging in a crowded region.
• This system is cost-effective since it uses less flow when pumping is necessary during disposal.	• Due to the need to maintain two sets of pipelines, this system has a higher maintenance cost.
Combined Sewerage System	
• There is no need to flush sewers in a location where rainfall occurs throughout the year since self-cleansing velocity will develop owing to increased quantity from the addition of stormwater.	• Unsuitable for regions with little annual rainfall since the flow during dry season will be little, and the sewers may not develop self-cleansing velocity, leading to silting.
• Plumbing in a house will only need one set of pipes.	• STPs must handle large flows of sewage before disposing of them, which raises the construction and operating costs of the facility.
• It is simpler to install one pipe in a crowded place than two, as required by other systems.	• This technology is not cost-effective when pumping is necessary and sewer overflows during rainy weather will compromise public hygiene.

laying out a separate sewerage system, the combined sewerage system is very useful since it requires less space (Butler et al., 2018). The schematic of the combined sewerage system has been presented in Figure 4.4.

The combined sewerage system is designed in such a way that the excess wastewater during the rainy season would overflow and reach a natural water body while the remaining portion is sent to the STP. On the other hand, during dry weather or normal conditions, there is no overflow, and all of the wastewater generated reaches the STP. The combined sewerage system also comes with its advantages and disadvantages. The various advantages and disadvantages of a combined sewerage system are presented in Table 4.2.

4.3.1.3 Partial Separate Sewerage System

Apart from the separate and combined, there is another type of sewerage system called the partial sewerage system. The partial separate system is a type of sewerage system that combines the characteristics of a separate and combined sewer system. The system employs two separate sewers, one only for stormwater collection and

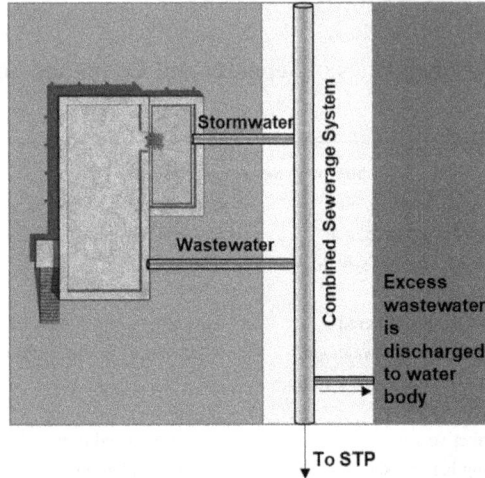

FIGURE 4.4 Combined sewerage system.

conveyance, and the other occasionally accommodate stormwater from roofs and paved portions of buildings along with sanitary sewage. These systems are suitable for those locations which receive uniform rains with occasionally heavy rainfall. However, a major drawback of this system is that self-cleansing velocity may not be achieved in dry weather due to less flow. Similarly, there may be a possibility of overflow during the rainy season if the system is not properly designed (Butler and Davies, 2004).

4.3.1.4 Types of Sewerage Systems Based on Sewage Driving Force

The sewerage system may be further classified into a gravity sewerage system or pressure sewerage system based on the driving force behind the conveyance of the wastewater from its source to the STP. The gravity sewerage system is the most common sewerage system, where the conduits are placed at a specified downward gradient to make sure the self-cleansing velocity. The system conveys wastewater primarily by gravity, and it should be designed to maintain adequate flow toward the discharge point without surcharging manholes or pressuring the system. The conveyance capacity compensates for groundwater percolation and inevitable inflow must be made in the case of gravity sewerage systems. This kind of sewer system largely depends on the topography of the area (Babbitt, 2020; Mackenzie, 2010).

Pressure sewerage system utilizes a network of pumps instead of gravity to convey wastewater through small diameter pipes that are placed in shallow trenches to a treatment system. They consist of a network of fully enclosed pipes that are serviced by pumping units at each connected community. They are commonly installed in such areas where either the landscape is hilly or very flat, or have high water tables, or where other types of sewerage systems are not feasible (Babbitt, 2020; Parcher, 1997).

4.3.2 Sewer Materials

There are different kinds of sewer materials are available, including brick, stoneware, cement concrete, asbestos cement (AC), cast iron (CI), corrugated iron, ductile iron (DI), steel, fiber-reinforced polymer (FRP), glass reinforced plastic (GRP), and polyvinyl chloride (PVC). Each of the sewer materials has certain advantages and drawbacks and is suitable for different scenarios. Hence, the selection of the sewer material should be carried out by taking into several considerations, which are as follows:

Cost of material: In centralized wastewater treatment systems, the treatment plant is often located far away from the source of wastewater generation. Hence, the cost of the sewer material should be low, such that the entire sewerage scheme is economical.

Weight of material: It is preferable for sewer materials to be lightweight since a significant amount of energy is lost in laying and transporting heavy pipes.

Strength and durability: The sewer pipes are subjected to a considerable amount of loads from the backfill material and traffic load. There may be an internal load from the water pressure. Hence, the sewer material should have high strength to overcome such loads. Also, they should be durable enough to overcome the prolonged natural weathering process.

Imperviousness: The wastewater may seep through the walls of the sewer and further contaminate the surrounding soil and, eventually, the groundwater. In order to prevent this, the sewer material should be impervious.

Abrasion resistance: Wastewater contains a number of suspended solids, sand, and grit. When the wastewater flows at high speed, these solids may strike against the walls of the sewer and cause internal wear and tear of the pipe. The thickness of the pipe may gradually get reduced. Furthermore, the pipes may lose their hydraulic efficiency if the inner wall of the pipe becomes rough after continuous abrasion.

Resistance to corrosion: During the transportation of wastewater, various gases are evolved. One such gas is H_2S, which can come in contact with the moisture present in the pipes to form H_2SO_4. The formed acid can corrode the sewer pipes. Hence, the sewer material should be corrosion-resistant.

Hydraulically efficient: The inner wall of the sewer should be smooth so that there is a low frictional coefficient.

Different sewer materials have different properties. Some sewers may be corrosion-resistant but are not light in weight. On the other hand, some sewers may have low weight but are not durable enough. Sewer materials are therefore chosen in accordance with the characteristics of wastewater to be conveyed, the distance to be traveled, and the topography of the place. The different types of available sewer materials are discussed in the following sections.

4.3.2.1 Brick Sewers

The earliest form of sewers were made of bricks. These sewers are particularly larger in diameter. The advantage of brick sewers is that they can be constructed to any required shape and size. However, brick sewers are not favored at present because they have a comparatively higher cost than few other materials, they often get deformed, and leakage may occur. Furthermore, they require large spaces and require intensive labor for construction. Sewer bricks must be complete, brand-new bricks of the highest quality, of a consistent standard size, with straight and parallel edges and square corners (Babbitt, 2020; Parcher, 1997). They must also be free from harmful cracks and faults, sturdy and tough, and have a distinct ring when struck together. All bricks must have parallel, right-angled planar surfaces on their sides, ends, and faces. The size of the bricks used for any sewer should be as uniform as possible because the strength of the sewer is diminished by the additional mortar filling required to make up for undersized bricks (Babbitt, 2020; Parcher, 1997). Small bricks are not preferred since they cost the same to lay as large bricks but result in a thinner completed sewer. With the exception of a particular paving brick designed to stop erosion at the invert, sewer brick should not absorb more than 10–20 % moisture by volume in a 24 h period. When vitrified bricks are needed for the sewer invert or bottom of the sewer, they must be smooth, hard, durable, and of a quality that makes them suitable for the job. They must be the same size, properly and consistently burned, fully vitrified, and free of warps, fissures, and other flaws. The surfaces, edges, and corners must all be true, straight, and square (Babbitt, 2020; Parcher, 1997).

4.3.2.2 Concrete Sewers

Concrete sewers are constructed using cement concrete with or without reinforcement. They are heavy and strong and are available in different pipe sizes that can be precast or cast in situ. However, various disadvantages, such as crown corrosion by H_2S gas, mid-depth water line corrosion by sulfate, and external wall corrosion by sulfate in groundwater, limit its application. Precast concrete sewers are generally employed as branch or main sewers. They are mainly suitable for small stormwater sewers. Pressure pipes and non-pressure pipes are the two forms of precast concrete pipes (Babbitt, 2020; Goyns and Krüger, 2009).. For gravity sewers, non-pressure pipes are often utilized, while pressure pipes are used for inverted siphons and force mains. Inverted siphons are employed to transport sewage or stormwater under streams, highway cuts, or other depressions in the ground. Unlike typical sanitary or storm sewers with open channels for gravity flow, an inverted siphon has liquid that completely fills the pipe and runs under pressure. Sewer force mains are employed at places where gravity is insufficient to transport sewage or stormwater runoff down a sewer line. At lifting stations, force mains with pumps are installed to assist in moving wastewater from a lower elevation to a higher level. They are used at treatment facilities are located at an elevation that necessitates pumping to move sewage.

Cast in-situ reinforced concrete sewers are large sewers constructed at the site with cement concrete reinforcement. They are mainly suitable for combined sewer and stormwater sewer. The main advantage of these sewers is that they are cost-effective when non-standard sections or a special shape is required, and the headroom and working area are restricted (Babbitt, 2020; Goyns and Krüger, 2009).

4.3.2.3   Stoneware or Vitrified Clay Sewers

The stoneware sewers are made from vitrified clay and shales by molding and burning. These pipes are available in various sizes. These sewers are highly resistant to corrosion caused by most acids, as well as erosion due to grits and high flow rates. They have a very smooth interior and are hydraulically efficient. These pipes are commonly used for laterals. However, skilled labor is required to seal the joints with cement mortar-soaked yarn and pack the bell and spigot joints. Also, these sewers are brittle in nature. Hence, they are likely to get damaged during transport and handling operations (Babbitt, 2020; Parcher, 1997).

When making clay pipes, the dried clay from the excavation is brought to a mill and processed as finely as possible. Afterwards, it is placed in storage containers, from which it is later removed for wet grinding and tempering. In this procedure, the right amount of water is added to the clay to make it plastic. Failure will result from a moisture content fluctuation of between 1% and 12%. A combination that is too wet won't have enough strength to keep its shape in the kiln. A combination that is too dry will exhibit laminations when squeezed through the discs. The pipes are then brought to a steam-heated drying room after being bent, where a consistent temperature is maintained to avoid the splitting of pipes. They spend three to ten days drying in the room before being transferred to the kilns. When placed in the kilns when wet, blisters will form. In order to distribute heat and weight as equally as possible, the dry pipes are carefully arranged inside the kiln before the fire is lit (Babbitt, 2020).

4.3.2.4   Asbestos Cement Sewers

Asbestos cement sewers are made from a pressure-combined mixture of asbestos fibers and cement. They are light in weight, easy to handle, cut, threaded, drilled, and fitted easily, and have good flow characteristics. These pipes have plain ends and are linked together using a unique joint or coupling known as a slip-type sleeve joint. Slip-type expansion joints are best suited for lines with significant axial (straight-line) motions. They are unable to tolerate lateral offset since these conditions would lead to packing distortion, binding, galling, and possibly leaking. Therefore, suitable pipe guides and slide expansion joints are typically employed to ensure alignment. Asbestos cement sewers offer enough resistance against corrosion caused by most natural soils, acids, salts, and other corrosive materials. However, they cannot resist corrosion caused by highly septic sewage and highly acidic or high sulfate soils. Also, these sewers are brittle, thus cannot withstand heavy external stresses, and can be easily broken (Babbitt, 2020; Radlinski and Wolf, 2016).

4.3.2.5   Iron and Steel Sewers

Cast iron (CI) pipes have cement mortar as an inner coating, whereas the outer coating is of coal tar. These pipes are longer, have tighter connections, and, when appropriately designed, can withstand relatively high internal and external pressures and resist corrosion against a wide range of natural soils. However, CI pipes are costlier than other sewer materials and cannot resist corrosion against highly septic sewage and acidic soils (Babbitt, 2020; CPHEEO, 2013).

Ductile iron (DI) pipes are made by mixing magnesium into low-sulfur molten iron. DI pipes are resistant to impact, wear and tear, and corrosion. They also have high tensile strength and can handle water hammers. Additionally, they are 30% lighter in weight as compared to CI pipes. Internal and external protection systems, on the other hand, should be provided for prolonged longevity (CPHEEO, 2013).

Steel sewers are made up of steel pipes manufactured by rolling mild steel plates to a certain diameter by riveting or welding the joints. Welded pipes are most commonly used as steel sewers as they are smoother and stronger than riveted pipes. They are typically preferred for main sewers, underwater river crossings, bridge crossings, pumping station connections, self-supporting spans, railway crossings, and penstocks. Steel pipes are more resistant to internal pressure, impact load, and vibrations. Also, they are more ductile and can endure higher water hammer pressures. The major drawback of steel pipes is that they cannot resist high external stresses and are susceptible to collapse when subjected to negative pressure. (CPHEEO, 2013).

4.3.2.6   Plastic Sewers

Unplasticized polyvinylchloride (UPVC) pipes are resistant to corrosion, tough, rigid, less weight, and easy to fabricate. UPVC pipes are easier to lay and maintain. They are mainly used for internal drainage purposes in buildings. However, UPVC pipes are rigid and hence cannot be laid on uneven terrains (CPHEEO, 2013).

High-density polyethylene (HDPE) pipes are used for sewers because the inner side has a smooth surface. As a result, they are resistant to corrosion from the suspended particles in the sewage. This is the reason why they can be used for a prolonged duration. The joints are usually fusion welded, or flange joined depending on the straight runs or fittings. They have all of the benefits of UPVC pipes but are less rigid. Due to their flexibility, they are best suited for laying on hilly or uneven terrains (CPHEEO, 2013).

Glass fiber-reinforced plastic (GRP) pipes are widely used in places where corrosion-resistant pipes are required at a low cost. GRP can be used as a corrosion-resistant lining material for traditional pipes. The glass fiber can withstand external and internal corrosion. On the other hand, fiber glass reinforced plastic (FRP) is made of a mixture of polyester resin and glass fiber. FRP pipes have high strength and durability. They are also lightweight because of the low density of the material. Furthermore, they possess high tensile strength. FRP pipes are also resistant to corrosion (CPHEEO, 2013).

4.3.3   SHAPE OF THE SEWER

The sewers may be of varying shapes, but usually, for sewage discharge, a circular sewer is preferred. As compared to other shapes, the perimeter of the circular section sewer is the least. Also, in the circular section, the inner walls are more uniform, which would facilitate the easy flow of wastewater and prevent the deposition of suspended particles. Furthermore, circular sewers are easier to construct. Hence, circular section sewers are preferred over other sewers, such as egg-shaped sewers, horseshoe-shaped sewers, parabolic-shaped sewers, U-shaped sewers, semi-circular-shaped sewers, basket handled-shaped sewers, rectangular-shaped sewers, and semi-elliptical shaped sewer (Ghangrekar, 2022). The characteristics of the different shapes of sewers are provided in Table 4.3.

The shape of sewer and sewer material are only the preliminary stages of a sewerage network. However, proper designing and planning is necessary for the sewerage system to work efficiently over a long period of time. If the discharge estimated is more than the actual, then the pipe sizes will be inadequately more, thereby increasing the project cost and making the system hydraulically inefficient. Other factors, such as leakage and unaccounted overflow, must be considered while the total discharge is estimated. Additionally, the population increase over the design period of the sewer network also needs to be considered while evaluating the discharge. The

TABLE 4.3
Characteristics of Different Shapes of Sewers

Shapes of Sewer	Salient Features
Circular-shaped sewer	They are ideal for carrying sewage with varying characteristics and flow. They are simple to build, requires fewer building resources, and is cost-effective.
Egg-shaped sewer	A typical egg-shaped sewer provides a higher velocity than circular-shaped sewage. It is hence appropriate for low-flow situations. Such sewers are challenging to build and demand a substantial amount of building supplies. Thus, it is expensive. For combined sewer, a standard egg-shaped sewer is preferred.
Horseshoe-shaped sewer	Large sewers in the shape of a horseshoe are frequently used for locations where substantial discharge is necessary. The inversion section could be flat, parabolic, or circular. Such sewers have a greater height than width. Due to the size of the sewer, maintenance is also quite simple.
Parabolic-shaped sewers	Smaller sewage discharges are handled with parabolic-shaped sewers. This sewer is quite small, and the top border is shaped like a parabola. The geometry of invert of the sewer portion could be parabolic or elliptical. It is also cost-effective.
U-shaped sewer	Heavy sewage discharge takes place through a U-shaped sewer. Due to its size, it is also employed for the transportation of stormwater. Such sewer types have an invert part that is semi-circular in shape. Additionally, it serves as a combined sewer. The large size of the sewer facilitates its maintenance tasks.
Semi-circular-shaped sewer	The area where a large sewer is required is suitable for a semi-circular-shaped sewer. The upper arch of the sewer is shaped like a semicircle. It is used for massive sewage discharge and has a big section.
Rectangular-shaped sewer	A huge sewer with a rectangular shape is used mostly to carry a lot of sewage, while it can also be utilized to discharge rainfall. With this kind of sewer, construction and maintenance tasks are quite simple. In comparison to other sections, this one is fairly stable. They have a rectangular shape, and the upper and lower portions are both flat.

different steps in design and planning of a sewerage system have been discussed in the following section.

4.3.4   DESIGN AND PLANNING OF SEWERAGE SYSTEM

4.3.4.1   Estimation of Sewage Discharge

Usually, it is estimated that 80% of the water supplied is discharged as sewage (CPHEEO, 2013). Apart from that, the water used from private wells or tube wells also contribute to the amount of sewage generation. This discharge quantity, due to the unaccounted private water supplies, should be added to the total sewage discharge. Furthermore, groundwater seepage into the sewers can occur if the sewers have faulty joints or cracks. This is only possible when groundwater levels are higher than the invert level of sewer line. If it is below, exfiltration can happen. This excess inflow due to infiltration should also be added while estimating total sewage discharge. In contrast, the water loss through leakages in the pipe needs to be subtracted while estimating sewage discharge. Additionally, there are certain activities that consume water but do not produce sewage, such as water for boiler feed, water for washing roads or lawns, and others. This factor is also required to be subtracted to estimate the sewage discharge amount more accurately. Therefore, the total amount of sewage generated can be represented by Equation 4.1 (Garg, 2012).

Net quantity of sewage = Sewage due to accounted water supplied from water works + Sewage due to unaccounted water supplied from private supplies + Infliltration – sewage loss through leaks – Sewage not entering the system

$$(4.1)$$

4.3.4.2   Variation in Sewage Flow

In order to efficiently design the sewerage system, the variation in sewage flow should also be taken into account. The amount of flow in the sewer can be varied with season, month, or even daily. The peak flow in the sewer systems may be different at different points in the sewer. Near the origin of the sewer network, the peak flow will be pronounced. However, the peak will be delayed as the distance increases because of the time accounted for collecting sewage from different points and the time required for traveling. Furthermore, the time at which peak flow will occur also depend upon the length of the sewer and the kind of population served. By taking into consideration all these factors, the maximum design discharge can be estimated using Equations 4.2 and 4.3 (Qasim and Zhu, 2017).

$$\log (M_{max}) = -0.19 \log (P') + 0.74 \qquad (4.2)$$

$$\text{Hourly maximum flow} = M_{max} \times \text{daily average wastewater flow} \qquad (4.3)$$

where M_{max} is the ratio of hourly maximum to daily average flows and P' is the population in thousands, provided that the value of P' is less than 10,000.

On the other hand, in order to prevent silting or settling of solids in the sewer lines, the minimum flow should provide the self-cleansing velocity. Hence, the minimum flow criteria also need to be adhered. The minimum design discharge can be estimated using Equations 4.4. and 4.5 (Qasim and Zhu, 2017).

$$\log (M_{min}) = 0.142 \log (P') - 0.682 \tag{4.4}$$

$$\text{Hourly minimum flow} = M_{min} \times \text{daily average wastewater flow} \tag{4.5}$$

where M_{max} is the ratio of hourly maximum to daily average flows and P' is the population in thousands, provided that the value of P' is less than 10,000.

4.3.4.3 Population Forecasting

In the design of a sewerage system, the amount of sewage generated is an essential parameter that needs to be considered. The sewerage network design not only considers the present population but also future density of population of a particular area. The time or the number of years into the future, the capacity of the sewer adequate to serve the targeted community is called the design period. Usually, the design period of sewerage networks are around 30 years (CPHEEO, 2013). Hence, the design should also take into consideration the increase in sewage generation with the increase in population over the design period. The different ways to estimate the population are listed below (CPHEEO, 2013).

Arithmetic Increase Method
The population of a city or community after n decades is given by Equation 4.6 (Sincero and Sincero, 2003).

$$P_n = P + nC \tag{4.6}$$

where P_n is the population after n decades, P is the present population, and C is the average of the rate of change in population over the past few decades and is a constant. This method is suitable for established and large cities. This is because the rate of growth in old cities usually become constant over time.

Geometrical Increase Method
In this method, the increase in percentage of the population is assumed to be constant. The population of a city or community after "n" decades is given by Equation 4.7 (Sincero and Sincero, 2003).

$$P_n = P\left(1+\frac{I_G}{100}\right)^n \tag{4.7}$$

where P_n is the population after n decades, P is the present population, and I_G is the geometric mean (%). This method usually overestimates the predicted value. Hence, it is suitable for newly developed towns.

Incremental Increase Method

The population of a city or community after n decades is given by Equation 4.8 (Sincero and Sincero, 2003).

$$P_n = P + nC + \left[\frac{n(n+1)}{2}\right]D \qquad (4.8)$$

where P_n is the population after n decades, P is the present population, C is the average of the rate of change in population, and D is the average incremental increase. This method is suitable for towns or cities of average sizes where the rate of growth is of an increasing order.

Example 1: Estimate the population of a town for the year 2050 using arithmetic increment method, geometrical increase method, and incremental increment method.

Year	Population (P)
1990	85000
2000	105000
2010	130000
2020	170000

Solution

Year	Population (P)	Increment (I)	Geometrical Increase Rate of Growth (R)	Incremental Increase (II)
1990	85000			
2000	105000	20000	23.5	
2010	130000	25000	23.8	5000
2020	170000	40000	30.7	15000
		C=28333	I_G= 25.8	D=1000

Using arithmetic increment method

C= Average of $(I_{2000}, I_{2010}, I_{2020})$

$I = P_n - P_{n-1}$

Therefore, population for the year 2050 $(P_{2050}) = 170000 + 3 \times 28333 = 255000$

Using geometrical increase method

$I_G = (R_{2000} \times R_{2010} \times R_{2020})^{1/3}$

$R = (I_n/P_{n-1}) \times 100$

Therefore, population for the year 2050 $(P_{2050}) = 170000\left(1 + \frac{25.8}{100}\right)^3 = 338448$

Using incremental increment method

D= Average of $(II_{2000}, II_{2010}, II_{2020})$

$II = I_n - I_{n-1}$

Therefore, population for the year 2050

$$P_{2050} = 170000 + 3 \times 28333 + \left[\frac{3(3+1)}{2}\right] \times 1000 = 261000$$

Logistic Curve Method

This method is applicable when the growth rate of the population is not significantly affected by any unexpected events, such as war, pandemic, earthquake, or others. Using this method, the population after t years can be calculated using Equation 4.9. In Equation 4.9, P_S is the saturation population, a and b are constants. The formula for deriving P_S, a, and b have been provided in Equations 4.10, 4.11, and 4.12, respectively (Sincero and Sincero, 2003). P_0, P_1, and P_2 are the known population at time $= 0$ (t_0), time $= t_1$, and time $= 2t_1 (t_2)$, respectively.

$$P = \frac{P_S}{1 + a \ln^{(-bt)}} \tag{4.9}$$

$$P_S = \frac{2P_0 P_1 P_2 - P_1^2 (P_0 + P_2)}{(P_0 P_2 - P_1^2)} \tag{4.10}$$

$$a = \frac{P_S - P_0}{P_0} \tag{4.11}$$

$$b = \frac{1}{t_1} \ln \frac{P_0 (P_S - P_1)}{P_1 (P_S - P_0)} \tag{4.12}$$

4.3.4.4 Quantity Estimation for Stormwater

In comparison to the amount of sewage generated from the household, the total amount of stormwater that reaches the sewers is much larger during rainy seasons. Hence, the sewers should be designed to tackle this stormwater. The amount of stormwater generated and the amount of it reaching the sewers depend on a number of factors and are as follows:

Area and type of catchment: Catchments which are large and fan-shaped will produce more runoff as compared to small and fern-shaped catchments. Similarly, if the terrain has more slope, the amount of runoff will be more.

Porosity of the soil: Rocky terrain will produce more stormwater as compared to catchments with sandy soil.

Existing soil moisture: If the rainfall occurs during dry season, when the soil is dry, most of the water will be absorbed into the soil. Hence, the amount of stormwater generated will be less.

Vegetative cover: The presence of dense vegetation cover increases the infiltration capacity of the soil, thereby producing less stormwater.

The quantity of stormwater may be estimated by various methods, but the rational method is most prominent (Equation 4.13), where Q is the quantity of stormwater (m³/s), C is the coefficient of runoff, I is the intensity of rainfall (mm/h), and A is the drainage area (hectares) (CPHEEO, 2013).

Rational Method

$$Q = C.I. \frac{A}{360} \qquad (4.13)$$

4.3.5 Hydraulic Design of Sewers

There are a number of factors that need to be taken into consideration while designing the sewerage network. The diameter of the sewers needs to be optimized, such that neither excessive excavation is required, nor is there a possibility of solid deposition. Furthermore, it should be noted to design the sewerage network and the connected treatment plant in such a way so that the wastewater flows under gravity.

The sewerage network is different from water supply network in many ways. Firstly, the wastewater contains a significant amount of organic and inorganic solids. Hence, there are chances that these solids may lead to abrasion of the inner wall of the pipeline. The entire sewerage network should be laid in a continuous falling gradient since the wastewater flows under gravity. Furthermore, if the velocity of the wastewater in the pipes is not higher than the self-cleansing velocity (Equation 4.14), then there may be deposition of solids occurs. The self-cleansing velocity can be calculated using the Equation 4.14. In this equation, n is the Manning's coefficient, R is the hydraulic mean radius (m), K_s is the dimensionless constant [0.04 to 0.08], S_s is the specific gravity of particle, and d_p is the particle size (mm) (CPHEEO, 2013).

$$\text{Self-cleansing velocity} \left(V_s \right) = \frac{1}{n \left[R^{\frac{1}{6}} \sqrt{(K_s \left(S_s - 1 \right) d_p}} \right]} \qquad (4.14)$$

The primary factors that affect the flow of sewage in sewers are as follows:

- Characteristics of sewage
- Conditions of flow
- Cross-sectional area of sewer
- Presence or absence of bends and obstruction
- Gradient of the sewer line
- Roughness of interior surface

The different formulas that are used to calculate the velocity in the sewers are given in Equations 4.15–4.18 (Mackenzie, 2010; Qasim, 2017; Rangwala, 2005; Tchobanoglous et al., 2003).

Manning's formula

$$V = \frac{1}{n} R^{2/3} S^{1/2} \qquad (4.15)$$

where V is the velocity of flow, n is the Manning's coefficient (0.012 to 0.018), R is the hydraulic mean radius (m) (ratio of wetter area to wetted perimeter), and S is the slope or hydraulic gradient (Mackenzie, 2010).

Chezy's formula

$$V = C(\sqrt{R.S})$$ (4.16)

where V is the velocity of flow in the channel (m/s), R is the hydraulic mean radius (m), S is the slope or hydraulic gradient, and C is the Chezy's constant. The value of C can be obtained using Kutter's formula (Equation 4.17):

$$C = \frac{\left[23 + \dfrac{0.00155}{S} + \dfrac{1}{n}\right]}{1 + \left(23 + \dfrac{0.00155}{S}\right)\dfrac{n}{\sqrt{R}}}$$ (4.17)

where n = roughness coefficient or rugosity factor same as Manning's coefficient.

Hazen–Williams equation

$$V = 0.849 \, H \, R^{0.63} S^{0.54}$$ (4.18)

where V is the velocity of flow in the channel (m/s), R is the hydraulic mean radius (m), S is the slope or hydraulic gradient, and H is the Hazen–Williams constant. H depends on the material and the age of the pipe. H has higher values when the pipe is new. However, for design purposes lesser value of H is considered since over the design period, the inner walls of the pipe may become rough, thereby affecting the flow.

While designing the sewer network, it should also be noted that there is a limit to the maximum velocity at which the wastewater should flow. This velocity is called the non-scouring velocity. Scouring occurs throughout in the sewer lines due to the presence of suspended solids. However, at higher flow rates the effect of scouring is magnified. The effect of scouring on the sewer mainly depends on the pipe material. Hence, the maximum or non-scouring velocity of the sewers also depends on the material of the pipe. Brick lined provides lower limiting or non-scouring velocity as compared to concrete sewers. Cast-iron sewers provide the highest non-scouring velocity. Additionally, free board in sewers is necessary because it acts as a safety factor against low average and maximum flow estimates based on erroneous data. It also considers the infiltration of stormwater and the unanticipated increase in sewage generation (CPHEEO, 2013).

Example 2

Design a sewer for the following condition:

Maximum discharge =0.5 m³/s running half full.
Manning's coefficient $n = 0.01$, and gradient = 0.0001.

Solution

$$\text{Hydralulic mean depth or radius } (R) = \frac{\text{Wetted Area}}{\text{Wetted Perimeter}}$$

$$R = \frac{(\pi \times \text{radius}^2)/2}{(2 \times \pi \times \text{radius})/2}$$

$$R = \frac{\text{radius}}{2}$$

$$Q = A \times V = (\pi \times \text{radius}^2)/2 \times V$$

$$0.5 = (\pi \times \text{radius}^2)/2 \times \frac{1}{n} R^{2/3} S^{1/2}$$

Solving, we get radius = 0.77 m.
Hence, the radius of the sewer should be 0.77 m.

4.3.6 SEWER APPURTENANCES

Structures built at appropriate intervals along a sewerage system to aid in efficient operation and maintenance work are known as sewer appurtenances. Manholes, street inlets, clean-outs, catch basins, and storm regulators are the most common sewer appurtenances.

Manholes are built at appropriate intervals along sewer lines to provide access into them for maintenance and cleaning. They have perforated covers that prevent gases from accumulating inside. Manholes are typically installed at every bend, junction, or location where the sewer diameter or gradient changes. However, when the sewer runs straight, manholes are spaced at regular intervals. The distance between two manholes usually increases as the diameter of the manholes increases. However, it usually does not exceed 90 to 180 m (Qasim, 2017). The depth of a manhole can be used to classify it. A detailed figure of a manhole has been provided in Figure 4.5.

Shallow manholes have a depth of 0.7–0.9 m. They are typically installed at the beginning of a branch sewer where there is little traffic. Normal or medium manholes have a depth of about 1.5 m and a heavy cover on top. On the other hand, the deep manholes have a depth of more than 1.5 m and are equipped with a heavy cover at the top (B. C. Punmia et al., 1998). Drop manholes are built to connect the high-level branch sewer to the lower-level main sewer. They are built where a branch sewer enters a manhole that is more than 0.5 m to 0.6 m above the main sewer. The installation of a drop manhole eliminates the need for a steep gradient in the branch sewer

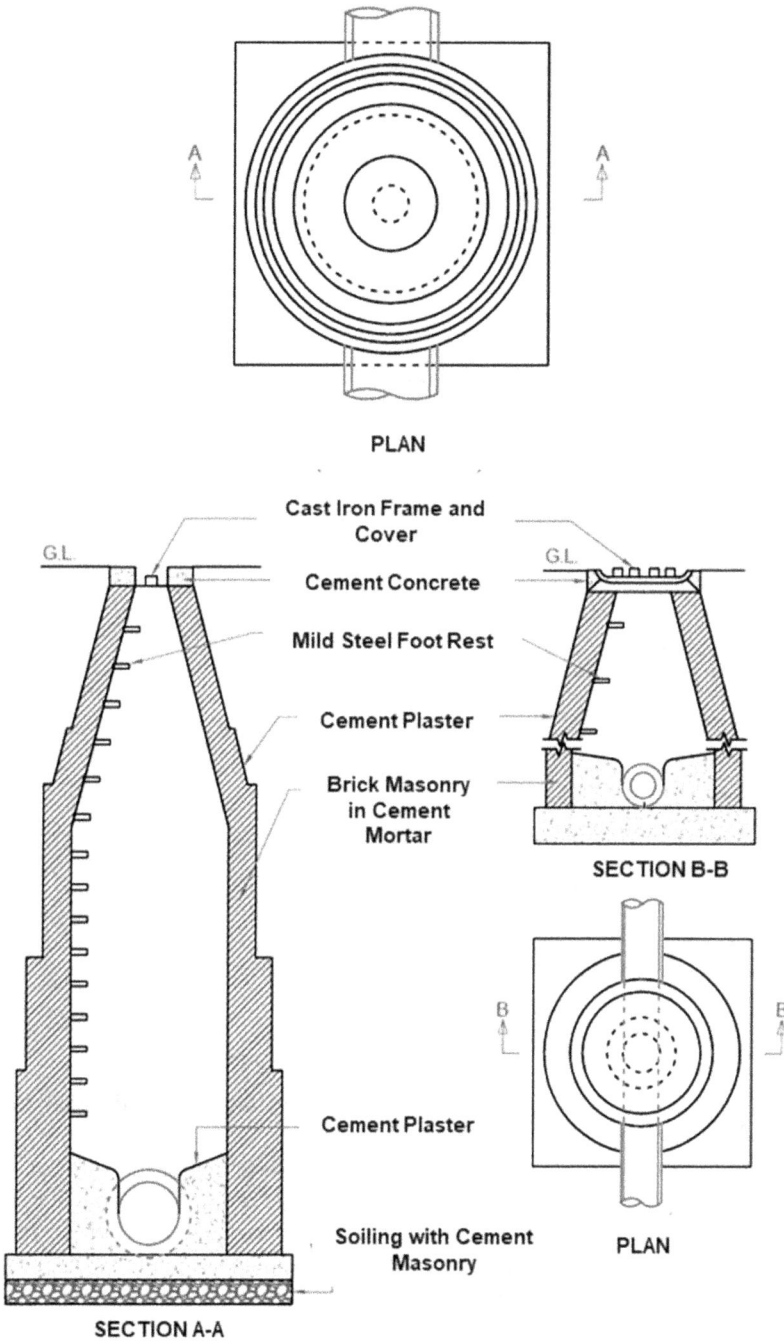

PLAN

Cast Iron Frame and Cover

Cement Concrete

Mild Steel Foot Rest

Cement Plaster

Brick Masonry in Cement Mortar

G.L.

G.L.

SECTION B-B

Cement Plaster

Soiling with Cement Masonry

PLAN

SECTION A-A

FIGURE 4.5 Schematic of a typical manhole.

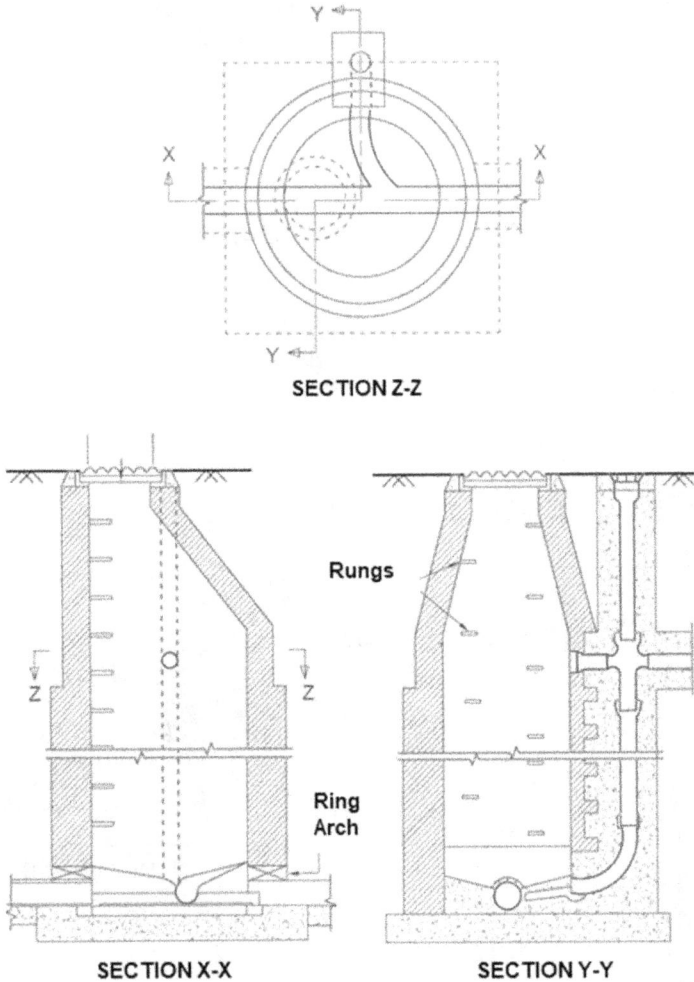

SECTION Z-Z

Rungs

Ring
Arch

SECTION X-X   **SECTION Y-Y**

FIGURE 4.6   Schematic of a typical drop manhole.

and reduces the amount of earth work required. Figure 4.6 shows a detailed illustration of a drop manhole.

Clean-outs are necessary to remove any obstacles in the sewer line. A clean-out is referred to as an inclined pipe connected from the ground level to the underground sewer. During clogging, water is passed through the inclined pipe at high force to clear the obstacles in its path desirable (Babbitt, 2020; Punmia et al., 1998). They are usually provided in place of manholes at the upper end of lateral sewers. A typical clean-out pipe is represented in Figure 4.7a.

Stormwater or water from street washing is sent to the stormwater sewer or combined sewer through a street inlet. A street inlet is typically a concrete box with vertical or horizontal openings. The inlets are positioned along the sides of the roads. The height and length of the entrance, the type of grating, and the position are crucial

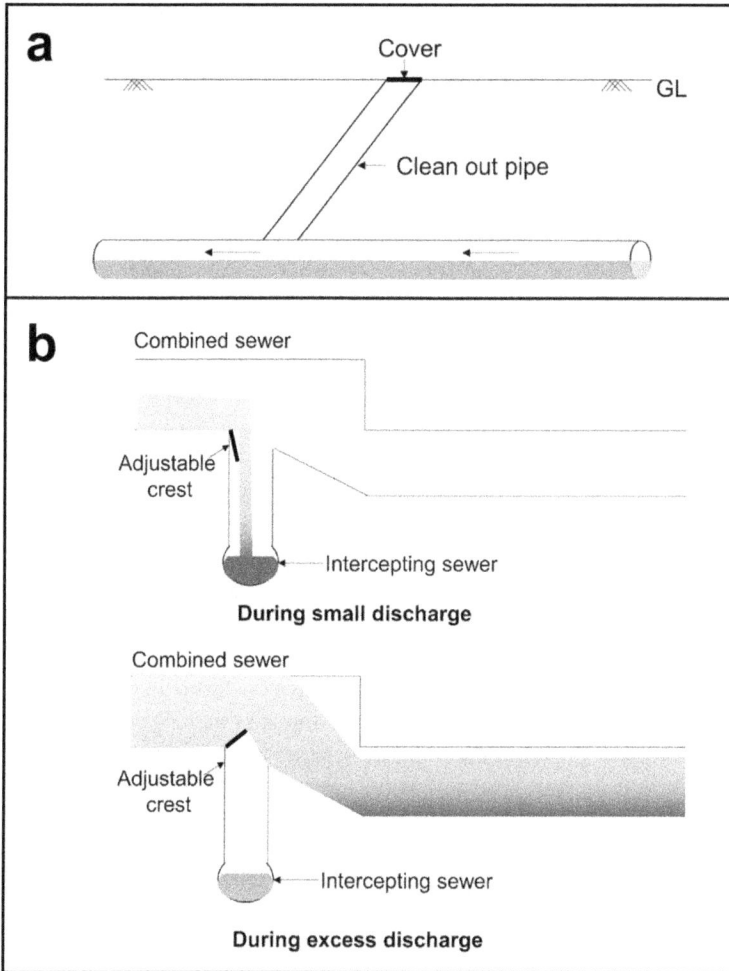

FIGURE 4.7 Schematic of (a) clean-out pipe and (b) leap weir during small discharge and excess discharge.

considerations when designing a street inlet. The typical clear height of an entrance is between 12 and 15 cm, with a clear length of at least 60–75 cm. On paved roadways with moderate slopes and drainage areas of little more than 1,000 m² of pavement, inlets of such size is desirable (Babbitt, 2020; Punmia et al., 1998). Street inlets with a small settling basin are known as catch basins or catch pits. The purpose of catch basins is to keep grit, sand, and debris out of the sewer system. They perform the function of a sedimentation chamber. A hood is also provided to keep foul gases from the sewer line from escaping.

Stormwater regulators are built to allow excess stormwater to be diverted into a nearby stream. The three different types of storm regulators have been described in the following paragraphs. Leaping weir consists of an opening in the invert of the storm drain through which the normal flow is diverted into the intercepting sewer, and the excess flow leaps over the combined flow into the nearby stream. The sewage falls directly into the intercepting sewer through the opening when the discharge is small. On the other hand, if the sewage discharge exceeds a certain threshold, it leaps across the weir and into the stream desirable (Babbitt, 2020; Punmia et al., 1998). A schematic of a leap weir during small discharge and excess discharge has been shown in Figure 4.7b.

In the overflow weir, there are connecting pipes between the combined sewer and the stormwater sewer. These connecting pipes or the channel are usually made in the manholes. During the event of an excess discharge, the combined sewer or the main sewer may get overloaded. The system helps to transfer the excess sewage into the stormwater sewer. Adjustable plates may also be used to prevent the floating materials from escaping from the main sewer into the stormwater sewer (Babbitt, 2020; Ghangrekar, 2022; Punmia et al., 1998). A schematic representation of an overflow pipe has been shown in Figure 4.8a.

Siphon spillway is considered as the most effective arrangement for dealing with stormwater. It is an automatic process and works on the principle of siphon action. The siphonic action starts when the sewage in the combined sewer rises above a fixed level and stops as soon as the sewage falls below this level desirable (Babbitt, 2020; Punmia et al., 1998). A schematic representation of an overflow pipe has been shown in Figure 4.8b.

The sewer appurtenances, along with the sewers, form the sewerage network. They are an integral part of the wastewater treatment system since they deal with the collection of wastewater from the point of generation and transport to the treatment

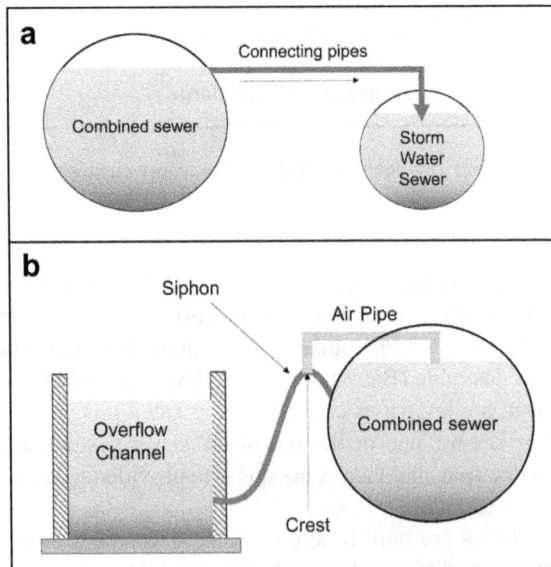

FIGURE 4.8 Schematic of (a) an overflow weir and (b) a siphon spillway.

unit. Furthermore, a proper sewerage network is vital to prevent water clogging in the streets during the rainy season. Adequate sewer systems play important roles in sanitation and disease prevention. Without a proper sewer network, there may be infiltration and leakage of wastewater, leading to contamination of the entire environment. Furthermore, water from unwanted places may also find its way into the sewer network. All these factors significantly affect the quantity and quality of the wastewater reaching the STPs. Variation in the hydraulic loading and organic loading may cause the STP to fail. Hence, designing and maintaining a proper sewerage system is vital for the efficient working of a wastewater treatment system.

4.4   CHAPTER SUMMARY

- The public sector typically owns centralized wastewater systems, which collect and treat a large amount of wastewater from a specific community.
- In order to efficiently collect and transport wastewater to treatment units, the centralized wastewater system necessitates several high-capacity pumps and sewerage networks. As a result, the cost of maintenance is significantly high.
- Decentralized STPs are typically owned and operated by non-professional organizations and have a number of limitations.
- Decentralized STPs typically treat a small amount of water generated by a group of nearby households, as well as commercial, institutional, and recreational facilities. As a result, they have a small sewerage network, and the costs associated with wastewater collection, transportation, and treatment are much lower than in centralized STPs.
- The separate sewerage system employs two separate sewer lines, one for collecting and transporting dry weather flow (sanitary sewage) to the treatment plant and the other for collecting and transporting wet weather flow (stormwater) to be disposed of in various surface water matrices without treatment. They are appropriate for locations that experience heavy rains.
- A combined sewerage system uses a single sewer to collect and transport both sanitary sewage and stormwater to the treatment plant. As the stormwater provides the self-cleaning velocity, this type of system is appropriate for locations that receive average and uniform rainfall throughout the year.
- Steps involved in design of sewerage system are as follows:
  a. *Estimation of design discharge*: This step involves the estimation of the design period, forecasting of the population based on the design period, and the amount of sewage that is likely to be generated based on the kind of infrastructure in the place. The infiltration of groundwater into the sewers and the stormwater runoff is also required to be considered.
  b. *Selecting the type of sewer system*: In this step, the kind of sewerage system, separate or combined, is to be decided. Also, the type of sewer materials is required to be finalized based on the quantity and quality of the sewage.
  c. *Sewer hydraulics*: The shape of the sewer needs to be finalized after which the minimum velocity, maximum velocity, depth of flow, and slope of the sewer are required to be calculated.
  d. Laying of manholes and other sewer appurtenances

4.5 CONCLUDING REMARKS

The sewerage network is an essential component of wastewater management. However, its function is only to convey the sewage from the point of origin to the STP. The treatment of the wastewater is only carried out in the STPs. The different types of treatment involved in a conventional STP are primary treatment, secondary treatment, and tertiary treatment. The different processes involved in each of the treatment step has been discussed in the subsequent chapters.

REFERENCES

Babbitt, H.E., 2020. *Sewerage and Sewage Treatment*. John Wiley & Sons.

Bassan, M., Koné, D., Mbéguéré, M., Holliger, C., Strande, L., 2015. Success and failure assessment methodology for wastewater and faecal sludge treatment projects in low-income countries. *J. Environ. Plan. Manag.* 58, 1690–1710. https://doi.org/10.1080/09640568.2014.943343

Boavida, S., Pinto, M., Salvador, T., 2016. Centralized vs. decentralized wastewater systems – potential of water reuse within a transboundary context. *New Water Policy Pract.* 2, 54–75. https://doi.org/10.18278/nwpp.2.2.6

Bogusz, M., Marzec, M., Malik, A., Jóźwiakowski, K., 2020. The state and needs of the development of water supply and sewerage infrastructure in the Radzyń district. *J. Ecol. Eng.* 21(3), 171–179. https://doi.org/10.12911/22998993/118282

Bradley, B.R., Daigger, G.T., Rubin, R., Tchobanoglous, G., 2002. Evaluation of onsite wastewater treatment technologies using sustainable development criteria. *Clean Technol. Environ. Policy* 4, 87–99. https://doi.org/10.1007/s10098-001-0130-y

Butler, D., Davies, J.W., 2004. *Urban Drainage*, 2nd Edition. Springer.

Butler, D., Digman, C.J., Makropoulos, C., Davies, J.W., 2018. *Urban Drainage*, 4th Edition. Taylor & Francis.

Capodaglio, A.G., Callegari, A., Cecconet, D., Molognoni, D., 2017. Sustainability of decentralized wastewater treatment technologies. *Water Pract. Technol.* 12, 463–477. https://doi.org/10.2166/wpt.2017.055

Carroll, S., Goonetilleke, A., Thomas, E., Hargreaves, M., Frost, R., Dawes, L., 2006. Integrated risk framework for onsite wastewater treatment systems. *Environ. Manage.* 38, 286–303. https://doi.org/10.1007/s00267-005-0280-5

CPHEEO, 2013. *Manual on Sewerage and Sewage Treatment Systems: Part A Engineering*, Ministry of Urban Development, Government of India.

Eggimann, S., Truffer, B., Feldmann, U., Maurer, M., 2018. Screening European market potentials for small modular wastewater treatment systems – an inroad to sustainability transitions in urban water management? *Land Use Policy* 78, 711–725. https://doi.org/10.1016/j.landusepol.2018.07.031

Garg, S., 2012. *Sewage Disposal and Air Pollution Engineering*. Khanna Publishers.

Ghangrekar, M.M., 2022. *Wastewater to Water Principles, Technologies and Engineering Design*. Springer Nature Singapore. https://doi.org/10.1007/978-981-19-4048-4

Goyns, A., Krüger, J., 2009. *Design Manual for Concrete Pipe Outfall Sewers*. Concrete Manufacturers Association.

Gupta, A.K., Majumder, A., Ghosal, P.S., 2022. Introduction to modular wastewater treatment system and its significance. *Modul. Treat. Approach Drink. Water Wastewater* 1, 81–106. https://doi.org/10.1016/B978-0-323-85421-4.00010-3

Jung, Y.T., Narayanan, N.C., Cheng, Y.L., 2018. Cost comparison of centralized and decentralized wastewater management systems using optimization model. *J. Environ. Manage.* 213, 90–97. https://doi.org/10.1016/j.jenvman.2018.01.081

Kumar, M.D., Tortajada, C., 2020. Wastewater treatment technologies and costs, in: *Assessing Wastewater Management in India*. Springer, pp. 35–42. https://doi.org/10.1007/978-981-15-2396-0_7

Leong, J.Y.C., Oh, K.S., Poh, P.E., Chong, M.N., 2017. Prospects of hybrid rainwater-greywater decentralised system for water recycling and reuse: A review. *J. Clean. Prod.* 142, 3014–3027. https://doi.org/10.1016/j.jclepro.2016.10.167

Lienhoop, N., Al-Karablieh, E.K., Salman, A.Z., Cardona, J.A., 2014. Environmental cost-benefit analysis of decentralised wastewater treatment and re-use: A case study of rural Jordan. *Water Policy* 16, 323–339. https://doi.org/10.2166/wp.2013.026

Mackenzie, D.L., 2010. *Water and Wastewater Engineering: Design Principles and Practice*. McGraw-Hill Education.

Massoud, M.A., Tarhini, A., Nasr, J.A., 2009. Decentralized approaches to wastewater treatment and management: Applicability in developing countries. *J. Environ. Manage.* 90, 652–659. https://doi.org/10.1016/j.jenvman.2008.07.001

Parcher, M.J., 1997. *Wastewater Collection System Maintenance*. Taylor & Francis.

Punmia, B.C., Jain, A.K., 1998. *Waste Water Engineering*. Laxmi Publications.

Qasim, S.R., 2017. *Wastewater Treatment Plants: Planning, Design, and Operation*.

Qasim, S.R., Zhu, G., 2017. *Wastewater Treatment and Reuse, Theory and Design Examples*, Volume 1 – Google Books. CRC Press.

Radlinski, M., Wolf, J., 2016. Condition assessment and service life analysis of an asbestos-cement sewer pipe, in: Pipelines 2016: Out of Sight, Out of Mind, Not Out of Risk – Proceedings of the Pipelines 2016 Conference. https://doi.org/10.1061/978078 4479957.030

Rangwala, S., 2005. *Water Supply and Sanitary Engineering*. Charotar Publishing House.

Sincero, A.P., Sincero, G.A., 2003. *Physical-Chemical Treatment of Water and Wastewater*. IWA Pub.

Stoklosa, R.J., del Pilar Orjuela, A., da Costa Sousa, L., Uppugundla, N., Williams, D.L., Dale, B.E., Hodge, D.B., Balan, V., 2017. Techno-economic comparison of centralized versus decentralized biorefineries for two alkaline pretreatment processes. *Bioresour. Technol.* 226, 9–17. https://doi.org/10.1016/j.biortech.2016.11.092

Suhak, T., Chimyshenko, S., 2021. Prospects of innovative development of water supply and sewerage en-terprises as one of the Sustainable Development Goals. *Her. Kiev Inst. Bus. Technol.* 48(2): 31–38. https://doi.org/10.37203/kibit.2021.48.05

Tchobanoglous, G., Burton, F.L., Stensel, H.D., 2003. *Wastewater Engineering: Treatment, Disposal, and Reuse* [WWW Document]. Metcalf Eddy.

Wilderer, P.A., Schreff, D., 2000. Decentralized and centralized wastewater management: A challenge for technology developers. *Water Sci. Technol.* 41, 1–8. https://doi.org/10.2166/wst.2000.0001

5 Overview of Conventional Wastewater Treatment Processes
Primary Treatment

CHAPTER OBJECTIVES

The chapter covers the basic concepts behind the functioning of the different primary treatment units in a conventional wastewater treatment plant, such as screening, grit chamber, skimming tank, and primary sedimentation tank (PST). The design of the mentioned treatment units are discussed with examples.

5.1 INTRODUCTION

Wastewater treatment technologies have been developing since the late 1800s. In the 1920s, the more conventional treatment processes we use today came into practice. Activated sludge process (ASP), constructed wetlands (CW), and rotating biological contractors (RBC) came into practice around 1914, 1950, and 1960, respectively. Upflow anaerobic sludge blanket reactors (UASB), membrane bioreactors (MBR), sequencing batch reactors (SBR), and moving bed biofilm bioreactors (MBBR) were developed between 1970 and 1990 (Lofrano and Brown, 2010). In addition, significant advances have been made in the last 30 to 40 years to develop technologies for the reuse of wastewater.

The typical flow diagram of a conventional sewage treatment plant (STP) is depicted in Figure 5.1. An STP comprises different treatment systems, including preliminary, primary, secondary, and tertiary. Each of these treatment steps has its role in treating wastewater. For instance, large debris and inorganic solids are removed in the preliminary stage. A large portion of the suspended solids and some organic matter is removed during the primary treatment. Primary treatment is followed by secondary treatment, where the biological conversion of colloidal and dissolved organics occurs. During the secondary treatment process, the generated biomass can be separated by secondary sedimentation or clarifier. The sludge generated from primary and secondary clarifiers is called primary and secondary sludge, respectively. The sludge generated from these two clarifiers is taken together for further treatment before its disposal. Sludge refers to any kind of substance formed during the primary and secondary treatment stages. After secondary treatment, tertiary treatment is

DOI: 10.1201/9781003364450-5

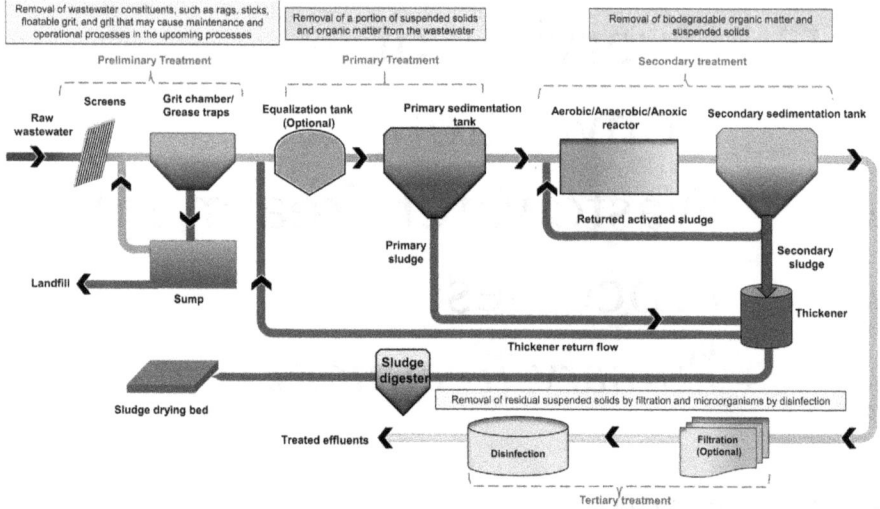

FIGURE 5.1   Wastewater treatment train in a convention sewage treatment plant.

required to either make the effluent suitable for discharge in surface water or can be reused. In tertiary treatment, the removal of any excess suspended solids, nutrients, microorganisms, and other organic compounds are targeted. Understanding the utility and mechanism of different units involved in the STP is essential for design purposes.

In this context, this book discusses the fundamental concept of wastewater treatment using different treatment methods and the steps involved in designing. This chapter focuses on the different treatment units involved in primary treatment. Primary treatment is the first stage of treatment in an STP, where the removal of large floating objects and suspended solids occur through physical forces. The most common units involved in primary treatment include screens, grit chamber, and primary sedimentation. Operations involved in eliminating large floating objects and grit are often referred to as preliminary treatment. The basic concepts of these treatment units and the steps involved in their design are discussed in the following sections.

5.2   SCREENING

Screening is the first operation unit in a conventional STP, where large floating substances, such as sachets, plastic sheet bits, leaves, fibers, rags, and others, are separated by employing different sizes of screens. The main purpose of the screens is to prevent the clogging of the valves and other types of equipment in STP and protect the pumps. A screen is a device that consists of a set of inclined bars with openings of different sizes. When the size of the openings is greater than 6 mm, this type of screen is called a coarse screen, while fine screens have an opening size ranging from 0.5 to 6 mm. Micro-screens (<0.5 mm) are also employed in different STPs to reduce the suspended solid load from the sedimentation tanks (Tchobanoglous et al., 2004). The selection of the type of screen depends on the characteristics of the wastewater, the degree of screening removal required, and the disposal options (Langford et al., 2015;

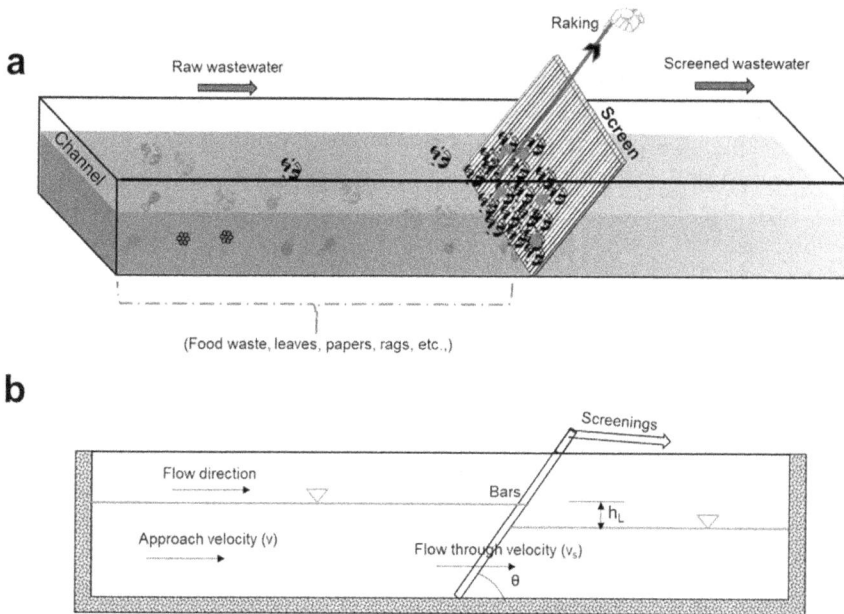

FIGURE 5.2   (a) Typical representation of screen chamber and (b) hydraulic flow diagram.

Mackenzie, 2010; Tchobanoglous et al., 2014). The typical diagram of the screen is shown in Figure 5.2.

It can be observed in Figure 5.2a and b that screens contain vertical bars (barracks), which are inclined (commonly 60° with the horizontal) away from the direction of the incoming wastewater (Mara, 2013). The cross-section of the screen chamber is always greater than the incoming channel. Providing screens having an equal or lower cross-sectional area than the incoming channel may lead to wastewater overflowing the screens. The length of the incoming channel should be sufficient to prevent the formation of eddies around the screen. The materials, such as papers, leaves, bottles, rags, and others retained on the screens, are called screenings. The screenings are removed either manually or mechanically. For instance, manual cleaning is preferred ahead of pumps in small wastewater pumping stations. In contrast, mechanical screening is used when raw wastewater contains a large amount of debris, where manual screening is difficult. The quantity of screenings collected depends on different factors, such as the type of screen used, sewer system, and geographic location. The collected screened solids must be disposed of safely to prevent environmental impacts. The most common disposal practices are sanitary landfills, incineration, and grinding (Tchobanoglous et al., 2004).

5.2.1   DESIGN OF COARSE SCREEN

The screen should be placed before the grit chamber because it removes the objects from wastewater that may damage or clog downstream equipment. The screens are essentially the first treatment unit in an STP. There are two or more screens that are cleaned automatically so that even if one unit is taken out of service, performance won't be affected. Each unit should be able to handle the peak hydraulic flow rate even if the biggest unit isn't working. Each unit should be able to work independently from the other units. In very small plants, a bypass channel may be built into a single unit. The screen on the bypass channel will have to be cleaned by hand. During the design, the approach velocity, flow-through velocity, spacing between each on the bars and head loss are the essential components that require to be considered (Mackenzie, 2010; Tchobanoglous et al., 2014).

> *Approach velocity*: The approaching velocity of wastewater in the screening channel should be maintained at least 0.4 m/s to minimize the solid deposition in the channel (Mackenzie, 2010).
> *Flow-through velocity:* The velocity through screens should be limited to 0.9 m/s at peak flow rates (Mackenzie, 2010).
> *Clear spacing:* For manually cleaned bar screens, the clear spacing ranges between 25 and 50 mm, while 15–20 mm of clear spacing is provided in the case of mechanically cleaned bar screens (Mackenzie, 2010).
> *Head loss through screens:* The head loss through mechanically cleaned coarse screens is typically limited to about 150 mm. The head loss through the bar rack when bar rack is 50% clogged should be less than 0.15 m (Tchobanoglous et al., 2004). The following equations (Equations (5.1) and (5.2)) can be used to calculate the hydraulic head loss through coarse and fine screens, respectively (Mackenzie, 2010; Tchobanoglous et al., 2014).

$$h_L = \frac{1}{C}\left(\frac{v_s^2 - v^2}{2g}\right) \qquad (5.1)$$

$$h_L = \frac{1}{2g}\left(\frac{Q}{CA}\right)^2 \qquad (5.2)$$

where
h_L = head loss (m),
v_s = flow-through velocity (m/s),
v = approach velocity (m/s),
C = empirical discharge coefficient (0.7 for clogged screen and 0.6 for clean screen) (Tchobanoglous et al., 2004),
A = effective open area of the submerged screen (m²),
Q = discharge through the screen (m³/s),
g = gravity (9.81 m/s²).

5.2.2   Design Example

Example 1
Design a bar rack and screen chamber for a plant with design flow of 40 MLD. The peak factor may be assumed to be 3. The diameter of the incoming sewer is 1.5 m and

depth of flow in sewer during peak condition (d_1) is 1.15 m. The velocity in the sewer at peak design flow (V_1) is 1.2 m/s. The drop of screen chamber flow or the difference between Z_1 and Z_2 is 0.08 m. Rectangular bars of width 10 mm and clear spacing of 25 mm may be assumed.

Solution

Step 1: Bar rack
Based on the provided data the peak design flow is = 40 MLD × 3 = 120 MLD = 1.39 m³/s.

The depth of flow in screen chamber (d_2) is 1.15 m (assumed), and the velocity of flow-through rack openings (V_3) has been assumed to be 0.9 m/s (should be in the range of 0.6 to 1.2 m/s).

Clear area of openings through the rack $(A) = \dfrac{Q}{V_3} = \dfrac{1.39 \text{ m}^3/\text{s}}{0.9 \text{ m/s}} = 1.54 \text{ m}^2$.

Clear width of openings through the rack $(B) = \dfrac{A}{d_2} = \dfrac{1.543 \text{ m}^2}{1.15 \text{ m}} = 1.34 \text{ m}$.

Number of clear spacing provided $(N_s) = \dfrac{B}{s} = \dfrac{1.34 \text{ m}}{25 \text{ mm}} = 54$

So, number of bars $(N_b) = (N_s - 1) = 54 - 1 = 53$
Total width of the screen chamber $(W) = [(N_s \times s) + (N_b \times b)]$
$= [(54 \times 0.025) + (53 \times 0.010)] = 1.88 \text{ m}$

Step 2: Calculation of the depth of flow in screen chamber during the peak flow
Applying Bernoulli's theorem (Equation 5.3) between section 1 and 2 (Figure 5.3), we get

$$Z_1 + d_1 + \frac{V_1^2}{2g} = Z_2 + d_2 + \frac{V_2^2}{2g} + h_L \tag{5.3}$$

where
Z_1 and Z_2 are datum heads (m)
d_1 and d_2 are depths of flow at sections 1 and 2 (m)
V_1 and V_2 are velocities of flow at section 1 and 2 (m)
h_L = head loss due to sudden expansion when wastewater is entering the screen chamber from the sewer (m)

$$h_L = K_e \left[\frac{V_1^2}{2g} - \frac{V_2^2}{2g} \right] \tag{5.4}$$

(K_e = coefficient of expansion (0.3))
Velocity of flow at section 2 (V_2) is given by Equation (5.5):

$$V_2 = \left[\frac{Q}{W \times d_2} \right] = \frac{1.736}{2.545 \times d_2} \tag{5.5}$$

Depth of flow at section 2 (d_2) is unknown.

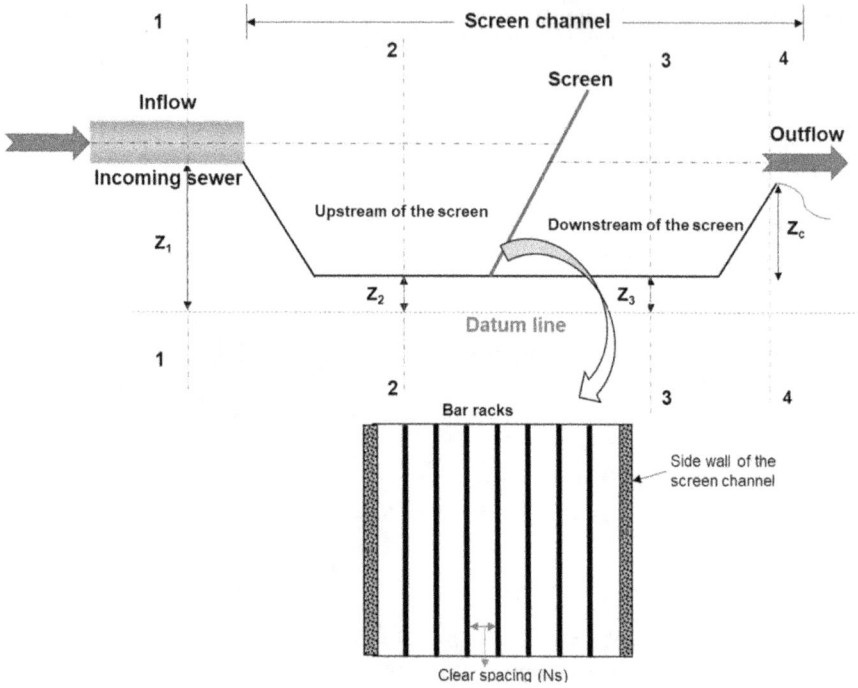

FIGURE 5.3   The longitudinal section of the screen chamber.

Putting all the known values in the following equation:

$$Z_1 + d_1 + \frac{V_1^2}{2g} = Z_2 + d_2 + \frac{V_2^2}{2g} + h_L$$

we get,

$$0.08 + 1.15 + \frac{1.2^2}{2g} = d_2 + \frac{\left(\dfrac{1.39}{1.88 \times d_2}\right)^2}{2g} + 0.3 \left[\frac{1.2^2}{2g} - \frac{\left(\dfrac{1.39}{1.88 \times d_2}\right)^2}{2g}\right]$$

$$d_2^3 - 1.28\, d_2^2 + 0.019 = 0$$

Solving by trial and error; $d_2 = 1.25$ m.

Therefore, velocity of flow at section 2 $(V_2) = \dfrac{1.39}{1.88 \times 1.25} = 0.59$ m/s

Step 3: Calculation of velocity through clear opening

$$V = \frac{\text{Flow}}{\text{Net area of opening through rack}} \qquad (5.6)$$

$$V = \frac{Q}{\text{Number of clear spacings} \times \text{Clear spacing between bars} \times \text{depth of flow at section 2}}$$

$$V = \frac{1.39}{54 \times 0.025 \times 1.25} = 0.823 \text{ m/s}$$

The velocity through the rack had been previously assumed to be 0.9 m/s. However, upon calculation, it was found to be 0.823 m/s. Since the velocity through the rack should lie between 0.6 and 1.2 m/s, the obtained value can be accepted (CPHEEO, 2013).

Step 4: Calculation of the head loss
The head loss through clean flat bar screens is calculated from the following formula:

$$h_L = \frac{1}{C}\left(\frac{V^2 - V_2^2}{2g}\right) (C = 0.6 \text{ for clean screen})$$

$$h_L = \frac{1}{0.6}\left(\frac{0.823^2 - 0.59^2}{2 \times 9.81}\right) = 0.028 \text{ m}$$

Step 5: Calculation of depth and velocity of flow
Applying Bernoulli's theorem between section 2 and 3, we get

$$Z_2 + d_2 + \frac{V_2^2}{2g} = Z_3 + d_3 + \frac{V_3^2}{2g} + h$$

When the bar rack is clean,

$$0 + 1.25 + \frac{0.59^2}{2g} = d_3 + \frac{\left(\frac{1.39}{1.88 \times d_3}\right)^2}{2g} + 0.028$$

$$d_3^3 - 1.236 \, d_3^2 + 0.0278 = 0$$

Solving by trial and error, $d_3 = 1.21$ m.

Therefore, velocity at section 3, $V_3 = \frac{1.39}{1.88 \times 1.21} = 0.611$ m/s

Step 6: Calculation of head loss assuming there is 50% clogging
During this process d' has been assumed to be the depth and V_2' has been assumed to be the velocity of flow at section 2 when bar rack is 50% clogged.

$$h_L = \frac{1}{C}\left(\frac{\text{Velocity through clogged rack}^2 - V_2^2}{2g}\right) (C = 0.7)$$

$$h_L = \frac{1}{0.7 \times 2 \times 9.81}\left(\left(\frac{1.39}{54 \times 0.025 \times 0.5\, d_2'}\right)^2 - \left(\frac{1.39}{1.88 \times d_2'}\right)^2\right)$$

$$h_{50\%} = \frac{0.26}{\left(d_2'\right)^2}$$

Now, applying Bernoulli's theorem between sections 2 and 3,

$$Z_2 + d_2' + \frac{(V_2')^2}{2g} = Z_3 + d_3 + \frac{V_3^2}{2g} + h_{50\%}$$

$$0 + d_2' + \frac{\left(\dfrac{1.39}{1.88 \times d_2'}\right)^2}{2 \times 9.81} = 1.21 + \frac{0.611^2}{2 \times 9.81} + \frac{0.26}{\left(d_2'\right)^2}$$

$$(d_2')^3 - 1.23 \times (d_2')^2 - 0.232 = 0$$

Solving by trial and error, $d_2' = 1.35$ m
   So, during the rack being 50% clogged, the velocity of flow at section 2 can be calculated.

$$V_2' = \frac{1.39}{1.88 \times 1.35} = 0.574 \text{ m/s}$$

Head loss through the bar rack when bar rack is 50% clogged

$$h_{50\%} = \frac{0.26}{\left(d_2'\right)^2} = \frac{0.26}{\left(1.35\right)^2} = 0.142\text{m}$$

Step 7: Calculation of floor raising required

Critical flow conditions are required near the outflow. The depth at the critical flow conditions is provided by d_c:

$$d_c = \left[\frac{Q^2}{g \times W^2} \right]^{\frac{1}{3}}$$

(5.7)

$$d_c = \left[\frac{1.39^2}{9.81 \times 1.88^2} \right]^{\frac{1}{3}} = 0.38 \text{ m}$$

Hence, the critical velocity can be found out, which is $V_c = \dfrac{1.39}{1.88 \times 0.38} = 1.94$ m/s

The floor of the screen chamber needs to be elevated by an amount Z_c, which can be estimated by applying Bernoulli's theorem between sections 3 and 4. This is necessary to ensure that the existing hydraulic profile at section 3 and beyond is not disrupted in any way.

$$Z_3 + d_3 + \frac{(V_3)^2}{2g} = Z_4 + Z_c + d_4 + \frac{V_4^2}{2g} + h$$

where Z_3 is equal to the value of Z_4 or the datum, d_4 is equal to the value of d_c and V_4 is equal to the value of V_c.

Neglecting the head loss, we get $Z_c = 0.657$ m.

5.3 GRIT REMOVAL

The physical separation of grit, such as sand, gravel, glass, pieces of metals, and other heavy inorganic solids, occurs in the grit chamber. The specific gravity of grit is assumed to be around 2.65 (Mackenzie, 2010). Most of the aforementioned substances are abrasive in nature and have specific gravity higher than organic solids in wastewater. Therefore, the main purpose of the grit chamber is to reduce the deposition in the sedimentation tanks. Also, by removing these substances, clogging and abrasion of different treatment units in the upcoming processes in the STP can be prevented. The inorganic solids which are collected in this stage are suitable for landfill disposal. Generally, there are different types of grit chamber, such as horizontal-flow grit chamber (rectangular or square) and aerated grit chamber. The schematic representation of horizontal and aerated grit chambers has been depicted in Figure 5.4a and b.

The horizontal-flow grit chamber is a velocity-controlled type long narrow channel. The velocity control can be achieved by providing the proportional weir or Parshall flume at the outlet of the grit chamber. At different flow conditions, the constant velocity can be achieved by varying the cross-sectional area of flow. There is a head loss in the grit chamber due to the installation of velocity control devices.

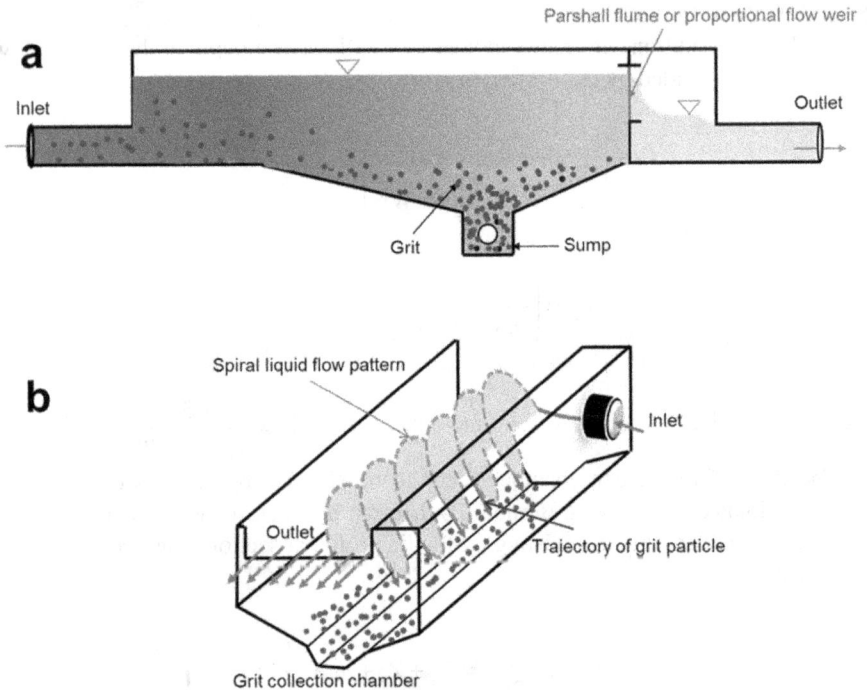

FIGURE 5.4   (a) Horizontal-flow grit chamber with velocity control device at the effluent and (b) aerated grit chamber.

A horizontal-flow grit chamber is designed to maintain the horizontal velocity (flow-through velocity) as close to 0.3 m/s, and a very short detention time, that is, 45–90 s is sufficient to remove the targeted grit particle (Tchobanoglous et al., 2004).

In the context of an aerated grit chamber, the movement of the wastewater through the chamber takes the form of a spiral while an air supply is present. It is necessary for the air supply and control system to be able to deliver air at a rate ranging from 0.0019 to 0.0125 m³/s per meter of the tank's length. (Mackenzie, 2010). The rate of air diffusion and the shape of the tank govern the rate of spiral velocity. Typically, 3 min of detention time is provided in the aerated grit chamber (Tchobanoglous et al., 2004). The heavier solids are deposited at the bottom of the tank, and lighter solids remain in suspension. Compared to a horizontal-flow grit chamber, an aerated grit chamber has several advantages, like efficiency is constant over a wide range of flow, organic content can be controlled by air rate, and the head loss is minimal. Generally, for every 1,000 m³ of wastewater treated, about 0.05–0.1 m³ of grit is collected (Mara, 2013). The deposited grit is removed by using mechanical scrappers. The collected grit is either taken to the landfill or buried. Sometimes, a sump is provided after the grit chamber, where the grit is collected, and the water is recirculated (Figure 5.1). Often, the grits are washed thoroughly and used on sludge drying beds or as a cover for screenings. They may also be used as a surfacing material for roadway. The grit chamber is placed downstream of the screens.

5.3.1 DESIGN OF GRIT CHAMBER

The shape of the grit chamber makes it easier for the water to roll in a spiral as it goes through the chamber. Even though some grit chambers have been made in the shape of a bulb to provide this geometry, the shape seems like it would be very hard and expensive to make. So, there are chambers with depths between 2 and 5 m and width-to-depth ratios between 1:1 and 5:1, with a typical value of 2:1. The ratio of length to width is between 2.5:1 and 5:1. Lengths are between 7.5 m and 27.5 m (Mackenzie, 2010). As the wastewater moves through the chamber, the heavier grit particles fall to the bottom of the tank. Less dense, mostly organic particles that are lighter stay in suspension and move out of the tank. The size of the particles that will settle out depends on how fast the water moves across the bottom of the tank. The velocity of the flow of water should also be high such that the settled particles do not come into suspension. Hence, the different parameters that are required to be considered for the design of grit chamber are the settling velocity of the grit particles, scouring velocity, specific gravity, and size of the particles (Mackenzie, 2010; Tchobanoglous et al., 2014).

Settling velocity of the targeted particles (V_s): All the particles are assumed to settle as per Newton's Law. Hence, Equation 5.8 can be used to calculate the settling velocity of grit particles (Mackenzie, 2010; Tchobanoglous et al., 2014).

$$V_s = \sqrt{\frac{4g(G_s - 1)d}{3C_D}} \qquad (5.8)$$

where V_s is the settling velocity of particle (m/s), G_s is the specific gravity of the particle, d is the diameter of particle (m), g is the acceleration due to gravity (9.81 m/s^2), and C_D is the drag coefficient is given by Equations 5.9 and 5.10 (Mackenzie, 2010; Tchobanoglous et al., 2014).

$$C_D = \frac{24}{R_e} + \frac{3}{\sqrt{R_e}} + 0.34 \rightarrow \quad \text{for} \quad 1 < R_e < 10^3 \, (\text{transition flow}) \qquad (5.9)$$

$$C_D = \frac{24}{R_e} \rightarrow \quad \text{for} \quad R_e < 1 \, (\text{laminar flow}) \qquad (5.10)$$

where R_e = Reynolds number.

When Reynolds number is between 1 and 1,000 or during transition flow, the value of C_D can be approximated to be by Equation 5.11 (Mackenzie, 2010; Tchobanoglous et al., 2014).

$$C_D = \frac{18.5}{R_e^{0.6}} \qquad (5.11)$$

Consequently, upon substituting the value of C_D in Equation 5.8, the settling velocity can be calculated by Equation 5.12 (Mackenzie, 2010; Tchobanoglous et al., 2014).

$$V_s = \left[0.707 \times \left(G_s - 1 \right) \times d^{1.6} \times \gamma^{-0.6} \right]^{0.714} \qquad (5.12)$$

where γ = kinematic viscosity of wastewater (m²/s).

Under laminar conditions, Stokes' law can be valid. Hence, V_s can be calculated by using Equation 5.13 (Mackenzie, 2010; Tchobanoglous et al., 2014):

$$V_s = \frac{g}{18} \frac{\left(G_s - 1 \right) \times d^2}{\gamma} \qquad (5.13)$$

where γ = kinematic viscosity of wastewater (m²/s).

> *Note*: In order to settle the smallest particle in the grit chamber, the settling velocity (V_s) should be equal to or higher than the overflow velocity (V_o).
>
> *Scouring velocity (V_c):* In order to prevent scouring of already deposited particles, the magnitude of horizontal velocity should not exceed critical horizontal velocity (V_c) (beyond this velocity settled particles start scouring). The following Equation 5.14 can be used to calculate the scouring or critical velocity (Mackenzie, 2010).

$$V_c = \sqrt{\frac{8\beta \left(G_s - 1 \right) g \times d}{f}} \qquad (5.14)$$

where
V_c = critical or scouring velocity (m/s)
β = constant that depends on type of material being scoured (0.04 for unigranular and 0.06 for non-uniform sticky material)
G_s = Specific gravity of particle
d = diameter of particle (m)
f = Darcy–Weisbach friction factor (Typical range 0.02 – 0.03)
g = Acceleration due to gravity (9.81 m/s²)

5.3.2 DESIGN EXAMPLES

Example 2.
In order to remove grit particles with a size of up to 0.15 mm and a specific gravity of 2.65, you will need to construct a grit chamber that can treat an average design flow of 40 MLD of wastewater at a peak factor of 3. The temperature never drops below 15°C. At the temperature that was specified, the kinematic viscosity was 1.14×10^{-6} m²/s. As a means of controlling the amount of grit that enters the chamber, a proportional flow weir has been installed.

Solution

$$\text{The peak discharge} = \frac{3 \times 40 \times 1000}{24 \times 60 \times 60} = 1.39 \text{ m}^3/\text{s}$$

Step 1: Calculation of settling velocity

Stokes' law: $V_s = \frac{g}{18} \times (G_s - 1) \times \frac{d^2}{3}$ (If $R_e < 1$)

Transitions law: $V_s = \left[0.707 \times (G_s - 1) \times d^{1.6} \times 3^{-0.6} \right]^{0.714}$ ($1 < R_e < 1000$)

Let us assume that Reynold's number $R_e < 1$, then by applying Stokes' law,

$$V_s = \frac{9.81}{18} \times (2.65 - 1) \times \frac{(0.15 \times 10^{-3})^2}{1.14 \times 10^{-6}} = 0.018 \text{m/s}$$

Check for Reynold's number

$$R_e = \frac{V_s d}{v} = \frac{0.018 \times 0.15 \times 10^{-3}}{1.14 \times 10^{-6}} = 2.37 > 1.0$$

Hence Stokes' law does not apply, apply transitions law for $1 < R_e < 10^3$:

$$V_s = \left[0.707 \times (2.65 - 1) \times (0.15 \times 10^{-3})^{1.6} \times (1.14 \times 10^{-6})^{-0.6} \right]^{0.714} = 0.0168 \text{ m/s}$$

Again, check for Reynold's number

$$R_e = \frac{V_s d}{\gamma} = \frac{0.0168 \times 0.15 \times 10^{-3}}{1.14 \times 10^{-6}} = 2.2$$

$1 < R_e < 10^3$. Hence OK.

Step 2: Calculation of surface overflow rate
The surface overflow rate (SOR) for 100% removal efficiency should be equal to the settling velocity of the minimum size of the particle to be removed:

$$= 0.0168 \text{ m/s}$$
$$= 0.0168 \times 60 \times 60 \times 24 \text{ m}^3/\text{m}^2/\text{day} = 1451.5 \text{ m}^3/\text{m}^2/\text{day}$$

However, due to turbulence and short-circuiting caused by various factors such as eddy, wind, and density currents, the actual value to be adopted must be reduced in order to account for the performance of the basin and the desired particle removal efficiency. This is necessary in order to ensure that the particles are removed as effectively as possible. Equation 5.15 can be used to calculate the actual overflow rate.

$$\eta = 1 - \left[1 + \frac{nV_s}{(Q/A)}\right]^{-\frac{1}{n}}$$ (5.15)

where

η = efficiency of removal of desired particles

n = measure of settling basin performance [range: 1/8 to 1]

  = 1/8 for very good performance

Assuming $\eta = 0.75$ and $n = 1/8$, we get

$$0.75 = 1 - \left[1 + \frac{(1/8) \times (1451.5)}{(Q/A)}\right]^{-\frac{1}{1/8}}$$

SOR or $\left(\dfrac{Q}{A}\right)$ = 959 m³/m²/day

Step 3: Calculation of dimensions of the grit chamber

$$\text{Plan area of grit chamber} (A) = \frac{\text{Discharge}}{\text{SOR}} = \frac{3 \times 40 \times 1000 \text{ m}^3/d}{959 \text{ m}^3/\text{m}^2/\text{day}} = 125.1 \text{ m}^2$$

Assumptions:

Number of channels (N) to be provided = 5

Length of each channel (L) = 23 m

Width of each channel (B) = 1.08 – 1.1 m

So, surface area of each channel = 23 × 1.1 = 25.3 m²

Total surface area provided = 5 × 25.3 m² = 126.5 m²

Step 4: Scouring velocity

Assuming a depth (D) of 1.0 m,

$$\text{Horizontal Velocity} = \frac{\text{Discharge}}{\text{cross-sectional area of grit chamber}}$$
$$= \frac{1.39 \text{ m}^3/\text{s}}{5 \times 1.1 \times 1} = 0.25 \text{ m/s}$$

$$\text{Scouring velocity } v_c \left(\text{Eq.} (5.14)\right) = \sqrt{\frac{8\beta(G_s - 1)g \times d}{f}}$$ (5.14)

$$= \sqrt{\frac{8 \times 0.06 \ (2.65 - 1) \times 9.81 \times 0.15 \times 0.001}{0.02}}$$

$$= 0.240 \text{ m/s}$$

Therefore, horizontal velocity is less than the scouring velocity. Hence design is OK.

$$\text{HRT of grit chamber} = \frac{V}{Q} = \frac{5 \times 25.3 \times 1}{1.39} = 91 \text{ s}$$

Assume free board (FB) = 0.25 m

Assume grit storage space (GSS) = 0.25 m

Total depth of grit chamber (h) = FB + GSS + D = 1 + 0.25 + 0.25 = 1.5 m

Hence, 5 channels (dimension: $L \times B \times D = 23 \times 1.1 \times 1$) and one standby channel should be provided.

Step 5: Proportional flow weir

$$\text{Peak flow for each weir} \left(Q_\text{w}\right) = \frac{Q}{N} = \frac{1.39 \text{ m}^3/\text{s}}{5} = 0.278 \text{ m}^3/\text{s}$$

Flow through a proportional flow weir is given by Equation 5.16.

$$Q = cb\sqrt{2ag}\left(H - \frac{a}{3}\right) \tag{5.16}$$

where

c = coefficient, which is assumed 0.61 for symmetrical sharp-edge weirs

a = dimension of weir usually assumed between 25 mm and 50 mm

b = base width of the weir

H = depth of flow

Assuming the value of a = 40 mm = 0.04 m; H = 1.0 m at peak flow; c = 0.61

$$0.278 = 0.61 \times b \times \sqrt{2 \times 0.04 \times 9.81} \times \left(1 - \frac{0.04}{3}\right)$$

$$b = 0.52 \text{ m}$$

So, base width of weir (b) = 0.52 m

Example 3.

The purpose of a grit chamber is to remove particles with a diameter of 0.2 mm and a specific gravity of 2.65. The settling velocity of these particles ranges between 0.016 and 0.022 m/s, depending on their form factor. A proportioning weir will keep the flow-through velocity at 0.3 m/s. Determine the channel size for a maximum flow of 10,000 m³/d of wastewater.

Solution

$$\text{Flow} = \frac{10000}{60 \times 60 \times 24} = 0.116 \text{ m/s}$$

Step 1: Calculation of dimensions

Assuming a rectangular cross-section with depth (D) and width (B) which is 1.5 times the depth at maximum flow.

$$A_s = B \times 1.5B = 1.5B^2$$
$$= \frac{Q}{v_h} = \frac{0.116}{0.3}$$
$$= 0.39 \text{ m}^2$$
$$B = 0.51 \text{ m}$$
$$D = 0.76 \text{ m}$$

Step 2: Calculation of detention time

Assuming a settling velocity of 0.02 m/s (given in range 0.016 to 0.022 m/s), the detention time is

$$t_d = D / v_t$$
$$= \frac{0.76}{0.02} = 38 \text{ s}$$

Step 3: Calculation of length

$$L = t_d \times v_h$$
$$= 38 \text{ s} \times 0.3 \text{ m}/\text{s} = 11.4 \text{ m}$$

Tank dimension are length (L) = 11.4 m; width (B) = 0.51 m; depth (D) = 0.76 m.

Example 4

Create a grit chamber that is aerated to treat municipal wastewater. Average flow rate is 0.5 m³/s, with a peaking factor of 2.75 times the average.

Solution

$$\text{Peak flowrate} = 0.5 \text{ m}^3/s \times 2.75 = 1.38 \text{ m}^3/s$$

Step 1: Determination of dimensions of grit chamber

In order to provide routine maintenance, two chambers should be used. The average detention time at the peak flowrate can be assumed to be 3 min or 180 s (Tchobanoglous et al., 2004).

$$\text{Volume of each grit chamber } (\text{m}^3) = \left(\frac{1}{2}\right) \times (1.38 \text{ m}^3/s) \times 180 \text{ s} = 124.2 \text{ m}^3$$

A width (B)-to-depth (D) ratio of 1.2:1 was used and the depth was assumed to be 3 m.
Width (B) = 1.2 × 3 m = 3.6 m

$$\text{Length}(L) = \frac{\text{volume}}{\text{width} \times \text{depth}} = \frac{124.2 \text{ m}^3}{3 \text{ m} \times 3.6 \text{ m}} = 11.5 \text{ m}$$

Step 2: Determination of detention time of grit chamber

The detention time in each grit chamber at average flow:

$$\text{Detention time} = \frac{124.2 \text{ m}^3}{0.25 \text{ m}^3/\text{s}} = 8.28 \text{ min}$$

Step 3: Determination of air supply requirement

0.3 m³/min per 1 m length is assumed to be adequate (Tchobanoglous et al., 2004).

Then air required (length basis) $= 11.5 \times 0.3 = 3.45$ m³/min for each grit chamber
Total air supply requied $= 3.45 \times 2 = 6.9$ m³/min

5.4   SEDIMENTATION

5.4.1   PARTICLE SETTLING THEORY

The settling velocity of particles to be removed is one of the controlling parameters during the design of an ideal sedimentation tank. In general, settling properties of particles are categorized into four categories: (1) discrete settling (type I settling); (2) flocculant settling (type II settling); (3) hindered or zone settling (type III settling); and (4) compression settling (type IV settling) (Peavy et al., 1985; Tchobanoglous et al., 2014).

Type I settling: The particle will settle as an individual and do not flocculate during settling. For example, settling of sand follows this type of settling during plain sedimentation in water treatment or settling in a grit chamber. Stokes' law can be valid for calculating the settling velocity of a particle (Peavy et al., 1985; Tchobanoglous et al., 2014).

Type II settling: This type of settling will occur in coagulation tanks in water treatment or in PSTs in wastewater treatment. The Stokes' equation cannot be used because the flocculating particles are continually changing in size and shape. This type of settling occurs in the sedimentation tank with coagulation/flocculation processes (Peavy et al., 1985; Tchobanoglous et al., 2014).

Type III settling: When a high concentration of particles is present in the water, they are close enough together so that inter-particle forces lead to hinder the settling of neighboring particles resulting in hindered settling. This type of settling occurs in a secondary settling tank used in biological treatment methods. This type of settling is also called zone settling (Peavy et al., 1985; Tchobanoglous et al., 2014).

Type IV settling: When the particles settle at the bottom of the tank, further settling will occur due to compression of the particle layer due to squeezing out of the water molecules from the pores between solids. Compression settling usually occurs in the lower layers of a deep sludge mass, such as in the bottom of a secondary settling tank or secondary clarifier (Peavy et al., 1985; Tchobanoglous et al., 2014).

PST is designed to remove a significant fraction of organic suspended solids from wastewater. PST is an important unit because it reduces the significant quantity of suspended solids. It also removes some fraction of biochemical oxygen demand (BOD), thereby decreasing the oxygen demand and operational problems associated with biological treatment processes located downstream. Moreover, it also removes the inert particulate matter that escaped from the grit chamber. Generally, the suspended solids present in the wastewater are adhesive in nature. Hence, they flocculate naturally (type 2 settling) without adding chemical coagulants. The extent of flocculation will depend upon the concentration of particles and contact opportunity between the particles (Peavy et al., 1985; Tchobanoglous et al., 2014).

The above-mentioned factors influence hydraulic short-circuiting. Hydraulic short-circuiting is caused by the presence of dead zones, which can occur if sedimentation tanks are not designed properly. The presence of dead zones in sedimentation tanks indicates that the fluid has less volume for flow and deposition. As a result, the presence of dead zones can reduce tank performance. High flow mixing problems and decreased particle sedimentation are caused by dead zones in the settling tanks. Henceforth, it is recommended to consider all factors carefully while designing the PST. The efficient design of the PST leads to the removal of 50–70% of suspended solids and 25–40% of BOD (Peavy et al., 1985). The PST are commonly rectangular or circular shaped.

5.4.2 RECTANGULAR SEDIMENTATION TANK

Rectangular tanks are preferred where tank roofs or covers are required, and space is a constraint. The wastewater enters the tank through one or more inlet ports. A baffle is provided immediately downstream of the inlet for the uniform distribution of flow and solids across the cross-sectional area of the sedimentation tank. The sludge that is settled at the bottom of the tank is removed by chain and flight scrapers. Although the construction cost of rectangular tanks may be low, the chain and flight collector system require more maintenance than the rotating sludge collector used in circular tanks. The problems associated with the rectangular sedimentation tank include sludge removal, flow distribution, and scum removal.

5.4.3 CIRCULAR SEDIMENTATION TANK

In circular tanks, wastewater is introduced either in the center (most common feed point) or around the periphery, and the flow pattern is radial. The flow is distributed equally in all directions of the circular tank. The bottom of the tank is sloped at about 1 in 12 (V: H) to form an inverted cone, and settled solids are scrapped to the center of the tank through a rotating radical collector with a turning speed of 1.5–2 m/min (Tchobanoglous et al., 2004). The effluent from the sedimentation tank is collected at the outlet weir (launder) extended around the periphery of the tank. The different stages involved in the operation of a circular primary sedimentation tank have been depicted in Figure 5.5a–d.

The settled sludge should be removed from the primary sedimentation tank before it develops anaerobic conditions. The anaerobic conditions lead to the formation of

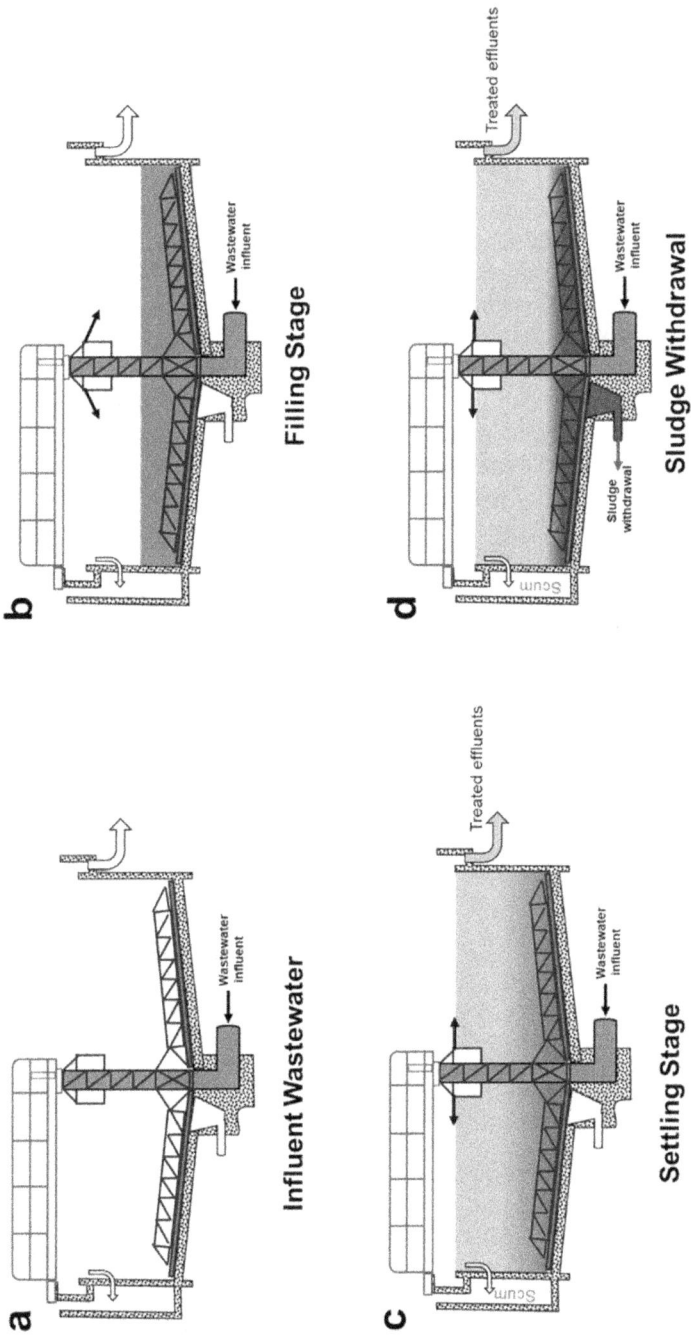

FIGURE 5.5   Different stages involved in the operation of PST (a) wastewater coming from the influent pipe; (b) filling stage; (c) settling of sludge stage; and (d) withdrawal of supernatant liquor.

gas bubbles, thereby making the removal of solids much less efficient. The quantity of sludge collected generally depends upon the strength of influent wastewater, the efficiency of the clarifier, and sludge conditions (i.e., water content and specific gravity).

5.4.4 Design of Primary Sedimentation Tank

All STPs with average design flows of more than 380 m³/d need to have more than one unit that can run on its own. If the goal of the design is to make the PST work efficiently so that it does not put too much stress on the biological processes that come after it, then the hydraulic design should be take into account the peak flow. This can be done with an equalization tank or by making the primary clarifier big enough to handle the peak flow. When estimating the hydraulic load, it is necessary to take into consideration not only the recycle streams coming from the waste activated sludge but also the supernatant from the thickening process, the supernatant from the digester, and the backwashing. It is likely that surges from these sources will affect how well the PST works. Hence, in order to achieve the best removal, the surface overflow rate and the amount of time the wastewater stays in the PST or the retention time are essential parameters that need to be considered. Furthermore, the settling velocity of the particles and the scouring velocity should also be considered to achieve the purpose of the PST (Mackenzie, 2010; Tchobanoglous et al., 2014).

> *Hydraulic retention time (HRT):* HRT (*h*) is the amount of time required for the wastewater to pass through the tank. The HRT is an important consideration while designing the PST because it influences the flocculation of suspended solids. The typical HRT of a PST ranges from 1.5 to 2.5 h for an average wastewater flow rate. In some instances, such as colder regions, high HRT is necessary to maintain good particle settling. Usually, there is an increase in water viscosity due to low temperatures (<20°C), which retard the particle settling in the clarifier, thereby reducing the performance of the PST in regions of cold climate.

Based on the HRT, the typical removal percentage of total suspended solids (TSS) and BOD can be estimated by using the following empirical Equation 5.17 (Tchobanoglous et al., 2004).

$$R = \frac{t}{a + bt} \tag{5.17}$$

where
R = expected removal efficiency (%)
t = hydraulic retention time (h)
a, b = empirical constants [For TSS: $a = 0.0075$, $b = 0.014$; for BOD: $a = 0.018$ and $b = 0.02$]

> *Surface loading rate (SLR):* SLR (m³/m². d) is defined as the amount of discharge that is being applied over the unit surface area of the tank (Equation 5.19).

TABLE 5.1
Typical Design Parameters of Primary Sedimentation Tank

Parameter	Range	Typical Value
Design Information		
HRT (h)	1.5–2.5	2
SLR (m³/m².d)		
At average flow	30–50	40
At peak flow	80–120	100
WLR (m³/d.m)	125–500	250
Typical Dimensions		
For rectangular		
Depth (m)	3–4.9	4.3
Length (m)	15–90	24 – 40
Width (m)	3–24	4.9 – 9.8
For circular		
Depth (m)	3–4.9	4.3
Diameter (m)	3–60	12 – 45
Bottom slope	1 in 16 to 1 in 6	1 in 12

Source: Tchobanoglous et al., 2014.

The typical ranges of SLR for average and peak flows have been provided in Table 5.1. The effect of SLR on the removal of suspended solids widely depends on the characteristics of wastewater, the concentration of solids, and the proportion of settleable solids (Mackenzie, 2010; Tchobanoglous et al., 2014).

$$SLR = \frac{\text{Design discharge}}{\text{Surface area of the tank}} \tag{5.18}$$

Settling velocity (V_s): The settling velocity (V_s) or the terminal velocity of the particle is the velocity at which the particles settle. The type II or flocculant settling of the targeted particles occurs in the PST. However, Stokes' law cannot be used to calculate the V_s of the particles because flocculant particles change their shape and size continuously.

Scouring velocity (V_c): Scouring velocity or the critical velocity (V_c) is the maximum horizontal velocity of the water or wastewater, such that the settled particles are not scoured. To avoid the scouring of settled particles, the horizontal velocity must be kept less than V_c. Equation 5.14 from Section 5.3.1 can be used to calculate the scouring or critical velocity.

Weir loading rate (WLR): Outlet weirs or submerged orifices are provided to maintain settling velocities and avoid short-circuiting. Weirs must be adjustable and at least equal in length to the tank's perimeter. The most common type of weir plate is one made with 90° v-notches at 150 or 300 mm intervals. Generally, WLR

(m^3/d. m) has little effect on the performance of the sedimentation tank. The typical range of WLR has been provided in Table 5.1. The weir loading rate can be calculated using Equation 5.19 (Mackenzie, 2010; Tchobanoglous et al., 2014).

$$WLR = \frac{Q}{\pi \times D} \qquad (5.19)$$

where
Q = discharge (m^3/d)
D = diameter (m)

5.4.5 DESIGN EXAMPLES

Example 5

Design a rectangular primary sedimentation basin for an average flow rate of 40,000 m^3/d and peak daily flowrate is 100,000 m^3/d. A channel width of 8 m should be provided for the rectangular primary clarifier. A minimum of two clarifiers are to be used. Calculate the scour velocity to check whether the settled material will become suspended. The overflow rate is 40 m^3/m^2.d at average flow and side water depth of 4 m. Cohesion constant (β) = 0.05, specific gravity (G_s) = 1.25, acceleration due to gravity (g) = 9.81 m/s^2, diameter of particles (d) = 100 × 10^{-6} m, Darcy–Weisbach friction factor (f) = 0.025.

Solution

Step 1: Calculation of the required surface area for average flow condition

$$A = Q/SOR = 40,000/40 = 1000 \text{ m}^2$$

Since 2 clarifiers are to be provided, the area of each clarifier should be 500 m^2.

Step 2: Determination of the length of each tank

$$L = A/W = 500/8 = 62.5 \text{ m}$$

Surface dimensions will be rounded to 62.5 m × 8 m.

Step 3: Computation of the HRT at average flow
At average flow using the assumed side water depth of 4 m,
Total volume of two tanks = 2 × (4 m × 62.5 m × 8 m) = 4000 m^3
HRT = volume/Q = (4000 × 24)/40,000 = 2.4 h

Step 4: Determination of HRT and overflow rate at peak flow
Overflow rate = Q/A = (100000)/2(62.5 × 8) = 100 m^3/m^2.d
HRT = volume/Q = (4000 × 24)/100000 = 0.96 h.

Step 5: Calculation of the scour velocity using following values (Tchobanoglous et al., 2004)

$$V_c = \sqrt{\frac{8\beta(G_s - 1)g \times d}{f}} = \sqrt{\frac{8 \times 0.05(0.25)(9.81)(100 \times 10^{-6})}{0.025}} = 0.063 \, \text{m/s}$$

Step 6: Comparison of horizontal velocity and scouring velocity
The peak flow horizontal velocity through the settling tank is

$$V = Q/A = 100000/2(8 \times 4) = 1562 \, \text{m/d} = 0.018 \, \text{m/s}$$

The horizontal velocity value at peak flow (0.018 m/s) is lower than the scouring velocity (0.063 m/s). Hence, settled matter will not come into suspension.

Example 6
Design a circular primary sedimentation tank for 40 MLD of sewage.

Solution
In this design, two circular tanks are being provided. So, the average flow-through each of the circular PST is 20 MLD.

Let the diameter of each clarifier is 30 m. The depth of the tank, excluding a free-board of 0.5 m, can be assumed to be 3.5 m. The typical depth of a circular sedimentation tank is between 3 and 4.9 m (Tchobanoglous et al., 2004).

The area of the clarifier is $= \dfrac{\pi \times 30^2}{4} = 706.8 \, \text{m}^2$.

Volume of the clarifier is = area × depth = 706.8 × 3.5 = 2474 m³.

Average overflow rate $= Q_{avg}/\text{area} = \dfrac{20 \times 1000}{706.8} = 28.3 \, \text{m}^3/\text{m}^2/\text{d}$.

The value of overflow rate for average flow in circular tank should be between 24 and 32 m³/m²/d (Tchobanoglous et al., 2004). Hence, the design is okay.

Weir loading rate $(\text{WLR}) = \dfrac{Q}{\pi \times D} = \dfrac{20 \times 1000}{\pi \times 30} = 212.2 \, \text{m}^3/\text{m}^2/\text{d}$.

The value of WLR is less than the typical WLR value of 250 m³/m²/d (Tchobanoglous et al., 2004). Hence, the design is okay.

The HRT of the tanks can be between 1.5 and 4.2 h (CPHEEO, 2013; Tchobanoglous et al., 2004).

HRT = volume/Q = (2474 × 24)/20 × 1000 = 2.96 h. Hence, the design is okay.

5.5 CHAPTER SUMMARY

- In the screening chamber, the velocity should be maintained between 0.4 and 0.9 m/s.
- The head loss in a screening chamber should be limited to 150 mm.

- The detention time in grit chamber should be between 45 and 90 s in a horizontal-flow grit chamber, and in an aerated grit chamber the detention time should be 3 min.
- In order to settle the smallest particle in grit chamber, the settling velocity should be equal or higher than the overflow velocity.
- Settling may be of 4 types: (1) discrete settling (type I settling); (2) flocculant settling (type II settling); (3) hindered settling or zone (type III settling); and (4) compression settling (type IV settling).
- In the primary sedimentation tank in an STP, usually type II settling occurs.
- PST should be able to remove 50–70% of suspended solids and 25–40% of BOD.
- The typical HRT of a PST ranges from 1.5 to 2.5 h.

5.6   CONCLUDING REMARKS

The purpose of the primary treatment is to reduce the solids that can be settled. They significantly remove a large portion of the suspended solids, which may create complications for subsequent treatment systems. However, a major portion of the dissolved organic component still persists in the wastewater. This organic matter is removed using the secondary treatment or biological treatment methods. The different biological treatment methods have been discussed in the following chapter.

REFERENCES

CPHEEO, 2013. *Manual on Sewerage and Sewage Treatment Systems: Part A Enigneering.* Central Public Health and Environmental Engineering Organisation.

Langford, K.H., Reid, M.J., Fjeld, E., Øxnevad, S., Thomas, K.V., 2015. Environmental occurrence and risk of organic UV filters and stabilizers in multiple matrices in Norway. *Environ. Int.* 80, 1–7. https://doi.org/10.1016/j.envint.2015.03.012

Lofrano, G., Brown, J., 2010. Wastewater management through the ages: A history of mankind. *Sci. Total Environ.* 408(22), 5254–5264. https://doi.org/10.1016/j.scitotenv.2010.07.062

Mackenzie, D.L., 2010. *Water and Wastewater Engineering: Design Principles and Practice.* McGraw-Hill Education.

Mara, D., 2013. *Domestic Wastewater Treatment in Developing Countries.* Taylor & Francis.

Peavy, H.S., Rowe, D.R., Tchobanoglous, G., 1985. *Environmental Engineering.* McGraw-Hill.

Tchobanoglous, G., Burton, F.L., Stensel, H.D., 2004. *Wastewater Engineering: Treatment and Reuse (Book),* 4th Edition, Metcalf & Eddy, Inc. McGraw-Hill Education.

Tchobanoglous, G., Burton, F.L., Stensel, H.D., 2014. *Wastewater Engineering: Treatment and Resource Recovery,* Metcalf & Eddy, Inc. McGraw-Hill Education.

6 Overview of Conventional Wastewater Treatment Processes

Secondary Treatment

CHAPTER OBJECTIVES

The chapter seeks to provide an understanding of the basic concepts of biological treatment units. The theory and design aspects of different processes, such as attached growth, suspended growth, and hybrid growth processes, have been covered in this chapter. Furthermore, the application of different units after biological treatment, such as secondary clarifiers and post-aeration, has also been discussed.

6.1 INTRODUCTION

Secondary treatment or biological treatment is implemented to remove both the suspended and dissolved organic fraction that remains after primary treatment. Different biological processes, such as activated sludge process (ASP), trickling filters (TF), rotating biological contactors (RBC), moving bed biofilm reactor (MBBR), sequencing batch reactor (SBR), upflow anaerobic sludge blanket (UASB), and others having a combination of physical and biological processes may be implemented depending on the influent quality and quantity. The wastewater contains a large number of microorganisms capable of degrading organic matter in a natural purification process under aerobic (presence of free dissolved oxygen (DO)) or anaerobic conditions (absence of free DO).

The different processes by which the organic matter can be degraded are discussed in later sections. The most simple governing process involved in the degradation of organic matter in biological treatment has been provided in Equation 6.1 (Tchobanoglous et al., 2004).

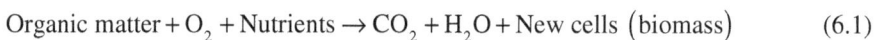

$$\text{Organic matter} + O_2 + \text{Nutrients} \rightarrow CO_2 + H_2O + \text{New cells (biomass)} \qquad (6.1)$$

The microorganisms convert the organic matter into carbon dioxide and biomass. The organic matter is the food, and the newly formed microorganisms are the biomass.

DOI: 10.1201/9781003364450-6

The newly formed cells or the biomass also starts to represent part of the total organic matter and contributes to the total organic content of the effluent. Hence, removing the newly formed biomass during the treatment process is important.

6.1.1 MICROBIAL METABOLISM

Different metabolic activities are carried out for the degradation of organic matter in a controlled or natural environment. The energy derived from the degradation of organic matter is utilized to synthesize new cells or biomass and maintain other cell functions. Catabolism is the process by which complex organic matter is oxidized or broken down into simpler compounds (primarily CO_2 and H_2O), resulting in the reduction of biological oxygen demand (BOD) and chemical oxygen demand (COD). In anabolism, organic matter is transformed into new cells or biomass. If the external substrate is not available, the microorganisms will use store food for their metabolic activities. This process is called endogenous catabolism. Anabolism requires energy, which is gained through catabolic reactions; hence, both processes depend on each other. During the biological treatment of wastewater, these two processes occur simultaneously in the system.

6.1.2 BIOMASS GROWTH CURVE AND MAJOR REQUIREMENTS FOR MICROBIAL GROWTH

As mentioned before, different types of microorganisms are present in the wastewater system, and they can be classified based on the energy and food sources required. For example, organisms that derive energy from inorganic (CO_2) sources are called autotrophs. Similarly, the organisms that obtain energy from organic (sugars, amino acids, and other organic compounds) sources are called heterotrophs. On the other hand, phototrophs use sunlight as an energy source. Among the different organisms mentioned, heterotrophs are the most important organisms for the degradation of organic matter. The factors for the growth and maintenance of these organisms are as follows:

- Appropriate environmental conditions (primarily pH and temperature)
- Macronutrients: carbon, nitrogen, and phosphorous
- Micronutrients: Trace elements and vitamins

The relationship between biomass growth and substrate utilization can be illustrated with the help of a batch reactor. The plot of the variation of biomass versus the amount of substrate present in the wastewater is normally referred to as the substrate utilization curve. Figure 6.1 represents the rate of utilization of food or substrate and biomass growth rate. As the substrate is consumed, four distinct phases are observed sequentially (Figure 6.1).

Lag phase: This is the time required for the acclimatization of microorganisms in the new environment. The length of the lag phase may vary based on the type of microorganisms.

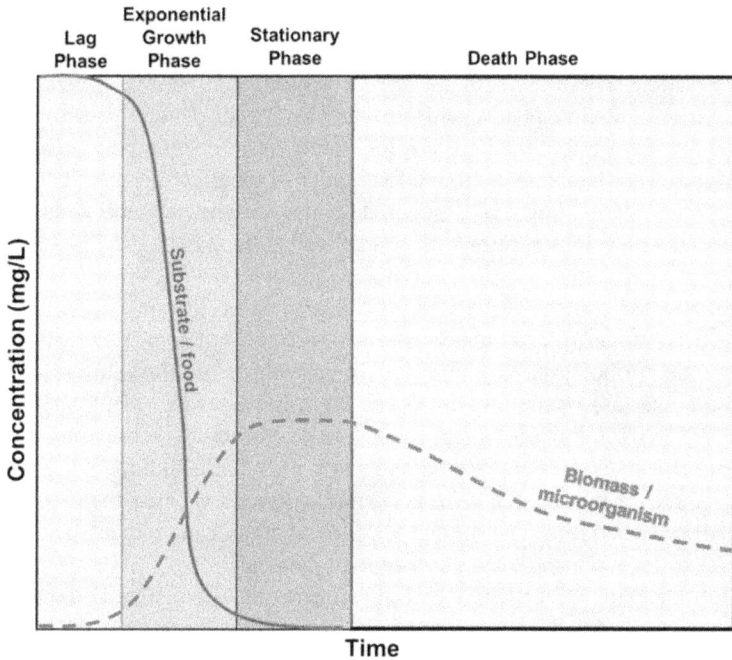

FIGURE 6.1 Biomass growth phase with changes in substrate and biomass versus time.

Exponential-growth phase: During this phase, bacterial cells multiply by the action of binary fission, that is, cells divide into segments, thereby forming two new independent cells. In this phase, the growth curve increases exponentially due to no limitation in substrate availability. However, the time required to regenerate new cells is highly influenced by temperature and other environmental factors.

Stationary-growth phase: In this phase, the biomass curve becomes relatively flat with time because the food supply may become limiting. The death of new cells roughly offsets the production of new cellular material.

Death/endogenous phase: The substrate has been depleted in this phase, so no growth is observed in biomass concentration. During this phase, biomass gets slowly depleted and approaches zero. The sludge generated in this phase has better settling properties and is more stable. The biomass yield (Y) is the ratio of the biomass produced to the amount of substrate consumed.

6.1.3 TYPES OF BIOLOGICAL TREATMENT

In biological treatment, the growth of microorganisms may be suspended in wastewater or attached to the media present. Based on the growth of the microorganisms, the biological treatment processes are termed suspended growth, attached growth, or a hybrid growth process. In hybrid processes, both suspended and attached growth are found.

In the suspended growth process, the microorganisms are maintained in suspension through appropriate methods, such as mixing with a mechanical mixer or by diffused aeration. There are different types of biological treatment methods, such as ASP, SBR, oxidation ponds, and others, where a suspended growth system is observed. These treatment methods are commonly used for removing organic matter and nutrients from municipal wastewater under aerobic conditions. The different reactor configurations, such as a completely mixed tank with or without sludge recycling and a plug-flow reactor with sludge recycling, may be used to maintain the suspended growth system.

In the attached growth system, microbial growth occurs on the surface of the media, such as stones, plastic media, and various others. The porosity of the media influences the surface area available for the growth of biofilm. Different attached growth processes are commonly employed for wastewater treatment, such as trickling filters (TFs), rotating biological contactors, and others. The microorganisms are attached to the surface of the media and grow into dense biofilms. The dissolved organics pass to the biofilm due to the concentration gradient between the surrounding fluid and the attached film. The soluble or colloidal organic matter may be retained on the biofilm surface, which decomposes in the process.

The biological treatment processes can be further classified based on the presence and absence of oxygen. In aerobic processes, the treatment occurs in the presence of free molecular oxygen. Oxygen serves as an electron acceptor. This process is mainly used for the removal of organic matter or nitrification.

In anaerobic processes, the organic waste is stabilized in the absence of oxygen. The electron acceptors are phosphate, nitrite, nitrate, sulfate, organic compounds, and carbon dioxide. This process is mainly used for the removal of organic matter or solids stabilization. Often, the anaerobic process is confused with the anoxic process. In the anoxic process, free oxygen is absent. However, bound oxygen may be present in the form of nitrate and nitrite. Both aerobic and anaerobic processes can have microorganisms with suspended growth, attached growth, and hybrid growth conditions. As a result, the biological treatment processes can be classified into different categories, as depicted in Figure 6.2.

6.2 AEROBIC TREATMENT PROCESS

6.2.1 ESTIMATION OF BIOMASS YIELD AND OXYGEN REQUIREMENTS

The exact chemical reactions during the biodegradation of organic compounds in wastewater treatment are very difficult to estimate. However, for the purpose of simplification, the organic matter is assumed to be glucose ($C_6H_{12}O_6$) and the newly formed biomass may be assumed to be $C_5H_7NO_2$. The equation involved in biological degradation can be represented as follows (Equation 6.2) (Tchobanoglous et al., 2014).

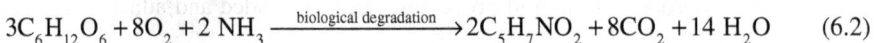

$$3C_6H_{12}O_6 + 8O_2 + 2\,NH_3 \xrightarrow{\text{biological degradation}} 2C_5H_7NO_2 + 8CO_2 + 14\,H_2O \qquad (6.2)$$

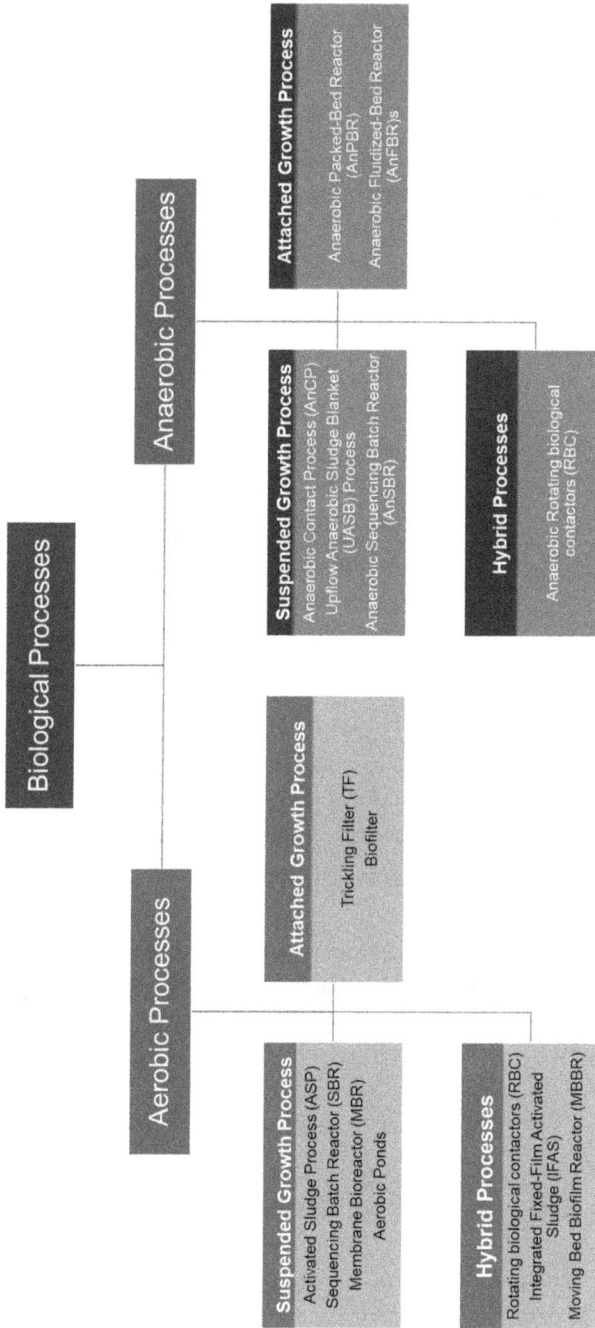

FIGURE 6.2 Classifications of different biological process.

From Equation 6.2, the biomass yield (Y) (g cells/g organic matter used) can be calculated. It is the ratio of the biomass produced to the amount of substrate consumed.

$$Y = \frac{\Delta C_5H_7NO_2}{\Delta C_6H_{12}O_6} = \frac{2(113 \text{ g/mole})}{3(180 \text{ g/mole})} = 0.42 \text{ g cells/g organic matter used}$$

In wastewater treatment, the organic matter is usually represented by chemical oxygen demand (COD) and the newly formed biomass is represented by VSS (volatile suspended solids). The COD of glucose is the amount of oxygen required to oxidize glucose to CO_2 and H_2O. The oxidation process of glucose can be represented by Equation 6.3 (Tchobanoglous et al., 2014).

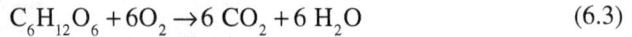

$$C_6H_{12}O_6 + 6O_2 \rightarrow 6\ CO_2 + 6\ H_2O \tag{6.3}$$

Hence, the COD of glucose can be calculated as follows:

$$COD = \frac{\Delta O_2}{\Delta C_6H_{12}O_6} = \frac{6\ (32 \text{ g/mole})}{(180 \text{ g/mole})} = 1.07 \text{ g } O_2/\text{g of glucose}$$

Similarly, the biomass yield in terms of COD can be calculated as follows:

$$Y = \frac{\Delta C_5H_7NO_2}{\Delta C_6H_{12}O_6 \text{ as COD}}$$

$$= \frac{2\ (113\text{g/mole})}{3\ (180\text{g glucose/mole})(1.07 \text{ g COD/g glucose})}$$

$$= 0.39 \text{ g ells/organic matter used}$$

During wastewater treatment, the biomass formed contributes to the total organic load of the effluent. Hence, the COD of the biomass needs to be calculated. The oxidation process of the biomass can be represented by Equation 6.4 (Tchobanoglous et al., 2014).

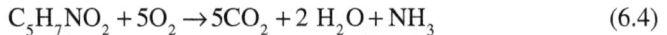

$$C_5H_7NO_2 + 5O_2 \rightarrow 5CO_2 + 2\ H_2O + NH_3 \tag{6.4}$$

Hence, the COD of the biomass can be calculated as follows:

$$COD = \frac{\Delta O_2}{\Delta C_5H_7NO_2} = \frac{5\ (32 \text{ g/mole})}{(113 \text{ g/mole})} = 1.42 \text{ g } O_2/\text{g of biomass}$$

The COD of the oxidized substrate is the total oxygen consumed and is provided by Equation 6.5 (Tchobanoglous et al., 2014).

$$\text{Oxygen consumed} = COD_{utilized} - COD_{cells} \tag{6.5}$$

Hence, the oxygen consumed can be calculated as follows:

1.07 g O_2/g of glucose $\times 3 \times 180$ g of glucose $- 1.42$ g O_2/g of biomass

$\times$ 2 $\times$ 113 g of biomass

$= 256.9$ g O_2

The oxygen consumed per unit of glucose as COD is

$$\frac{\text{Oxygen consumed}}{\Delta C_6 H_{12} O_6 \text{ as COD}} = \frac{256.9 \text{ g } O_2}{3(180\text{g glucose/mole})(1.07 \text{ g COD/g glucose})}$$

$$= 0.44\text{g } O_2/\text{g COD used.}$$

6.2.2 AEROBIC SUSPENDED GROWTH PROCESS

6.2.2.1 Activated Sludge Process

Activated sludge process (ASP) is a biological treatment process that converts organic matter in the wastewater into gases and cell tissue by aerobic microorganisms maintained in suspension in the presence of aeration. In 1914, the process was named activated sludge by Arden and Lockett because it involved the production of an activated mass of microorganisms capable of stabilizing the organic matter present in wastewater (Tchobanoglous et al., 2014). In the ASP, the biomass formed can be separated by gravity settling. The most commonly used configurations in ASP are plug-flow, step feed, and complete mixed reactors (Figure 6.3). The main differences between the configurations mentioned above are tank geometry and the mixing regime.

In the plug-flow reactor, the wastewater enters a long narrow tank from one end and escapes from the other end. The substrate concentration and the DO uptake or oxygen demand decrease along the length of the tank (Figure 6.3). The substrate concentration should ideally reach 0 mg/L if complete biodegradation is taking place. However, if the concentration of the substrate is high, the DO may be completely depleted. This may lead to fermentation or partial oxidation of the substrate, which may result in the production of organic acid and lower pH of the effluent.

In the step aeration feed, the influent enters the aeration chamber at different locations along its length. The substrate concentration is increased at every location where the wastewater is flowing in. Hence, the DO uptake rate or oxygen demand is also increased at these points (Figure 6.3). One benefit of step feeding is that there is mixing at continuous intervals, and hence there is an increase in DO at the inlet points. As a result, there are lower chances of DO deficit at the end of the tank.

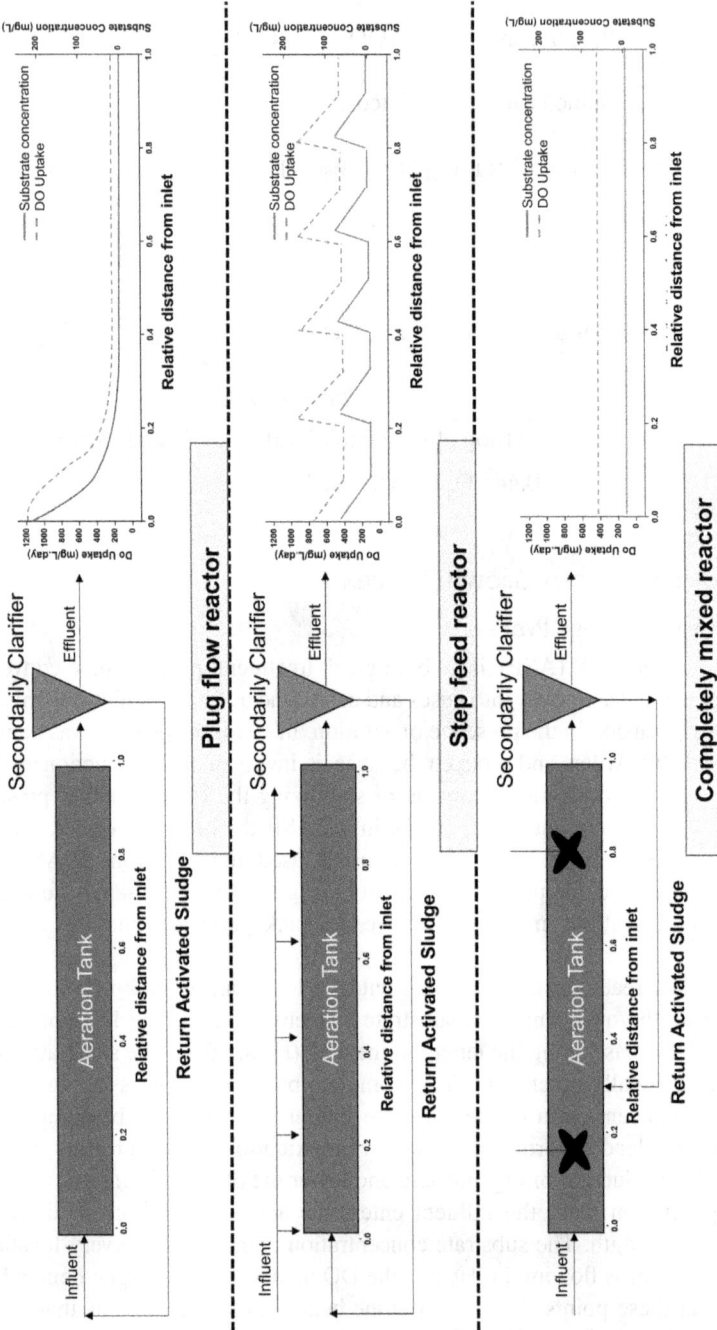

FIGURE 6.3 The dissolved oxygen uptake and substrate concentration in different configurations of activated sludge processes.

Source: Adapted from Mackenzie, 2010.

In the complete mixed reactor, the influent is diluted with the other contents of the tank as soon as it enters the aeration tank. As a result, the DO uptake rate and substrate concentration are uniform throughout the length of the tank.

The typical flow configuration of ASP has been depicted in Figure 6.4. The basic components of the ASP are as follows:

- An aeration tank in which microorganisms are responsible for treatment is kept in suspension and aerated
- Solid–liquid separation unit
- Recycle system for recirculation of sludge to maintain the microbial population

The wastewater coming out from preliminary and/or primary treatment flows into the aeration tank, where it gets mixed with the activated sludge. In the aeration tank, sufficient contact time is provided for mixing the influent wastewater with microbial suspension, that is, mixed liquor suspended solids (MLSS). Generally, the mixing is induced in the aeration tank by means of the diffused aeration system and/or mechanical aeration system. Sometimes a combination of diffused and mechanical aerators can also be used, and this type of combined ASPs equipped with such aeration systems requires less retention time. Typically, in conventional ASP, the wastewater is aerated for 6–8 h in long rectangular aeration tanks (Mackenzie, 2010). The aeration equipment must be designed to supply the oxygen demand rate of the microorganisms in the ASP under a wide range of loading rates. In general, DO concentration in the aeration tank must be maintained at about 1.5–2 mg/L (Tchobanoglous et al., 2014). If DO concentrations drop too low, the growth of filamentous organisms may predominate, thereby causing poor settling of the sludge in the secondary clarifier.

The sewage from the aeration tank flows to the secondary sedimentation tank (SST) or secondary clarifier, where solid–liquid separation takes place. The hindered settling (type 3) occurs in the SST, and the settling velocity of discrete particles may not govern its design as in the case of primary sedimentation tank (PST). Some amount of sludge that is settled in the SST is recycled to the aeration tank to maintain the desirable MLSS concentration in the aeration tank. However, the excess sludge from SST should be wasted (i.e., the proportion of sludge must be discarded from the system) periodically to maintain the required sludge retention time (SRT) in the tank.

6.2.2.1.1 Kinetic Model for a Completely Mixed Activated Sludge Process Reactor

The line diagram for a completely mixed reactor has been provided in Figure 6.4. By assuming the steady-state conditions within the controlled section of the treatment system, kinetic models can be developed. The mass balance in the controlled section for the biomass can be written as per Equations 6.6 and 6.7 (Mackenzie, 2010; Tchobanoglous et al., 2014).

$$\text{Biomass}_{in} - \text{Biomass}_{out} = \text{Net rate of change in biomass inside the system} \qquad (6.6)$$

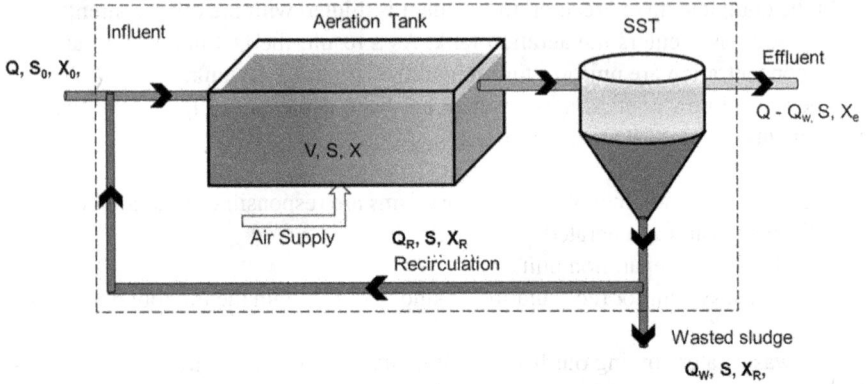

FIGURE 6.4 Schematic representation of activated sludge process.

Equation 6.6 can be simplified to Equation 6.7.

$$\text{Biomass}_{in} + \text{Biomass production} = \text{Biomass}_{out} \tag{6.7}$$

where
$\text{Biomass}_{in} = Q \times X_0$

$\text{Biomass}_{out} = \text{Biomass in the effluent} + \text{Biomass wasted} = (Q - Q_W) \times X_e + Q_W \times X_R$

Substituting the value of biomass$_{out}$ in Equation 6.7, we get

$\text{Biomass}_{in} + \text{Biomass production} = (Q - Q_W) \times X_e + Q_W \times X_R$

Biomass production as per Monod expression (Mackenzie, 2010) is provided in Equation 6.8:

$$= V \times \frac{dX}{dt} = V \times \left(\frac{\mu_m \times X \times S}{K_s + S} - K_d \times X \right) \tag{6.8}$$

Therefore, from Equation 6.7, we get

$$Q \times X_0 + V \times \left(\frac{\mu_m \times X \times S}{K_s + S} - K_d \times X \right) = \left[(Q - Q_W) \times X_e + Q_W \times X_R \right] \tag{6.9}$$

Similarly, the mass balance for the substrate

$$\text{Substrate}_{in} - \text{Substrate}_{out} = \text{Substrate utilized} \tag{6.10}$$

where

- Substrate$_{in}$ = $Q \times S_0$
- Substrate$_{out}$ = Substrate in effluent + Substrate wasted = $(Q - Q_W) \times S + Q_W \times S$
- Substrate utilized = $V \times \dfrac{dS}{dt}$

$$= V \times \left(\frac{\mu_m \times X \times S}{Y\,(K_s + S)} \right)$$

Therefore, from Equation 6.10, we get

$$Q \times S_0 - \left[(Q - Q_W) \times S + Q_W \times S \right] = V \times \left(\frac{\mu_m \times X \times S}{Y\,(K_s + S)} \right) \tag{6.11}$$

where
Q = Design flow rate (m³/d)
S_0 = Influent substrate concentration (mg/L)
S = Substrate concentration in the effluent (mg/L)
X_0 = Biomass concentration (MLSS) in influent (mg/L)
V = Volume of the aeration tank (m³)
X = MLSS concentration in aeration tank (mg/L)
X_R = MLSS concentration in the return line from clarifier (mg/L)
X_e = MLSS concentration in the effluent (mg/L)
Q_W = Waste discharge flow rate (m³/d)
k_s = Half saturation constant (mg/L)
k_d = Endogenous decay rate constant (d⁻¹)
Y = Synthesis yield (g cells/g organic matter used)
μ_m = Maximum biomass growth rate (d⁻¹)

The following assumptions can be made for the simplification of Equations 6.9 and 6.11:

- The biomass concentration in the influent and effluent of the system can be negligible, that is, X_0 and $X_e = 0$.
- All the reactions occur in the aeration tank only (the calculated volume (V) represents the aeration tank volume only).
- The influent substrate concentration S_0 is immediately diluted to concentration S due to complete mixed reactor.

Based on the assumptions, Equations 6.9 and 6.11 can be modified as follows:

$$V \times \left(\frac{\mu_m \times X \times S}{K_s + S} - K_d \times X \right) = \left[Q_W \times X_R \right] \quad \rightarrow \text{For biomass}$$

$$\left(\frac{\mu_m \times S}{K_s + S} \right) = \left[\frac{Q_W \times X_R}{V \times X} \right] + K_d \tag{6.12}$$

$$Q \times \left(S_0 - S \right) = V \times \left(\frac{\mu_m \times X \times S}{Y (K_s + S)} \right) \quad \rightarrow \text{For substrate}$$

$$\left(\frac{\mu_m \times S}{K_s + S} \right) = \frac{Q \times Y \times \left(S_0 - S \right)}{V \times X} \tag{6.13}$$

By equating, Equations 6.12 and 6.13, we get

$$\left[\frac{Q_W \times X_R}{V \times X} \right] + K_d = \frac{Q \times Y \times \left(S_0 - S \right)}{V \times X} \tag{6.14}$$

In Equation 6.14, the inverse of $\dfrac{Q_W \times X_R}{V \times X}$ represents the mean cell residence time

(MCRT) or sludge age, denoted by θ_c or solid retention time (SRT). The inverse of $\dfrac{Q}{V}$

represents the hydraulic retention time (HRT) denoted by θ. The MCRT ot SRT can be depicted by Equation 6.15:

$$\text{MCRT or SRT} \left(\theta_c \right) (d) = \frac{V \times X}{Q_W \times X_R} \tag{6.15}$$

However, if the effluent has non-negligible biomass concentration, the sludge age may be rewritten as per Equation 6.16:

$$\theta_c = \frac{V \times X}{Q_W \times X_R + \left(Q - Q_W \right) \times X_e} \tag{6.16}$$

$$\text{Hydraulic retention time } (\theta)(d) = \frac{V}{Q} \tag{6.17}$$

Therefore, by substituting θ_c and θ in Equation 6.14, we get

$$\frac{1}{\theta_c} = \frac{Y \times (S_0 - S)}{\theta \times X} - K_d = \frac{\mu_m \times S}{K_s + S} - K_d \tag{6.18}$$

The MLSS concentration can be calculated by using Equation 6.19, and it is correlated with θ_c and θ

$$X = \frac{\theta_c \times Y \times (S_0 - S)}{\theta \times (1 + K_d \theta_c)} \tag{6.19}$$

The volume of the aeration tank volume can be calculated from Equation 6.20 when kinetic coefficients are known.

$$V = \frac{Q \times \theta_c \times Y \times (S_0 - S)}{X \times (1 + K_d \theta_c)} \tag{6.20}$$

6.2.2.1.2 Operational Problems Associated with Activated Sludge Process

ASP is the most commonly used biological treatment technology for the treatment of wastewater. However, for the ASP to function properly, there are a few operational parameters that need to be carefully considered. The various operational problems associated with ASP are as follows:

Sludge bulking: A principal concern in ASP with clarifiers for solid–liquid separation is to maintain a good settling sludge. However, due to the presence of filamentous type bacteria, poor settling of sludge can be observed. For this reason, the sludge volume index (SVI) is used as an indicator for sludge settling properties. SVI is defined as volume occupied in ml by 1 g of sludge when allowed to settle for 30 min.

SVI can be calculated by taking 1,000 mL of sludge (usually from the aeration chamber) and putting it into an Imhoff cone. The sludge is allowed to settle for 30 min. After the sludge has settled, the volume of the settled sludge is measured. The MLSS (mg/L) of a corresponding sample of the unsettled sludge is then calculated.

In order to calculate MLSS, a certain volume of the sludge is taken and placed in a silica crucible and dried at 110°C. The weight difference between the empty crucible and the crucible after evaporation is the weight of mixed liquor total solids (MLTS).

Similarly, the same volume of sludge is allowed to pass through Whatman Filter paper No. 42, and the filtrate is taken in a silica crucible and dried at 110°C. The weight difference between the empty crucible and the crucible after evaporation is the weight of mixed liquor dissolved solids (MLDS). The difference between MLTS and MLDS is the MLSS, and it is denoted in mg/L. Standard methods as per the APHA manual can also be carried out for measuring MLSS concentration.

SVI can be calculated as per Equation 6.21 (APHA, 2017; Mackenzie, 2010):

$$\text{SVI} \left(\text{mL/g}\right) = \frac{\text{Volume of settled sludge} \left(\text{mL/L}\right)}{\text{MLSS} \left(\text{mg/L}\right)} \times 1000 \text{ mg/g} \qquad (6.21)$$

Typically, a good settling sludge has an SVI of 100 mL/g, while sludge having an SVI of more than 150 mL/g is associated with filamentous growth.

> *Foaming:* Another nuisance in ASP is foaming, which is related to the develop-ment of bacteria having hydrophobic cell surfaces (*Nocardia and Microthrix parvicella*) that attach to air bubbles. The foaming may be reduced by using surface sprays.
>
> *Rising of sludge:* The most common cause of this phenomenon is denitrification, in which nitrogen gas is trapped in the sludge mass, thereby making sludge mass buoyant.

6.2.2.1.3   Design of Activated Sludge Process

In conventional activated sludge systems, the wastewater is usually aerated for 6–8 h in long, rectangular aeration basins. There is enough air to keep the sludge in the suspension. Furthermore, other parameters, such as the organic loading rate, amount of microorganisms in the system, the hydraulic retention time (HRT), the solids, or sludge retention (SRT) time are important factors that need to be considered.

> *Organic loading rate (OLR):* The amount of BOD or COD applied to the aeration tank volume per day· (Kg BOD or COD/m³. d) (Equation 6.22):

$$\text{OLR} \left(\text{kg BOD or COD/m}^3 \cdot \text{d}\right) = \frac{\text{Sewage flow rate} \times \text{Influent BOD/COD}}{\text{Volume of the tank}}$$

$$= \frac{Q \times S_0}{V} \qquad (6.22)$$

> *Food to microorganism ratio (F/M):* F/M is defined as the rate of BOD or COD applied per unit volume of mixed liquor (Equation 6.23). F/M ratio is the main controlling factor of BOD removal. Lower F/M values give higher BOD removal and vice versa. F/M ratio can be controlled by varying the MLSS concentration in the aeration tank.

$$\text{F/M(mg/mg.d)} = \frac{\text{Sewage flow rate} \left(\text{Influent BOD/COD} - \text{Effluent BOD/COD}\right)}{\text{Volume of the tank} \times \text{MLSS}}$$

$$= \frac{Q \times (S_0 - S)}{V \times X} \qquad (6.23)$$

Mean cell residence time (θ_c) *or sludge retention time (SRT) (d):* The average period of time, in days, that solids or bacteria are maintained in the reactor (also called sludge age).

Recirculation ratio: It is the amount of sludge from the secondary clarifier that is returned to the sewage flow. It is also known as the return sludge ratio.

$$\left(Q_R/Q\right) = \frac{X}{X_{\mathrm{w}} - X} \tag{6.24}$$

The excess sludge produced is given as per Equation 6.25 (Mackenzie, 2010)

$$P_x = Y_{\mathrm{obs}} \times Q \times \left(S_0 - S\right)\left(10^{-3}\,\mathrm{kg/g}\right) \tag{6.25}$$

where $Y_{\mathrm{obs}} = Y/\left(1 + k_d \theta_c\right)$

Oxygen demand (R_o): It is the oxygen utilization rate for the decomposition of carbonaceous and nitrogenous organics.

$$\underset{(A)}{R_o\ (\mathrm{kg\ of\ O_2/h}) = [Q\ (S_0 - S)\ /\ f]} - \underset{(B)}{1.42\ P_{X,\,\mathrm{bio}}} + \underset{(C)}{4.57\ Q\ (NO_x)} \tag{6.26}$$

In Equation 6.26, the terms (A) and (B) are related to oxygen demand for carbonaceous matter only, whereas (C) represents the oxygen demand for nitrification. The term f represents the ratio of BOD_5 to BOD ultimate, and NO_x is the amount of total Kjeldahl nitrogen oxidized to nitrate.

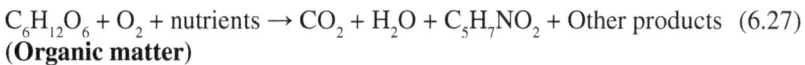

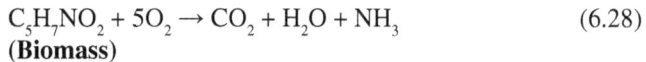

$$\underset{\textbf{(Organic matter)}}{C_6H_{12}O_6} + O_2 + \text{nutrients} \rightarrow CO_2 + H_2O + C_5H_7NO_2 + \text{Other products} \tag{6.27}$$

$$\underset{\textbf{(Biomass)}}{C_5H_7NO_2} + 5O_2 \rightarrow CO_2 + H_2O + NH_3 \tag{6.28}$$

Equation 6.27 represents the aerobic degradation of organic matter, and Equation 6.28 represents the oxidation of the biomass. The formed new cells $(C_5H_7NO_2)$ have an oxygen demand of approximately 1.42 units. The oxygen demand by the newly formed biomass has already been discussed in the earlier section. The extra oxygen requirement for nitrification is theoretically found to be 4.57 kg of O_2 per kg of ammonia nitrogen oxidized to nitrate (Tchobanoglous et al., 2004).

It is to be noted that there are different process arrangements available for the design of ASP. The variations in the design of ASP involve subtle differences, such as the rate of air or wastewater applications, reactor shape, method of introducing the air, and others. The operational parameter values for the different ASP processes (Mackenzie, 2010; Tchobanoglous et al., 2014) have been presented in Figure 6.5.

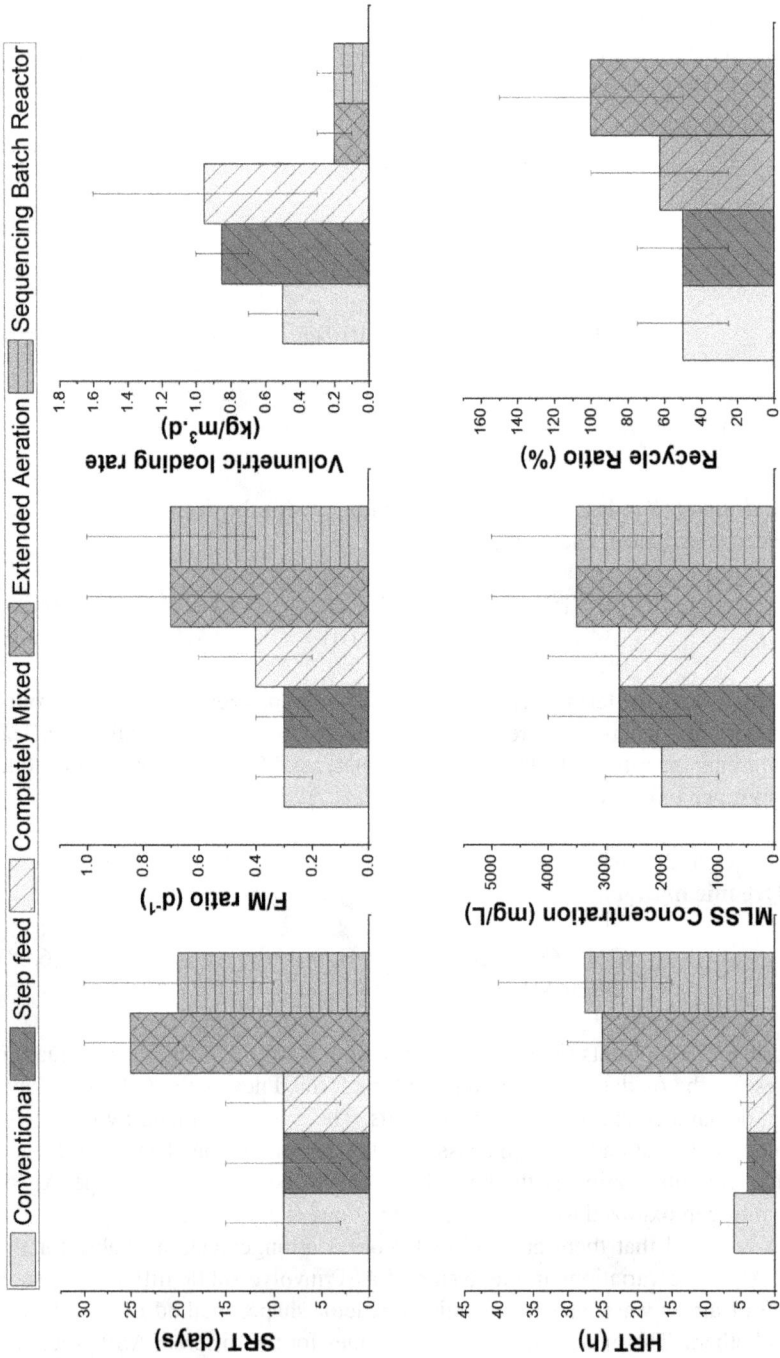

FIGURE 6.5 Typical design values for different activated sludge processes.

Source: Mackenzie, 2010; Tchobanoglous et al., 2014.

Example 1

Design an aeration tank for activated sludge: Secondary treatment of 12,000 m³/d of municipal wastewater will be accomplished using an activated sludge system. It is preferred that the BOD in the effluent be no higher than 5 mg/L after initial clarity, with the initial BOD being 150 mg/L. Following analysis in a pilot plant, the following kinetic values have been determined for usage in a totally mixed reactor: Assume $Y = 0.5$ kg/kg and $k_d = 0.05$ d⁻¹, $f = 0.68$, and SRT is 10 d. Determine (1) the volume of the reactor; (2) the mass and volume of solids that must be wasted each day; (3) the recycle ratio if the MLSS concentration is 3 kg/m³ and the underflow concentration is 10 kg/m³ from the secondary clarifier; (4) biomass produced; and (5) oxygen requirements.

Solution

Step 1: Volume of the reactor
MCRT or SRT $(\theta_c) = 10$ d,

$$V = \frac{Q \times \theta_c \times Y \times (S_0 - S)}{X \times (1 + K_d \theta_c)}$$

$$V = \frac{(12000 \text{ m}^3)/\text{d} \times 10 \times 0.5 \times (0.15 \text{ kg/m}^3 - 0.005 \text{ kg/m}^3)}{3.0 \text{ kg/m}^3 (1 + 0.05 \times 10)}$$

$$V = 1933.3 \text{ m}^3$$

Step 2: At equilibrium conditions

$$\theta_c = \frac{\text{mass of silids in reactor}}{\text{mass of solids wasted}} = \frac{VX}{Q_w X_R}$$

$$Q_w X_R = \frac{VX}{\theta_c}$$

$$= \frac{1933.3 \text{ m}^3 \times 3.0 \text{ kg/m}^3}{10 \text{ d}}$$

$$Q_w X_R = 579.9 \text{ kg/d}$$

If the concentration of solids in the underflow is 10000 mg/L

$$Q_w = \frac{579.9 \text{ kg/d}}{10 \text{ kg/m}^3} = 57.9 \text{ m}^3/\text{d}$$

Step 3: Applying mass balance
A mass balance around the secondary clarifier can be written as follows:

$$\left(Q+Q_r\right)X = \left(Q+Q_r - Q_w\right)X_e + (Q_r + Q_w)X_R$$

Assuming that the solids in the effluent are negligible compared to the influent and underflow,

$$QX + Q_r X = Q_r X_R + Q_w X_R$$

$$Q_r\left(X_R - X\right) = QX - Q_w X_R$$

$$Q_r = \frac{QX - Q_w X_R}{\left(X_R - X\right)}$$

$$= \frac{12000 \text{ m}^3/\text{d} \times \left(3.0 \text{ kg}\right)/\text{m}^3 - 579.9 \text{ kg/d}}{10 \text{ kg/m}^3 - 3 \text{ kg/m}^3}$$

$$Q_r = 5060 \text{ m}^3/\text{d}$$

The recirculation ratio is

$$\frac{Q_r}{Q} = \frac{5060}{12000} = 0.42$$

Step 4: Determination of biomass produced or excess sludge quantity

$$\text{Excess sludge quantity}\left(P_{X,bio}\right), \text{kg/d} = \frac{Q \times Y \times \left(S_0 - S\right)}{1 + K_d \times \theta_c}$$

$$\frac{12000 \times 0.5 \times \left(150 - 5\right)}{1 + 0.05 \times 10}$$

$$= 580 \text{ kg/d}$$

Step 5: Determination of oxygen demand

Oxygen demand (R_o), kg of O_2/d $= [Q\,(S_0 - S)\,/\!f)] - 1.42\,P_{X,bio}$

$$R_o = \frac{12000}{1000} \times \frac{150 - 5}{0.68} - 1.42 \times 580$$

$$= 1735 \text{ kg of } O_2\text{/d.}$$

Example 2

To meet the effluent criterion of 30 mg/L BOD and 30 mg/L TSS, design a fully mixed activated sludge system. Calculate how big an aeration tank will need to be if the BOD of the TSS may be assumed to be equal to 60% of the TSS concentration. The current principal facility has made the following information available: The current primary plant effluent features a flow $=10,000$ m^3/d and a BOD$_5 = 100$ mg/l.

Growth constants are assumed to be as follows: $K_s = 100$ mg/L BOD, $\mu_m = 2.5$ d^{-1}, $k_d = 0.050$ d^{-1}, and $Y = 0.50$ mg VSS/mg BOD eliminated. Assume that MLVSS $= 2,000$ mg/L and that the secondary clarifier can produce an effluent with 30 mg/L TSS.

Solution

Step 1: Estimate the allowable soluble BOD$_5$ in the effluent using the 63% assumption from above

Soluble BOD in the treated effluent = Total BOD in the influent– BOD in suspended solids

$$S = 30.0 \text{ mg/L} - (0.60)(30 \text{ mg/L}) = 12 \text{ mg/L}$$

Step 2: Calculation of SRT

The SRT can be estimated from Equation 6.18 and the assumed values for the growth constants.

$$\frac{1}{\theta_c} = \frac{\mu_m \times S}{K_s + S} - K_d$$

$$\left(\frac{2.5 \times 12}{100 + 12}\right) = \left[\frac{1}{\theta_c}\right] + 0.05$$

$$\theta_c = 4.59 \text{ d}$$

Step 3: Calculation of HRT

Using the provided value of 2,000 mg/L for the MLVSS, the HRT can be calculated using Equation 6.18.

$$\frac{1}{\theta_c} = \frac{Y \times (S_0 - S)}{\theta \times X} - K_d$$

$$\frac{1}{4.59} = \frac{0.5 \times (100 - 12)}{\theta \times 2000} - 0.05$$

$$\theta = 0.082 \text{ d or } 1.97 \text{ h}$$

The volume of the aeration tank is then estimated:

$$0.082 \text{ d} = \frac{V}{10000 \text{ m}^3/\text{d}}$$

$$V = 820 \text{ m}^3$$

Two tanks of this size are required to meet redundancy requirements.

6.2.2.2 Sequential Batch Reactor

A sequential batch reactor (SBR) is a complete mixed batch reactor in which wastewater is treated over a series of time steps, such as fill, react, settle, decant, and idle. In SBR, the aeration and setting occur in the same tank. Therefore, recirculation of sludge is not required. Moreover, due to the involvement of only one reactor, this process can be used where space limitation is the main problem (Dutta and Sarkar, 2015). The different stages in the operation of SBR have been depicted in Figure 6.6.

Fill: The raw wastewater or wastewater from primary treatment is added to the reactor. During this stage, the reactor is mixed to promote biological activities. The mixing can be achieved by means of mechanical mixers or diffused aerators.

React: Under the controlled environmental conditions, the biomass degrade the organic matter. Biological nitrification and denitrification also happen during this stage for the removal of nitrogen. The sludge wasting is carried out in this phase to ensure the desired MLSS is maintained.

Settle: During this stage, solids are allowed to settle under quiescent conditions. Typically, a duration of 0.5 h is provided for settling.

Decant: The supernatant fluid is removed during the decanting period by means of floating or adjustable weirs.

Idle: An idle time period is used in a multi-tank system to provide time for one reactor to complete its fill phase before switching into another unit.

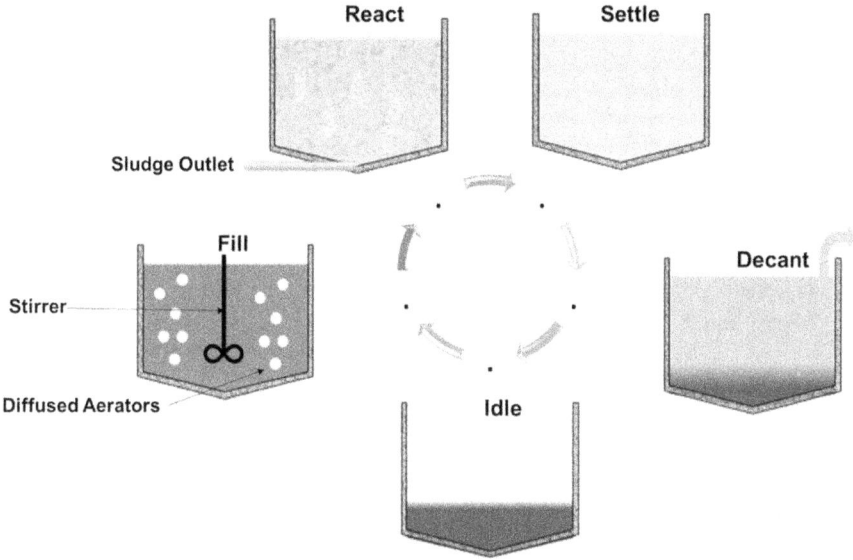

FIGURE 6.6 Different stages involved in the operation of SBR treatment unit.

The aforementioned cyclic steps are flexible, and they can be adjusted to undergo aerobic, anoxic, and anaerobic conditions, thereby, the treatment objectives, such as nitrification, denitrification, and phosphorous removal, can be achieved.

Apart from the hydraulic and organic loading rate, the performance of the SBR depends on several factors such as the total operating cycle time, volume of the tank, and the fill volume fraction. The different variables are discussed below.

SBR operating cycle time (T_C): This is the total length of cycle time (T_C), which contains fill time (t_f), reacting time (t_r), settling time (t_s), decanting time (t_d), and idle time (t_i) (Equation 6.29) (Mackenzie, 2010; Tchobanoglous et al., 2014).

$$T_C(h) = t_f + t_r + t_s + t_d + t_i \tag{6.29}$$

Therefore, the total number of cycles can be operated in a day $= \dfrac{24\text{ h}}{T_C}$

Fill volume (V_F): It is the amount of wastewater that needs to be filled in a tank per cycle. It can be calculated by Equation 6.30:

$$V_F(m^3) = \frac{Q\ (m^3/d)}{\text{Total number of cycles per day}} \tag{6.30}$$

Volume of the tank (V_T): If the full liquid depth in the tank is H, then decant depth (h) is generally considered 20% of full liquid depth (H) (Tchobanoglous et al., 2014).

$$\left(V_T\right)\left(m^3\right) = \frac{V_F}{\text{Ratio of fill volume to the tank volume}} \tag{6.31}$$

Fill volume fraction: Fill volume fraction is an important design parameter that is used to determine the liquid volume of the tank. It is the fraction of SBR tank volume occupied by settled solids. It can be determined by taking the product of MLSS and the SVI.

Example 3

If the total operating cycle time is 6 h and flow rate is 600 m³/d, find the fill volume and volume of the tank in a SBR system.

Solution

Step 1: Calculation of number of cycles

The total number of cycles can be operated in a day $= \dfrac{24\ h}{T_C} = \dfrac{24\ h}{6h} = 4$ cycles/d

Step 2: Calculation of fill volume

$$\text{The fill volume}\left(V_F\right) = \frac{Q\left(m^3/d\right)}{\text{Total number of cycles per day}} = \frac{600\left(m^3/d\right)}{4\left(cycles/d\right)} = 150\ m^3$$

Step 3: Calculation of the total volume of the tank

Assuming decant volume is 20% of full volume,

$$\text{Volume of the tank}\ \left(V_T\right) = \frac{V_F}{\text{Ratio of fill volume to the tank volume}}$$
$$= \frac{150\ m^3}{0.2} = 750\ m^3.$$

6.2.2.3   Membrane Bioreactor

Membrane bioreactor (MBR) is a type of ASP that involves suspended growth biological degradation followed by membrane separation by means of microfiltration (MF) (pore size: 0.1–1 μm) or ultrafiltration (UF) (pore size: 0.01–0.1 μm). Initially, the activated sludge process converts the organic matter into flocs in the presence of an air supply. Subsequently, the flocs are separated by the membrane and hence secondary clarifier is not required. Therefore, the land requirement is significantly reduced. Generally, two types of membranes, hollow fiber membranes and flat sheet membranes are used. These membranes consist of a thin surface layer of

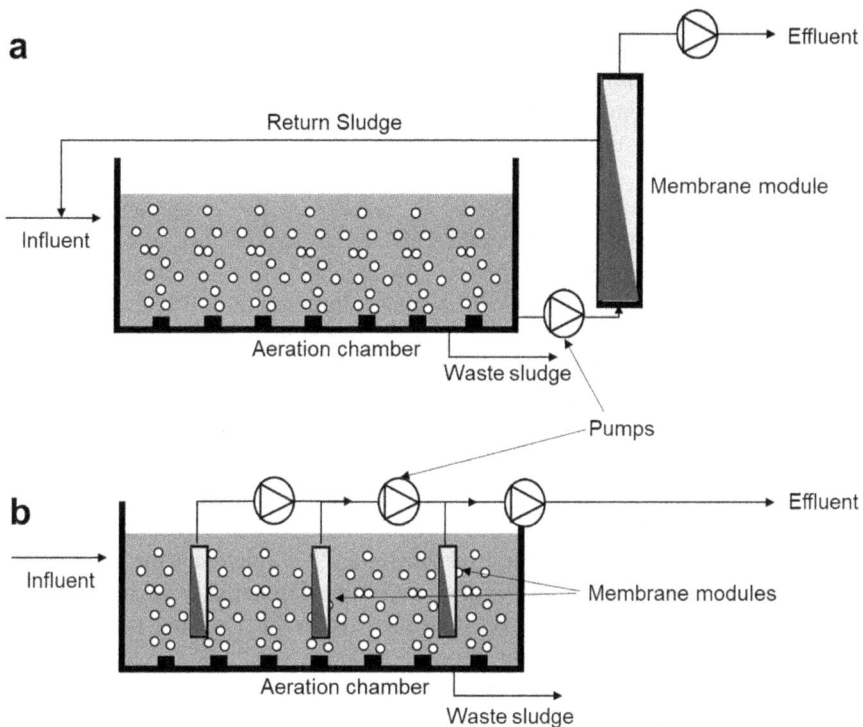

FIGURE 6.7 The typical representation of (a) external and (b) submerged membrane modules.

polymeric substances with high surface porosity. Polyethylene, polypropylene, and polyvinylidene difluoride are commonly used as polymeric materials for membranes. The MF can retain the bacteria, and UF can retain bacteria and viruses. However, permeate disinfection is preferred in both cases (Saidulu et al., 2021; Tchobanoglous et al., 2014). MBR has two basic configurations, such as an integrated system or submerged membrane module that have membranes immersed in the activated sludge reactor and a separate system or external membrane module with membranes placed outside the reactor. The typical representation of external and submerged membrane modules has been shown in Figure 6.7a and b.

In submerged MBR, the permeate is generated by means of transmembrane pressure (TMP) difference created by suction or vacuum pump. Recirculation is not required in the submerged membrane module, and the aeration creates the cross-flow. In the case of the external membrane module, the sludge is pumped back to the membrane. However, in most cases, submerged MBR is preferred due to lower energy requirement and high membrane packing density (Gukelberger et al., 2018).

Some of the important notations used in MBR are as follows.

- *Permeate:* The effluent from the membrane separation is called permeate.
- *Retentate:* The solids collected behind the membrane are called retentate.

- *Membrane flux:* Membrane flux is the flow rate per unit area of the membrane, and it is usually expressed as $L/m^2.d$.
- *Transmembrane pressure (TMP):* TMP is the pressure difference between the two sides of a membrane. TMP is used to estimate the excess pressure required to push the water through the membrane.

The effectiveness of the membrane is influenced by design and operating parameters, such as membrane flux and TMP. It can be said that, at a given TMP, the flux is inversely proportional to viscosity, indicating that flux can be increased at lower temperatures and higher MLSS concentrations. In MBR, the most cost-effective working range for MLSS is 8,000 to 10,000 mg/L (Tchobanoglous et al., 2004). Permeability (related to permeate and retentate) is a parameter that reflects the membrane flux and TMP. The decline in permeability through the membrane represents membrane fouling.

Membrane fouling is a phenomenon that represents the deposition of solids on the membrane surface or within the pores of the membrane, thereby reducing the membrane performance. The deposition of the solids occurs in two stages, that is, gradual deposition of extracellular polymeric substances (EPS), which includes proteins and polysaccharides on the membrane, followed by the formation of a cake layer. As a result, the decline in the permeability or increase in the TMP has been observed during membrane fouling. Therefore, special attention should be given for cleaning the membrane and performance (Iorhemen et al., 2016). The following methods can be applied to control the membrane fouling.

- *Pre-treatment:* Pre-treatment with fine screens (0.8–2 mm) is effective for the optimized performance of membranes. Fine screens will protect the membranes from hair and fibrous materials.
- *Air scouring:* Through coarse bubble aeration, scouring will be created at the surface of the membrane, thereby preventing the accumulation of solids. The typical air scouring rate is 3–12 L air/min m^2 of the membrane surface area.
- *Maintenance cleaning:* It involves backflushing (about 60–75 min) the membrane with NaOCl solution (200 mg/L) and/or citric acid (2000 mg/L) about one to two times per week (Tchobanoglous et al., 2004).
- *Recovery cleaning:* The recovery cleaning involves extensive chemical contact time, that is, 4 to 6 h, by using chlorine (1,000 mg/L) and citric acid (2,000 mg/L) (Tchobanoglous et al., 2004).

6.2.2.4 Oxidation Ponds

Oxidation ponds are another type of suspended growth biological system used for treating wastewater. The pond is typically a large, shallow earthen basin in which wastewater is retained for long periods, that is, days to weeks, for natural purification. Oxidation ponds are open to sun and air and they treat wastewater naturally using algae and bacteria. The symbiotic relationship between algae and bacteria leads to the purification of wastewater. In oxidation ponds, oxygen is added to the system through the photosynthesis process. However, some amount of oxygen is also provided by the air diffusion. In some cases, oxygen is supplied to the ponds through artificial aeration; these are called aerated ponds or aerated lagoons. Although oxidation ponds involve a slow rate of oxidation and high HRT, they are highly preferred in areas where the land availability is abundant and the temperature is favorable. In general, ponds are classified into aerobic, anaerobic, and facultative ponds.

Aerobic ponds are shallow depth ponds ranging from 0.15 to 0.45 m, and the DO is present at all depths of the pond. The symbiotic relation between microorganisms and algae leads to aerobic conditions in the pond. The microorganisms release the CO_2 and water molecules upon degradation of organic matter. The algae will uptake the CO_2 for their growth. The algae release the O_2 into the system, which helps in developing aerobic conditions. As mentioned earlier, the depth of tank is relatively shallow; therefore, the solar rays can easily penetrate through the entire depth of the pond to maintain the photosynthesis activity. Aerobic ponds are frequently used as a polishing or tertiary unit.

Anaerobic ponds have a relatively high depth, that is, ranging from 2 to 5 m, and the oxygen is absent in the entire depth of the pond except for the top surface layer. Anaerobic ponds are used to treat high organic loading rates (Mara, 2003). The removal of BOD is achieved by the sedimentation of settleable solids, and the settled solids undergo anaerobic decomposition. At the surface of the tank, bubbles can be observed due to the release of biogas (typically 70% CH_4 and 30% CO_2). Generally, the anaerobic ponds are used prior to aerobic or facultative ponds for improving the removal of organic matter and toxicity. For instance, many organic contaminants, such as phenols, pesticides, and others which are toxic to algae are degraded into non-toxicant form in anaerobic ponds. The accumulated digested solids are removed periodically, that is, once every 1 to 3 years. In recent times, the anaerobic ponds have been covered using polyethylene sheets to recover biogas and eliminate greenhouse gas emissions and odor problems.

The facultative ponds involve both aerobic and anaerobic conditions. Aerobic conditions are maintained in the upper portion of the tank by oxygen generated from the algae, while at the bottom of the pond, anaerobic conditions prevail. The intermediate zone, also called the facultative zone, involves both aerobic and anaerobic conditions. In the facultative zone, the DO fluctuates. The process that occurs in the facultative ponds has been presented in Figure 6.8. The biological solids produced in the aerobic zone get settled in the anaerobic zone and act as food for anaerobic microorganisms. The facultative ponds are most commonly used compared to others, and the performance of these ponds is comparable with other conventional treatment methods.

The oxidation ponds are similar to the completely mixed reactor without sludge recycling. Hence, the complete mixed reactor without sludge recycle model is commonly used for the design of ponds. Pre-treatment is an essential requirement for the oxidation pounds. The pre-treatment may include medium screens and grit chamber. The following mass balance equation can be used for the soluble substrate present in the system, where the amount of organic matter consumed by the microorganisms is equal to the influent organic matter and effluent organic matter (Peavy et al., 1985).

$$BOD\ cosumed = BOD\ in - BOD\ out$$

$$V(kS) = Q \times S_0 - Q \times S \tag{6.32}$$

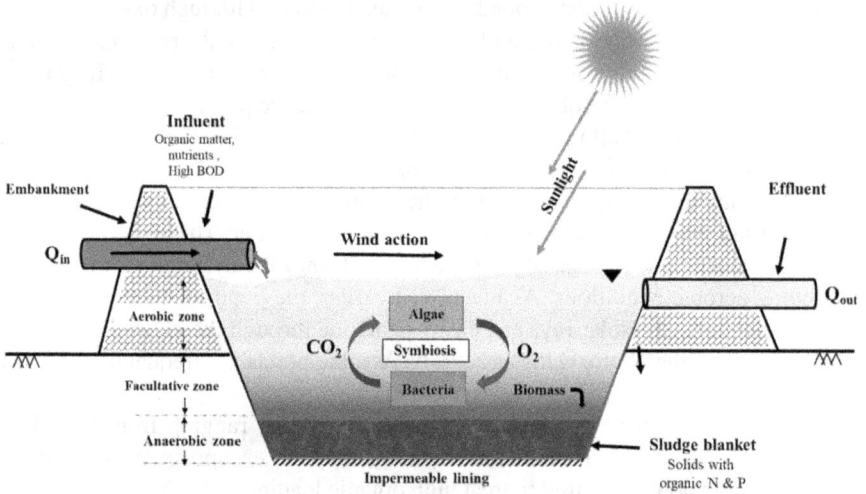

FIGURE 6.8   Schematic diagram of facultative ponds.

By rearranging the above equation, we get

$$\frac{S}{S_0} = \frac{1}{1 + \dfrac{K \times V}{Q}} = \frac{1}{1 + K \times \theta} \tag{6.33}$$

where
S/S_0 = Fraction of soluble BOD remaining
k = Reaction rate coefficient (d^{-1})
V = Volume of the pond (m^3)
Q = Flow rate (m^3/d)
θ = Hydraulic retention time (h)

If n ponds are arranged in the series, Equation 6.34 can be used for measuring the fraction of soluble BOD after n ponds (Peavy et al., 1985).

$$\frac{S}{S_0} = \frac{1}{\left(1 + \dfrac{K \times \theta}{n}\right)^n} \tag{6.34}$$

Temperature: The reaction rate constant k varies with the temperature (T, °C). Equation 6.35 gives the relation between k and T (Peavy et al., 1985).

$$\frac{k_T}{k_{20}} = \varnothing^{(T-20)} \tag{6.35}$$

6.2.3 Aerobic Attached Growth Processes

6.2.3.1 Trickling Filters

Trickling filters (TFs) are non-submerged fixed-film biological reactors in which media, such as stones or plastic material, is packed to support biofilm growth. For many years, the most popular design consisted of a 1–3 m deep bed of stones through which the wastewater traveled. A revolving arm normally distributes the wastewater over the surface of the rocks. Rock filter diameters can reach up to 60 meters. As their name implies, trickling filters are not primarily used for filtering or straining liquids. The 25–100 mm diameter rocks in a rock filter have too large of holes to effectively strain out particles. The rocks provide a wide surface area for microbes to attach to and grow as slime as they feed on organic substances. Although rock trickling filters have operated admirably for a number of years, they have certain restrictions. Under large organic loads, slime growths can become so abundant that they fill the vacuum areas between the rocks, leading to floods and system collapse. In addition, there are restricted vacuum spaces in a rock filter. This reduces air circulation and the amount of oxygen accessible to microorganisms. In turn, this barrier limits the amount of effluent that may be processed. To circumvent these constraints, other filling materials for the trickling filter have gained popularity. These materials include corrugated plastic sheet modules and plastic ring modules. Specific surface area (SSA) and porosity of the media play a key role in the effective performance of the system. The SSA refers to the amount of surface area available for biofilm growth, while porosity represents the void spaces available for the passage of air and wastewater. The typical flow diagram of the TFs is shown in Figure 6.9.

TFs involve three main components such as filter media, wastewater distribution system, and underdrain system.

Filter media: The selection of filter media depends on SSA availability, cost of the material, porosity, and durability. Initially, in TFs, crushed stones (size varies from 50 to 100 mm) were used as a filter media. The crushed stones offer good durability and provide a chemical-resistant surface. However, the volume of void spaces available in rock-type filters is limited. Henceforth, in recent times, stones are being replaced by plastic materials. The plastic carriers offer relatively large SSA for biofilm growth. Moreover, these materials are lighter in weight, thereby more structural height can be provided (up to 12 m) for TFs along with an increment in treatment capacity.

Wastewater distribution system: This rotatory distribution system distributes the wastewater uniformly over the surface of the media. The rotating distribution system contains two or more arms that contain nozzles, and through these nozzles, wastewater is discharged over the filter bed. This arrangement provides the intermittent wastewater supply and provides some resting period for air circulation. The rotation of the arms will take place through the jet action generated by nozzles. However, sometimes electric motor can also be used to drive the rotor.

Underdrain system: In TFs, the underdrain system serves two purposes: collecting the treated effluents and sloughed biomass, and another one is to provide an open area for the movement of the air.

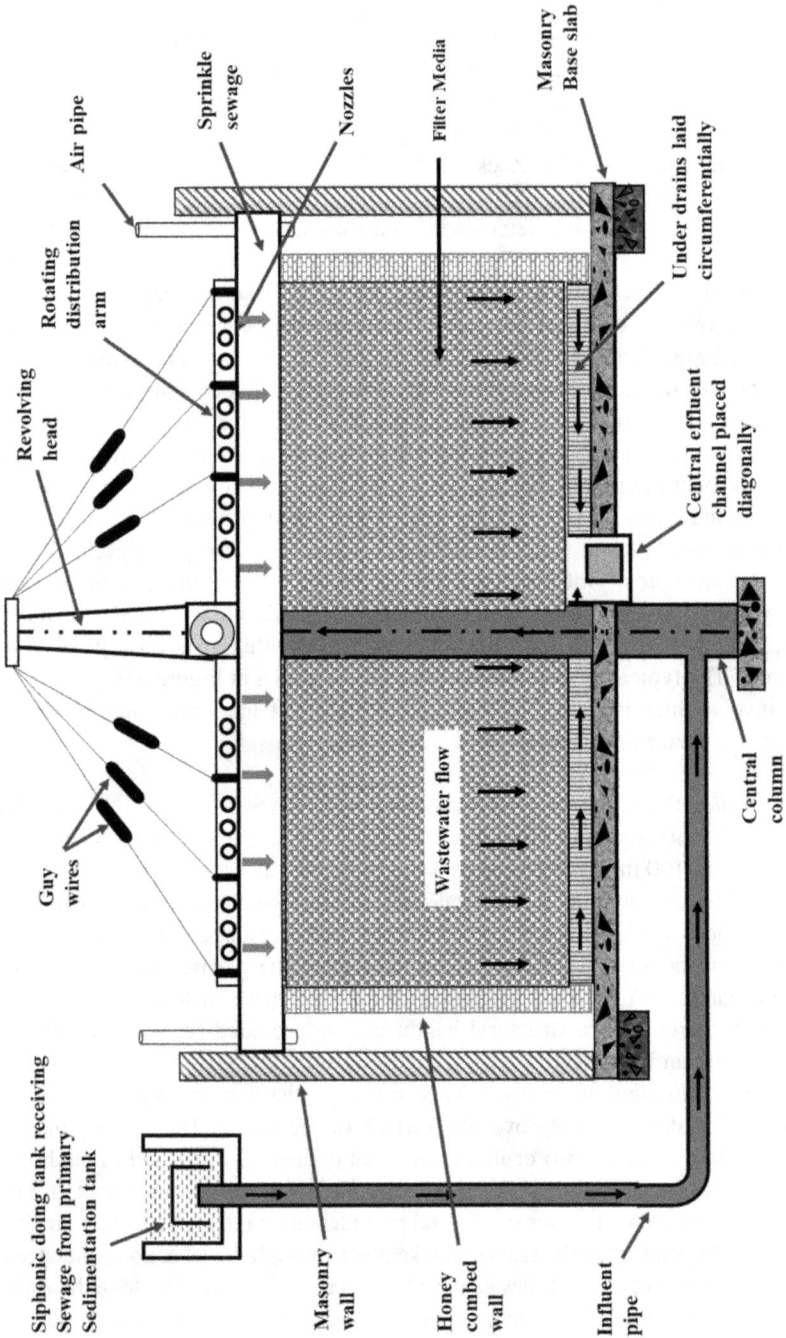

FIGURE 6.9 Typical representation of a trickling filter.

6.2.3.1.1   Working Principle of Trickling Filters

As mentioned before, when the wastewater is distributed over the media, it forms biofilm as it trickles down. This biofilm is often called slime layer or zoogloeal film. The thickness of the biofilm can reach up to 10 mm based on the operating conditions. As the thickness of biofilm increases, anaerobic conditions develop at the surface of the packing. During this stage, biofilm enters an endogenous respiration phase due to limitation in organic matter to the inner depth of the biofilm. Therefore, microorganisms lose their ability to stick to the packing material. Hence, biofilm is removed from the surface of the material by the shear force developed by influent wastewater. This phenomenon is called sloughing. The detached biofilm is collected at the secondary clarifier. In TFs, recirculation of the effluents from the clarifier is carried out to maintain the wetness of the packing materials during low-flow conditions.

Some of the limitations pertaining to the operation of trickling filters are ponding and nuisance due to odor and flies. The presence of large suspended material or excessive growth of biomass often leads to bridge formation in pore spaces (i.e., filling pore spaces between media particles), thereby leading to ponding. This process leads to poor water distribution and air in the filter media, which in turn leads to the formation of odor and mosquito breeding. This problem can be solved with proper pretreatment of the influent wastewater and increasing the HLR to induce sloughing. Furthermore, the undesirable growth of biofilm leads to anaerobic conditions, and these conditions lead to the generation of odor gases. Flies can also choke the filter bed, thereby causing interference with the working of filters. Chlorination or other appropriate methods can be used to control the flies and larvae developed at the TF site.

6.2.3.1.2   Design of Trickling Filters

Trickling filters (TFs) were the method of choice for secondary treatment of wastewater from cities because they worked well and were easy to use. Attached growth processes are often chosen because they require less skilled workers to run and use less energy than activated sludge processes. For redundancy, at least two parallel systems are set up. Based on the hydraulic loading rate (HLR), organic loading rate (OLR), recirculation ratio, etc., TFs can be classified into two categories: low-rate TFs and high-rate TFs. Low-rate TFs have higher depth of filter media (1.5–3 m), lower hydraulic loading rate (1–4 $m^3/m^2.d$), and lower organic loading rate (0.08–0.22 $kg/m^3.d$). Recirculation is not required and intermittent sloughing is carried out. On the other hand high-rate TFs have lower depth of filter media (1–2 m), around 10 times higher hydraulic loading rate (10–40 $m^3/m^2.d$), and higher organic loading rate (0.36–1.8 $Kg/m^3.d$). Recirculation is carried out and the recirculation ratio is between 1 and 3 and continuous sloughing is carried out. Hence, the dosing rate, recirculation ratio, amount of oxygen supply, HLR, OLR, and the sludge retention time are essential operational parameters for the proper functioning of the TFs.

> *Dosing rate (DR):* DR is the depth of the liquid discharged on the top of the packing for each pass of the distributor. The DR can affect the biofilm thickness. For instance, Low DR leads to excessive biofilm thickness.

$$\text{Dosing rate } (DR) = \frac{q \times (1+R)}{N_A \times n} \qquad (6.36)$$

where
DR = Dosing rate (mm per pass of distributor)
q = Hydraulic loading rate (m³/m².d)
R = Recirculation ratio
n = Rotational speed (rotations per minute)
N_A = number of arms in the rotary distributor

Recirculation ratio (R): The recirculation of the effluents is an important design consideration. The ratio of returned flow to the incoming flow is termed as recirculation ratio (R). As mentioned before, recirculation helps dilute the strong influent loads and prevents the drying out of biological slime during low-flow conditions.

$$\text{Recirculation ratio } (R) = \frac{Q_R}{Q} \qquad (6.37)$$

where
Q_R = Returned flow rate (m³/d)
Q = Incoming flow rate (m³/d)

Hydraulic loading rate (HLR): HLR influences the contact time of liquid with the biofilm. Equation 6.38 can be used to estimate the required contact time between liquid and biofilm:

$$t = \frac{C \times D}{THL^n} \qquad (6.38)$$

where
C = Constant depends on the type of packing material used
D = Depth of packing (m)
t = Contact time (d)
n = Hydraulic constant (typical range 0.5 to 0.67)

THL = Total hydraulic loading rate (m³/m².d) = $\dfrac{Q(1+R)}{A}$

A = Filter area (m²)
R = Recirculation ratio

The change in the BOD concentration in the filter with the time is described by the first-order reaction (Equation 6.39) (Mackenzie, 2010; Tchobanoglous et al., 2014).

$$\frac{dS}{dt} = -kS \tag{6.39}$$

By solving the above equation, we get

$$\frac{S_e}{S_i} = e^{-\frac{KD}{THL^n}} = e^{-\frac{KD}{(q(1+R))^n}} \tag{6.40}$$

where
S_e = Effluent BOD concentration, mg/L
S_i = Influent BOD concentration in total flow to TFs, mg/L
k = Filter treatability and packing coefficient (typical value = 0.69 d^{-1} for 20 °C)
q = Hydraulic loading rate (Q/A)

For solving Equation 6.40, we should know S_i, but this value is related to S_e. Therefore, S_i needs to be replaced with S_o (influent BOD based on primary effluent). By applying BOD mass balance of primary effluent and recirculation flows entering into the tank,

$$Q \times S_o + Q_R \times S_e = (Q + Q_R) \times S_i \tag{6.41}$$

$$\frac{S_i}{S_e} = \frac{R \times S_e + S_o}{(1+R) \times S_e} \tag{6.42}$$

Inverting Equation 6.40 and substituting it in Equation 6.42, we get

$$S_e = \frac{S_o}{(R+1) \times e^{\left(\frac{k_{20} \times A_s \times D \times \varnothing^{T-20}}{[q(1+R)]^n}\right)} - R} \tag{6.43}$$

where
S_o = Influent BOD based on primary effluent (mg/L)
S_e = Effluent BOD concentration (mg/L)
k_{20} = Filter treatability constant at 20 °C
A_s = SSA of the packing (m^2/m^3)
D = packing depth (m)
$\varnothing$ = Temperature coefficient (1.035) {$K_T = K_{20} \times \varnothing^{T-20}$}
q = Hydraulic loading rate based on the primary effluent flow (L/m^2.s)
R = Recirculation ratio (ratio of recirculation flow rate to primary effluent flow rate)
n = constant of packing

Sludge retention time (SRT): As compared to ASP, the SRT in TFs is typically very high, that is, around 100 d. Moreover, the sludge produced in TFs is less than 60–70% of sludge generated during ASP.

Efficiency of the TFs: The National Research Council has provided an empirical equation to calculate the efficiency of the single-stage or two-stage TFs. Generally, two-stage TFs are used for improving the effluent quality, where the secondary stage acts as a polishing unit (Mackenzie, 2010).

Removal efficiency for single-stage TFs (η_1)

$$\eta_1 = \frac{1}{1 + 4.12 \left(\dfrac{Q \times S_i}{V \times F} \right)^{0.5}} \tag{6.44}$$

where
η_1 = Fraction of BOD_5 removal at 20 °C, including recirculation
Q = Wastewater flow rate (m^3/s)
S_i = Influent BOD_5 concentration (mg/L)
V = Volume of filter media (m^3)
F = Recirculation factor = $\dfrac{1 + R}{(1 + 0.1 R)^2}$

Removal efficiency for single-stage TFs (η_2)

$$\eta_2 = \frac{1}{1 + \dfrac{4.12}{1 - \eta_1} \left(\dfrac{Q \times S_e}{V \times F} \right)^{0.5}} \tag{6.45}$$

η_1 = Fraction of BOD_5 removal after second-stage TF at 20°C, including
S_e = Effluent BOD concentration from first-stage of TF (mg/L)

Required oxygen supply in TFs (R_o): The oxygen transfer efficiency is about 5% for BOD removal applications:

$$R_o = 20 \left[0.8e^{-9 L_B} + 1.2e^{-0.17 L_B} \right] \times P.F \tag{6.46}$$

where
R_o = Oxygen supply (kg of O_2/kg of BOD applied)
L_B = BOD loading to filter (kg BOD/m^3.d)
P.F = Peaking factor

Example 4

Design flow rate: 10,000 m^3/d; BOD influent: 125 mg/L; TSS influent: 65 mg/L; targeted effluent BOD and TSS: 20 mg/L; Temperature: 14°C. Filter treatability constant at 20°C = 0.23. Consider the above-mentioned design parameters and provide two towers at 6.1 m depth. Plastic packing with an SSA of 90 m²/m³ and assume recirculation ratio of one- and two-arm distribution system. The required minimum wetting rate = 0.5 L/m². The packing coefficient value is 0.5. Design the following parameters:

1. Diameter of trickling filter
2. Volume of packing required
3. Oxygen required

Solution

Step 1: Calculation of hydraulic loading rate

$$S_e = \frac{S_o}{(R+1) \times e^{\frac{k_1 \times A_s \times D \times \varnothing^{T-20}}{[q(1+R)]^n}} - R}$$

Rearranging the above equation,

$$q(1+R) = \left\{ \frac{k_t \times D}{\ln\left[\frac{S_o + RS_e}{S_e(1+R)}\right]} \right\}^{1/n}$$

Recirculation ratio $(R) = 1$

$$K_{20} = 0.23$$

$$K_{14} = K_{20} \times \varnothing^{T-20} = 0.23 \times 1.035^{14-20} = 0.187$$

$$q(1+1) = \left\{ \frac{0.187 \times 6.1}{\ln\left[\frac{125 + 1 \times 20}{20(1+1)}\right]} \right\}^{1/n}$$

Therefore, hydraulic loading rate based on the primary effluent flow $(q) = 0.443$ L/m².s.

Step 2: Diameter of trickling filter

$$\text{Filter area} = \frac{Q}{q} = \frac{10000 \times 10^3}{0.443 \times 24 \times 60 \times 60} = 261.2 \text{ m}^2$$

Area of each tower = 261.2/2 = 130.6 m²

$$\text{Diameter} = \sqrt{\frac{4 \times 130.6}{\pi}} = 12.89 \text{ m} \text{ ` } 13 \text{ m}$$

Hence, provide two towers with 13 m diameter of each.

Step 3: Volume of packing required

Packing volume = 261.2 × 6.1 = 1593.3 m³

Surface area of the packing = 1593.3 m³ × 90 m²/m³ = 143398.8 m².

Step 4: Required oxygen supply
Using Equation 6.46, required oxygen for BOD removal can be calculated (Assume P.F =1.4)

$$R_o = 20\left[0.8e^{-9L_B} + 1.2e^{-0.17L_B}\right] \times \text{P.F}$$

$$L_B = \text{BOD loading to filter, kg BOD/m}^3.\text{d} = \frac{Q \times S_o}{V} = \frac{10000 \times 125 \times 10^{-3}}{2412}$$

$$= 0.52 \text{ kg BOD/m}^3.\text{d}$$

$$R_o = 20\left[0.8e^{-9 \times 0.52} + 1.2e^{-0.17 \times 0.52}\right] \times 1.4$$

$$= 30.9 \text{ kg of O}_2/\text{kg of BOD applied.}$$

6.2.3.2 Biofilters

Biofiltration is a kind of fixed-film attached growth process mostly used for odor control applications. Microorganisms present in the biofilters can biodegrade the compounds in their gas or vapor phase. In biofilters, a packed bed is used. Peat, wood chips, compost (rich in the microorganism and provides some mineral nutrients), or a mixture of compost with other materials are often used as a packing material. As mentioned before, packing media should have high SSA and porosity. Two processes occur simultaneously when the gases pass through the filter bed: absorption/adsorption and bioconversion. The microorganisms immobilized on the surface of the packed bed oxidize the adsorbed/absorbed gases. Also, some of the gases are adsorbed by the packing material. The design of the biofilter system is similar to that of TFs. The components present in the biofilter are similar to that of TFs. Moisture content and temperature are the two critical environmental parameters that affect the performance of biofilters.

6.2.4 AEROBIC HYBRID SYSTEMS

6.2.4.1 Rotating Biological Contactors

Rotating biological contactor (RBC) is a fixed-film media system in which a rotating disk is used to support biological growth. A set of closely spaced disks or series of modules that are mounted on a horizontal shaft are installed in contoured-bottomed tanks in which wastewater flows continuously. These disks are rotated by means of a power supply to the shaft. The rotational speed of the disk must generate sufficient turbulence to keep the solids in suspension. Moreover, shear velocity for the sloughing of biofilm is influenced by the rotational speed of the disk. Generally, rotating disks are immersed up to about 40% of their diameter in the wastewater. Typically, the disks rotate at 1–2 rpm. The surface area of the slowly rotating discs is alternatively exposed to wastewater and atmospheric air. It was reported that around 10,000 m^2 of the surface area is available for a disk having a diameter of 3.7 m and a length of 7.6 m (Peavy et al., 1985). When influent wastewater is applied, the microorganisms start to adhere to the surface of the rotating disk. As the disc rotates, the attached microbes start to absorb the organic matter present in the wastewater, following which oxidation occurs. A minimum HRT of 0.7–1.5 h is required for BOD removal to occur. The attached biofilm utilizes oxygen when it gets into contact with the atmospheric air. However, some modifications to RBC, including artificial aeration, have been implemented to overcome the oxygen limiting conditions in the bulk solution. The typical flow diagram of the RBC module has been shown in Figure 6.10a and b.

It can be observed that the working principle of RBC is similar to that of TF, except that the microorganisms are passing through the wastewater rather than the wastewater passing over microbes. The carbonaceous organic matter conversion takes place during the first stage of the series of modules; however, nitrification is usually completed after the fifth stage of the module. During the degradation process, the biofilm grows on the surface of the disk. The biofilm can grow up to a thickness of 2–4 mm, and it is highly influenced by the strength of influent wastewater and the rotational speed of the disc. The typical HLR and OLR applied in the RBC are 0.08 to 0.16 m^3/m^2.d and 4 to 10 g BOD/m^2.d, respectively (Tchobanoglous et al., 2004). As mentioned before, sufficient hydraulic loads must be applied for continuous sloughing of biofilm. The sloughed or detached biomass is relatively dense and settles well in the secondary clarifier. The recirculation of effluent through the tank is not necessary for RBC. The different physical units involved in RBCs are as follows:

- *Rotating media:* The rotating media and shaft are referred to as a drum. Approximately 40% of the drum is immersed in the wastewater. It supports biological activities.
- *Reactor basin:* A reactor basin is a physical unit where continuous wastewater flow occurs.

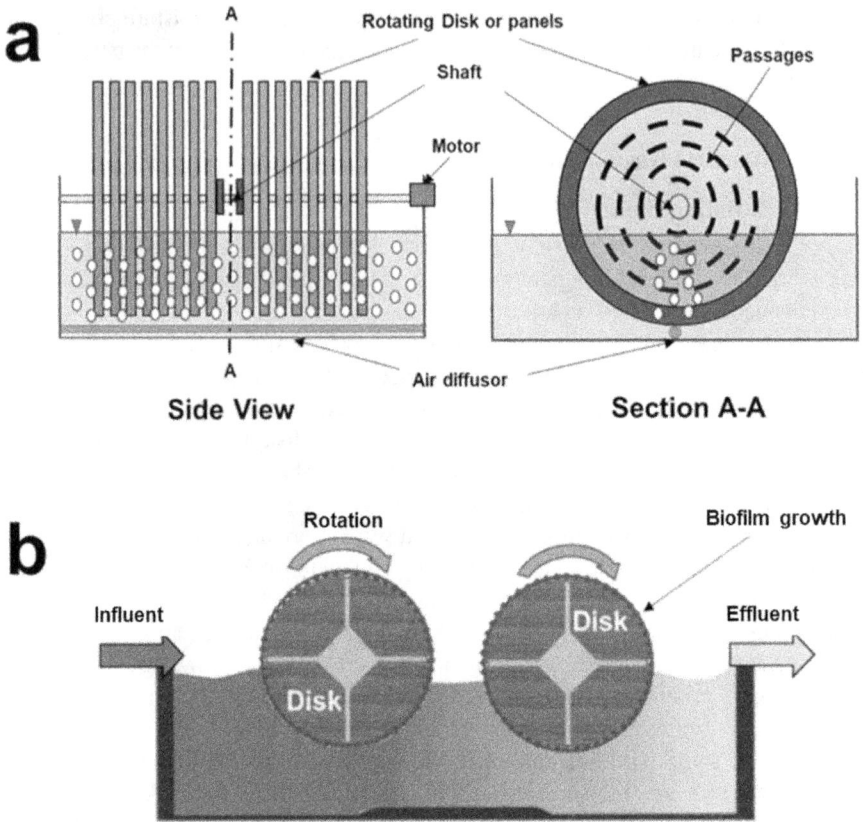

FIGURE 6.10 (a) Side view and component of rotating biological contactor and (b) process flow for rotating biological contactor.

- *Protective covers:* Covers that are made up of fiberglass are generally used to cover the RBC unit. These covers help in protecting the slime layer from different climatic conditions, such as rain, snow, and sunlight.

RBCs are easy to operate and require low energy and recirculation of sludge is not needed. It has a short HRT and sludge volume produced is low. Furthermore, the sludge has good settling properties. However, covers must be required to protect the media from damage and to stop excess algal growth. Furthermore, it requires high capital cost and is sensitive to temperature.

6.2.4.2 Integrated Fixed-Film Activated Sludge

The iintegrated fixed-film activated sludge process (IFAS) is a kind of hybrid system in which both suspended and attached biomass growth are involved. IFAS consists of an activated sludge system in which synthetic media has been added to support the attached growth of biomass in addition to suspended biomass growth. The media that is added to the reactor can be either dispersed type or fixed type. The addition of an attached growth system leads to an increase in effective MLSS concentration that may be around two times greater than the MLSS concentration present in ASP alone. This higher MLSS concentration can result in an increase in the effective SRT, thereby improving the overall performance of the system.

The IFAS system is preferred when a high level of treatment for biological nitrification–denitrification for nitrogen removal is required. Moreover, by converting the conventional ASP to an IFAS system, additional tank requirements can be eliminated. Higher energy demand due to operation elevated DO concentrations, media removal for diffuser maintenance, and additional hydraulic profile head loss due to flow through the media screening devices are some of the drawbacks associated with the IFAS systems.

There are different types of media that can be used in the IFAS. The floating or fixed media are called biofilm carriers and are typically either sponge-type polyurethane or specially designed polypropylene plastic.

Sponge-type biofilm carriers: Sponge-type biofilm carriers are foam cuboids with a specific gravity of about 0.95 g/cc. The filling ratio (FR) of this media typically ranges between 15 and 30% of the reactor. Screens or perforated plates can be used at the downstream end of the reactor to retain the carriers in the tank. Fine or coarse bubble size aeration can be applied to promote the mixing of the sponge carriers in addition to the oxygen supply.

Plastic-type biofilm carriers: The plastic carriers with the shape of wagon-wheel and a specific gravity of about 0.96–0.98 g/cc were originally developed by a Norwegian company. Thereafter, several modifications have been carried out to the plastic carriers in terms of shape and size. Plastic biofilm carriers have a bulk specific area of 500–700 m^2/m^3, and the FR of these type carriers can go up to a maximum of 70% (as these materials have less tendency to bind the downstream screens) of tank volume (Bassin and Dezotti, 2018).

Fixed material: Plastic sheets can also be used as supporting media in IFAS. Coarse bubble aeration is often used to ensure the proper oxygen transfer efficiency and to maintain the thickness of the biofilm attached to the fixed material (Tchobanoglous et al., 2004).

6.2.4.3 Design Parameters of Integrated Fixed-Film Activated Sludge

Critical design elements that need to be considered for the sponge or plastic media are the effluent sieves and forward velocity. Stainless steel effluent sieves are installed at the effluent of the tank to retain the media and collect the effluent flow. The sieves let the suspended media pass through and trap the foam in the aeration tank. The forward velocity influences the media movement in the tank. For example, higher forward velocity leads to pushing the media to the effluent end of the tank, where the accumulation of media can happen. Therefore, the recommended forward velocity is around 0.5–0.6 m/min (Tchobanoglous et al., 2004).

Sparged aeration is used to provide oxygen and circulate the floating media. Agitation and mixing caused by sparged aeration are capable of controlling excessive

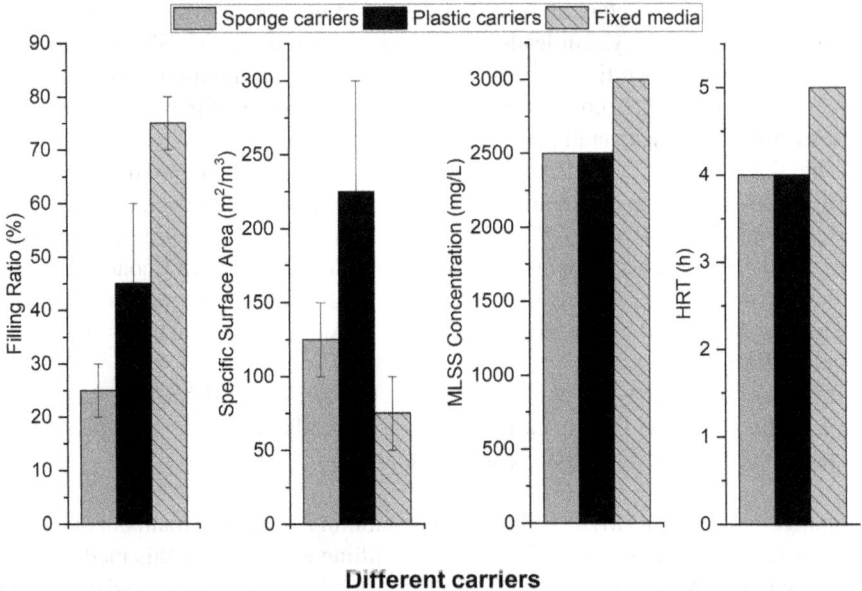

Different carriers

FIGURE 6.11 Typical design values for integrated fixed-film activated sludge system for different types of media.

Source: Tchobanoglous et al., 2004.

biofilm growth on the carrier. Coarse bubble aeration is generally used for IFAS applications; however, fine bubble aeration can also be used sometimes. The reason for applying coarse bubble aeration more often is that it avoids moving the media out of the basin for diffuser cleaning and maintenance. In the case of anoxic tanks, mechanical mixers are used to maintain the mixing of the carriers.

The solids loading to the secondary clarifier is the same as that of ASP. However, it was reported that the SVI value of mixed liquor from IFAS is slightly higher than from ASP alone. Furthermore, the filling ratio, SSA of the carriers, MLSS concentration, and the HRT are important parameters in the design of IFAS (Tchobanoglous et al., 2004). The typical design values for filling ratio, SSA of the carriers, MLSS concentration, and the HRT for different of media have been provided in Figure 6.11

6.2.4.4 Moving Bed Biofilm Reactor

The moving bed biofilm reactor (MBBR) process works similar to that of IFAS system, with the exception that there is no recirculation of activated sludge. MBBR process involves both suspended and attached growth processes, in which biofilms are grown on a sponge type or plastic carriers. These carriers keep moving in the reactor through aeration in aerobic reactors or by mechanical stirring in anoxic or anaerobic reactors. The media fill volume fraction can go up to 70%. The addition of attached growth in the reactor leads to an increase in effective SRT, thereby increasing the systems treatment capacity. A variety of biofilm carriers are investigated in MBBR, however, most of the present installations involved plastic-based biofilm carriers.

6.2.4.4.1   Moving Bed Biofilm Reactor Process Applications

The different configurations of MBBR have been investigated for BOD removal and nitrification. Most of the treatment schemes require pre-treatment that involves screening and grit removal. The physical facilities involved in MBBR are the same as that of IFAS system. Usually, BOD removal takes place in a single aerobic tank. For BOD removal and nitrification, a staged reactor can be used, where the first stage is primarily for removal of soluble BOD followed by nitrification stages. The reason could be to minimize the heterotrophic bacteria growth competing with nitrifying bacteria for the surface area on biofilm carriers in the downstream nitrification stages. Moreover, the staged reactor improves the volumetric efficiency of MBBR. Several elements, including as the HRT, the filling ratio of carriers, DO, pH, the biofilm thickness, temperature, reactor configuration, the kind of microbe, and others, all have an effect on the removal of organic matter and nutrients in an MBBR. Moreover, liquid flow diffusion and nutrient penetration through the biofilm influence the performance of MBBR. It is to be noted that the amount of plastic biofilm carriers added is defined in terms of its bulk volume filling fraction or percent of tank volume. Furthermore, media replacement has not been a significant issue in MBBR process. (Tchobanoglous et al., 2014). Also, recycling of solids is not required due to relatively low suspended solid concentration (typically 100–250 mg/L) from the effluent of MBBR process (Tchobanoglous et al., 2014).

The major advantages of MBBR compared to ASP are as follows.

- Requirement of less area
- Simplicity of operation with no need for manual sludge wasting and SRT control and sludge recycle
- Elimination of concerns of sludge bulking
- Ability to withstand peak wet weather flow variations
- Compared to other attached growth processes, such as TFs and RBCs, the MBBR process is much more versatile and adaptable for biological nitrogen removal.

However, the MBBR process has some disadvantages like higher energy demand due to operation of MBBR in elevated DO concentrations, need to use priority media, media removal for diffuser maintenance, additional hydraulic profile head loss, etc.

6.2.4.4.2   Types of MBBR Process

The MBBR may be used in different ways for the removal of COD and nitrogen. MBBR may be used as a single-stage treatment unit for COD removal (Figure 6.12a). It may be used as a stand-alone secondary treatment unit or a roughing treatment before other secondary treatment processes. In this process, the MBBR basin receives primary treated wastewater, and aeration is provided. The effluent from the MBBR basin goes to the secondary clarifier. Screens are provided at the outlet of the MBBR basin so that the carriers do not escape from the MBBR basin. In a two-staged MBBR designed for COD removal, the primary treated wastewater is passed through two MBBR basins (Figure 6.12b). The secondary clarifier is only provided at the end

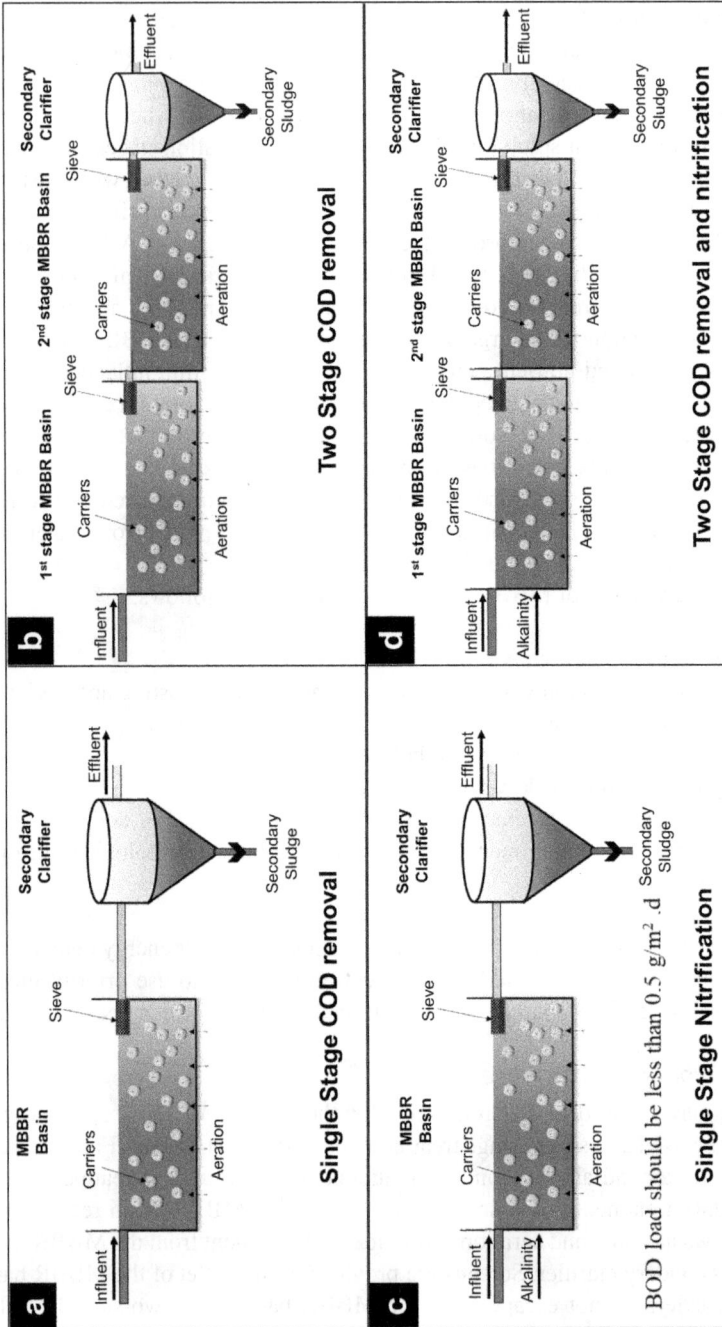

BOD load should be less than 0.5 g/m² . d

FIGURE 6.12 Different MBBR processes: (a) single-stage COD removal; (b) two-stage COD removal; (c) single-stage nitrification; and (d) two-stage COD removal and nitrification.

of the second MBBR basin. In the two-staged MBBR process, a high surface area loading rate (SALR) is provided for the first stage, and a low SALR is provided for the second.

A single-stage MBBR may be used to remove nitrogen from secondary treated wastewater. In order to achieve enhanced nitrification, the BOD loading should be less than 0.5 g/m².d (Figure 6.12c). During the nitrification process, alkalinity is consumed. As a result, the pH of the wastewater turns acidic, resulting in decreased removal of ammonia. Hence, alkalinity is required to be added on a regular basis to maintain the pH of the wastewater. A two-stage MBBR is also used for the removal of COD and nitrification (Figure 6.12d). In this process, the first MBBR basin is used for COD removal. In this basin, the organic matter is greatly removed. The next stage is used for nitrification. Alkalinity is added periodically to maintain the pH of the wastewater (Ødegaard, 2014; Odegaard et al., 1994).

6.2.4.4.3 Design for Moving Bed Biofilm Reactor

The surface area loading rate (SALR) (g/m²/d) is the most important empirical design parameter that is used to figure out the size of the MBBR tank. SALR is the ratio of the grams per day of the BOD/COD being removed and the surface area of the carrier. A single-stage MBBR can be used as a secondary treatment process on its own, as a roughing treatment before another secondary treatment process, or sometimes as an additional secondary treatment unit. In either case, the SALR is the most important design parameter for determining the size of the MBBR tank. This is the amount of BOD that enters the MBBR tank per m² of carrier surface area per day.

Total surface area of biofilm carrier (S_A): The amount of substrate removal by biofilm is a function of available surface area and substrate flux. The biofilm surface area is a product of reactor volume, media fill volume fraction, and specific surface area of the media. The substrate flux or surface area loading rate (SALR) is a key process design parameter. The required total surface area of biofilm carrier in the MBBR can be calculated by Equation 6.47.

$$S_A = \frac{\text{Rate of substrate removal } (\text{g/d})}{\text{SALR } \left(\text{g/}\left(\text{m}^2 \cdot \text{d}\right)\right)} \tag{6.47}$$

The SALR for BOD removal (partial) is typically between 15 and 20 g/m².d and 4.5–6 kg BOD/m³.d. On the other hand, the SALR for secondary treatment for complete BOD removal is typically between 5 and 15 g/m².d and 1.7–5 kg BOD/m³.d. The SALR values are significantly higher for MBBR as compared to ASP (Tchobanoglous et al., 2014).

Media volume (V_M): The media volume can be calculated by using Equation 6.48.

$$V_M = \frac{\text{Surface area of biofilm carrier } \left(\text{m}^2\right)}{\text{Specific surface area}\left(\text{m}^2/\text{m}^3\right)} \tag{6.48}$$

Volume of the tank (V_T): The volume of the tank can be denoted by Equation 6.49:

$$V_T = \frac{\text{Volume of the media } (m^3)}{\text{Filling ratio } (\%)} \tag{6.49}$$

Volume of the liquid (V_L): The volume of the liquid is the difference between the volume of the tank and volume occupied by the carrier .

$$V_L = V_T - (1 - V\%)V_M \tag{6.50}$$

where V% represents the void ratio of the carrier.

Biomass produced ($P_{x,bio}$): The biomass production per day can be calculated by using Equation 6.50 (Tchobanoglous et al., 2014):

$$P_{X,bio} = \frac{Q \times Y_H \times (S_0 - S)}{1 + K_d \times \theta_c} + \frac{f_d \times Q \times Y_H \times (S_0 - S) \times K_d \times \theta_c}{1 + K_d \times \theta_c} \tag{6.51}$$

where
  θ_c = SRT or mean cell residence time (d)
  Y_H or Y = Yield coefficient (g cells/g organic matter used)
  f_d = Kinetic coefficient (d^{-1})
  K_d = Endogenous decay coefficient (d^{-1})

Oxygen demand in actual conditions (R_o): The oxygen requirement is similar to that of ASP. It can be represented by Equation 6.52:

$$R_o \text{ (kg of O}_2\text{/h)} = \underbrace{[Q\ (S_0 - S) / f)]}_{(A)} - \underbrace{1.42\ P_{X,bio}}_{(B)} + \underbrace{4.57\ Q\ (NO_x)}_{(C)} \tag{6.52}$$

In Equation 6.52, the terms (A) and (B) are related to carbonaceous demand only, whereas (C) represents nitrification.

Example 5
Design a MBBR for 0.2 MLD flow. The BOD of the wastewater is 200 mg/L and effluent BOD should be 20 mg/L. Assume SALR to be 9 g/m^2.d. Specific surface area of media is 500 m^2/m^3, fill ratio is 40% and void ratio of carrier is 60%.

Solution

Step 1: Calculation of rate of substrate removal

BOD loading rate = 200000 L/d × (200 – 20) mg/L = 36000 g BOD/d

Step 2: Calculation of media surface area

$$S_A = \frac{\text{Rate of substrate removal (g/d)}}{\text{SALR (g/(m}^2.\text{d))}} = \frac{36000 \text{ (g/d)}}{9 \text{ (g/(m}^2.\text{d))}} = 4000 \text{ m}^2.$$

Step 3: Calculation of volume of media surface area

$$V_M = \frac{\text{Surface area of biofilm carrier (m}^2)}{\text{Specific surface area (m}^2/\text{m}^3)} = \frac{4000 \text{ (m}^2)}{500 \text{ (m}^2/\text{m}^3)} = 8 \text{ m}^3.$$

Step 4: Calculation of volume of tank

$$V_T = \frac{\text{Volume of the media (m}^3)}{\text{Filling ratio (\%)}} = \frac{8 \text{ (m}^3)}{0.4} = 20 \text{ m}^3.$$

Step 5: Calculation of volume of liquid

$$V_L = V_T - (1 - V\%)V_M = 20 - (1 - 0.6) \times 8 = 16.8 \text{ m}^3.$$

Step 5: Calculation of HRT

$$\text{HRT} = V_L/Q = \frac{16.8 \text{ (m}^3)}{200 \text{ (m}^3/\text{d)}} = 0.084 \text{ d} = 2.01 \text{ h}.$$

6.3   ANAEROBIC WASTEWATER TREATMENT

Over the years, aerobic treatment processes have been favored over anaerobic treatment processes. However, rapid urbanization in recent times has led to the rise in demand for processes that can treat high-strength wastewater at a low cost. As a result, anaerobic treatment is being used as a substitute for aerobic treatment because it has lower energy demand, lower biomass yield, lower nutrient requirement, and the ability to handle high organic loadings. Anaerobic wastewater treatment is provided when the wastewater contains a high organic load. Typically, high-strength municipal wastewater having a COD of 1,200 mg/L and BOD of 560 mg/L is treated using an anaerobic process (Tchobanoglous et al., 2014). Anaerobic wastewater treatment requires the involvement of specialized bacteria that can function in the absence of molecular oxygen. Similar to aerobic treatment, the anaerobic treatment also involves suspended and attached growth processes. Anaerobic treatment has been used as an alternative to aerobic treatment because of various advantages it holds over aerobic treatment processes. The various advantages of anaerobic treatment processes are as follows.

- *Energy generation*: In anaerobic treatment, there is a huge potential for energy production and resource recovery. The amount of energy production in the form of methane gas depends on the strength of the wastewater.
- *Lower sludge production*: The amount of biomass produced is around six to eight times lower than that in aerobic processes. In anoxic or anaerobic conditions, the microorganisms rely on nitrate, nitrite, sulfur, and sulfate for energy. In the absence of oxygen, the anabolism is reduced, resulting in lower biomass yield.
- *Low requirement of nutrients*: Since there is lower production of biomass, the amount of nutrient required is less.
- Low volume of the reactor is required.

However, anaerobic treatment has various drawbacks, which have made the implementation of anaerobic treatment processes a significant challenge. The different drawbacks associated with anaerobic treatment are as follows.

- *May require the addition of alkalinity*: The optimum pH for carrying out the anaerobic digestion process is between 6.6 and 7.4 since that is the pH range where the methanogens are most active. In anaerobic treatment, due to the production of volatile fatty acids, carbonic acid, and other organic acids, the pH of the wastewater is often reduced. However, for the efficient performance of the treatment, a pH between 6.6 and 7.4 is required to be maintained. As a result, excess alkalinity is required to be added. Typically, an alkalinity concentration of 2,000–3,000 mg/L as $CaCO_3$ is required to maintain the desired pH.
- *Requires aerobic treatment as a post treatment in order to meet discharge standards*: Due to the production of volatile fatty acids and other dispersed solids during anaerobic treatment, the effluent may not be suitable for discharge. Hence, an aerobic post treatment process is provided to remove the volatile fatty acids and other organics produced in the anaerobic treatment.
- Susceptible to temperature change, toxic substances, and variation in feed rate.
- Generation of gases, which may be foul smelling and corrosive.

6.3.1 Processes Involved in Anaerobic Treatment

In anaerobic treatment, the breakdown of complex organic molecules to biogas requires several groups of microorganisms. These microorganisms primarily take part in four different steps involved in the conversion of organic matter to biogas. The steps are depicted in Figure 6.13.

Hydrolysis: In this step, bigger complex compounds are converted to simpler smaller compounds, which are suitable source of energy and cell carbon. Suspended organic matter, proteins, lipids, and carbohydrates are converted to fatty acids, amino acids, and sugar with the help of hydrolytic bacteria.

Acidogenesis: In this step, the compounds formed in hydrolysis is converted to identifiable lower-molecular-mass intermediates, such as propionate, butyrate,

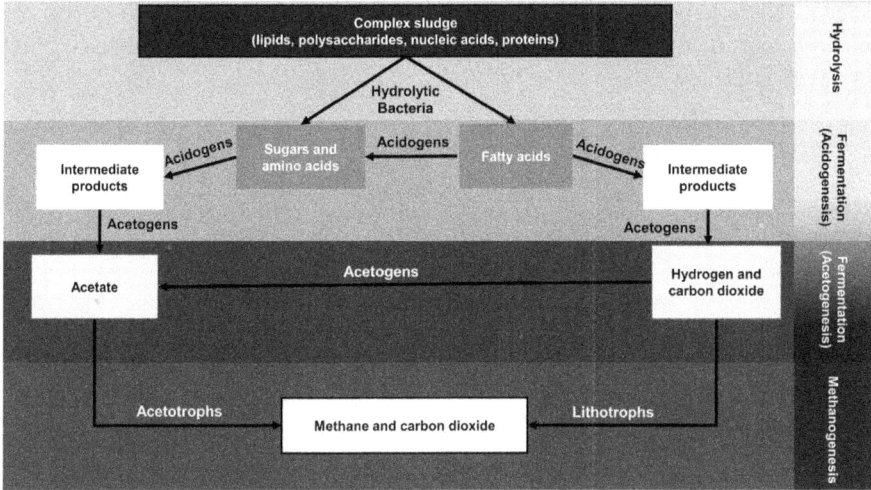

FIGURE 6.13   Anaerobic digestion of sludge.

lactate, formate, ethanol, valeric acid, isovaleric acid, and caproic acid compounds. Microorganisms, known as acidogens, are responsible for this conversion.

Acetogenesis: In this step, the compounds formed during acidogenesis, such as propionate and butyrate are utilized by a group of bacteria (acetogens) to produce acetate and hydrogen.

Methanogenesis: In this step, the intermediate products formed are converted to simpler end products, such as carbon dioxide and methane. Methanogenic bacteria, such as lithotrophs and acetotrophs are responsible for the conversion. Lithotrophs (hydrogen utilizing bacteria) convert carbon dioxide and hydrogen gas to methane, while acetotrophs convert acetic acid to methane and carbon dioxide.

The overall combined microbial synthesis yield during fermentation and methanogenesis is typically around 0.08 g VSS/g COD. On the other hand, the overall combined decay coefficient is typically around 0.03 g VSS/g COD (Tchobanoglous et al., 2013).

Methane production: The COD of methane is the amount of oxygen required to oxidize methane to carbon dioxide and water. This process can be represented by Equation 6.53:

$$CH_4 + 2O_2 \rightarrow CO_2 + 2H_2O \tag{6.53}$$

It can be seen from Equation 6.53 that 1 mole of methane requires 2 moles of oxygen to convert it to carbon dioxide and water. So, the COD of 1 mole of methane is

64 g O_2. The volume of 1 mole of methane at standard condition is 22.4 L. Hence, the methane equivalent of COD which gets converted under anaerobic conditions is 22.4/ 64 = 0.35 L CH_4/g COD.

6.3.2 FACTORS AFFECTING ANAEROBIC TREATMENT

pH
Methanogenic bacteria are sensitive to pH. The optimum pH for the methanogenic bacteria is between 6.6 and 7.4. Non-methanogenic bacteria are not as sensitive to pH change as methanogenic bacteria, but they also do not function in acidic conditions. The optimum pH for the proper functioning of non-methanogenic bacteria is between 5 and 8.5. Due to the continuous production of various acids during anaerobic treatment, the pH of the wastewater often shifts to acidic. In order to maintain near neutral pH, sodium bicarbonate, sodium carbonate, calcium carbonate, calcium bicarbonate, and calcium hydroxide are added to provide a buffering action (Qasim and Zhu, 2017; Tchobanoglous et al., 2014).

Temperature
Anaerobic digestion can occur over a wide range of temperatures. Conventional anaerobic digestion is usually carried out under mesophilic conditions (35°C to 37°C). However, at thermophilic conditions (55°C to 60°C), higher reaction rates may be achieved. Anaerobic digestion can also occur at a low temperature of 12°C to 15°C (Kim et al., 2006).

Inhibitors
The presence of volatile acids, ammonia nitrogen, sulfides, and heavy metals significantly decrease the performance of anaerobic systems. The presence of volatile acids, such as acetate and propionate, can significantly lower the pH of the system and may pose toxicity to the microorganisms. This may lead to the failure of the systems. A high concentration of ammonia-nitrogen may also favor the generation of volatile fatty acids. The presence of volatile fatty acids leads to a lowering of pH, thereby affecting the performance of the anaerobic process. Soluble heavy metals in the wastewater are often toxic to microorganisms. Hence, the presence of soluble heavy metals can also lead to the malfunction of anaerobic processes (Qasim and Zhu, 2017; Tchobanoglous et al., 2014).

6.3.3 ANAEROBIC WASTEWATER TREATMENT TECHNOLOGIES

6.3.3.1 Upflow Anaerobic Sludge Blanket
Upflow anaerobic sludge blanket (UASB) is an anaerobic suspended growth process. In UASB, the wastewater enters the reactor through the bottom of the reactor and flows upward. The UASB is filled with a granular sludge blanket. The wastewater passes through the blanket. The microorganisms in the sludge break down the organic matter through anaerobic digestion. The granules also release small biogas bubbles. The biogas is collected at the top and guided through a gas dome at the top of the

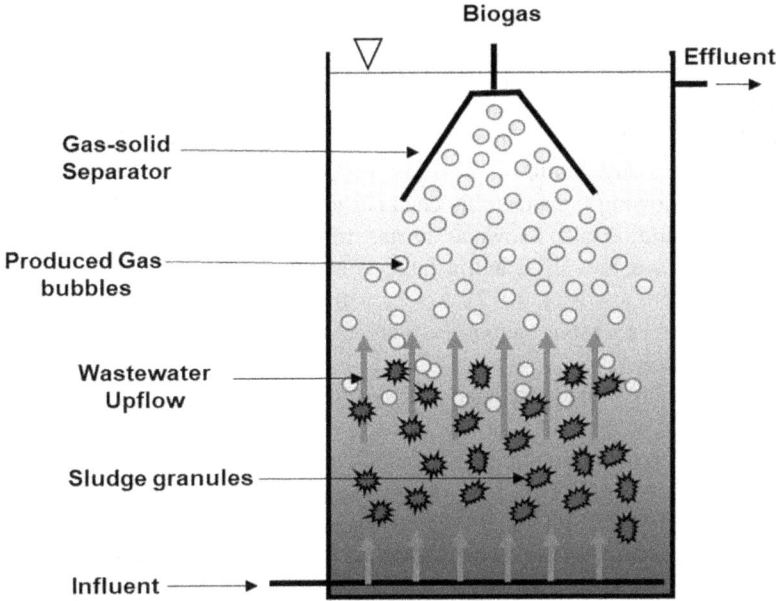

FIGURE 6.14 Schematic of upflow anaerobic sludge blanket.

reactor. Small granules may tend to escape with the flow of the wastewater. However, settling plates are provided after the reactor collects the escaped granules. The treated water is collected from the top. The UASB is efficient in removing BOD, COD, and TSS, but is not efficient in terms of removal of nutrients. A schematic representation of an UASB has been provided in Figure 6.14.

The effective sludge concentration in the UASB is typically 35–40 kg/m³. The SRT of the sludge blanket is usually above 30 days, and HRT varies from 4 to 16 h. The UASB is designed for COD loading of 5–20 kg/m³.d. The upflow velocity may vary from 1 to 6 m/h, and the height of the reactor may vary from 5 to 20 m (Qasim and Zhu, 2017; Tchobanoglous et al., 2014).

The expanded granular sludge blanket (EGSB) is a modification of the conventional UASB. Hence it is also an anaerobic suspended growth process. Higher upflow velocity (4–10 m/h) is provided along with the recirculation of effluent. The height of the reactor may reach up to 25 m. Furthermore, EGSB has a greater height-to-diameter ratio as compared to UASB. The higher upflow velocity in the EGSB facilitates better mixing, reduced sludge production, and increased diffusion rates from bulk liquid to the sludge blanket.

6.3.3.2 Anaerobic Contact Process

The anaerobic contact process (AnCP) is similar to the activated sludge process. It is also a suspended growth process. Typically, an airtight, heated, mixed reactor is followed by solid separation equipment. In order to maintain a high MLVSS, the solids are added back to the reactor. The solid separation apparatus could be a gas flotation machine, filtration system, lamella or tube settler, or sedimentation basin. Gas production and the uplift of biological floc might cause gravity settling to behave poorly. Hence, gas stripping or degasification is carried out before settling. Normally, the volumetric organic loading of the AnCP is around 2–5 kg COD/m³d. The HRT of AnCP (12–120 h) is significantly more than UASB, while the SRT of AnCP (15–30 days) is lower than that of UASB. The surface overflow rate in the clarifier following the anaerobic chamber is around 0.5–1 m/h. The AnCP with membrane separation, also known as the anaerobic membrane bioreactor (AnMBR), has a long SRT, a low HRT, and a large volumetric organic loading of 5–15 kg COD/m³d (Qasim and Zhu, 2017). A schematic diagram of anaerobic contact process reactor and anaerobic membrane bioreactor have been shown in Figure 6.15a and b, respectively.

Often baffles are provided to anaerobic systems. This system is called an anaerobic baffled reactor. The process is similar to a series of UASB. In a continuously stirred anaerobic reactor, there is complete mixing of the semisolid wastes and the suspended anaerobic matter. The SRT of the system is equal to that of the HRT of the system. Typically, the HRT or SRT of the system ranges between 15 and 30 days, and the system is designed to handle COD loading of less than 4 kg/m³.d (Qasim and Zhu, 2017; Tchobanoglous et al., 2014).

6.3.3.3 Other Anaerobic Processes

Anaerobic fluidized bed reactor (AnFBR) (Figure 6.16) involves the development of anaerobic biomass on small inert particles (0.1–0.3 mm), such as fine sand or plastic. This is an attached growth process. The inert particles containing the biomass are kept in suspension by passing wastewater through a high upward velocity. High upward velocity leads to a bed expansion of more than 25%, while low upward velocity leads to a bed expansion of around 15–25%. In AnFBR, the upward velocity ranges from 10 to 20 m/h, and the COD loading varies from 20 to 40 kg/m³.d. Recirculation is carried out in this process (Qasim and Zhu, 2017; Tchobanoglous et al., 2014).

In anaerobic filters, the water is allowed to pass through a bed of media on which anaerobic biomass is grown. The anaerobic filter is an attached growth process; hence there is no mixing. The removal mechanism involved in anaerobic filter is filtration and anaerobic digestion. Anaerobic filters usually have an HRT of 1–3 days and are designed for a COD loading of 5–20 kg/m³.d. The flow may be provided in an upward and downward manner. The anaerobic filters should be checked from time to time to ensure that the system is watertight. The flow should be gradually increased with time, and the filters must be cleaned if the removal efficiency starts to decrease (Qasim and Zhu, 2017; Tchobanoglous et al., 2014). Anaerobic rotating biological contactor (AnRBC) is a hybrid anaerobic process. The AnRBC reactors use fixed covers, have submerged discs, and are subjected to a little pressure. The generated gases are gathered. After settling, the solids are added back to the effluent (Qasim and Zhu, 2017; Tchobanoglous et al., 2014)..

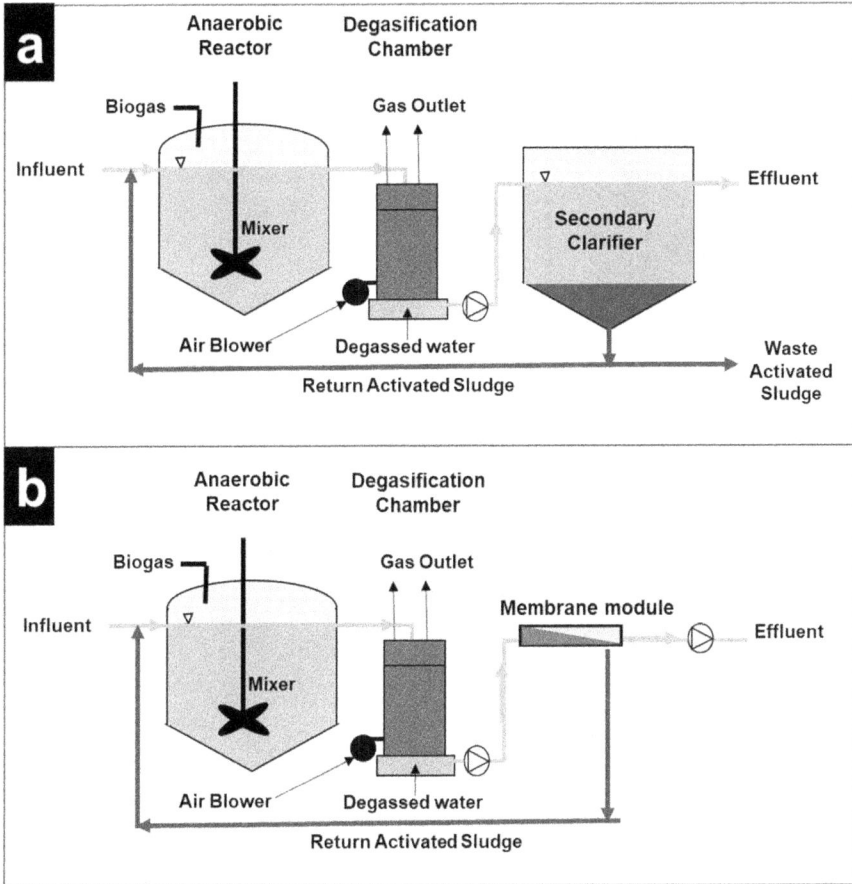

FIGURE 6.15   Schematic diagram of (a) Anaerobic contact process reactor and (b) Anaerobic membrane bioreactor.

Example 6

Design an UASB reactor to treat wastewater having a COD of 1,000 mg/L (1 kg/m³), TSS of 400 mg/L, VSS of 320 mg/L. The flowrate is 4,000 m³/d. The design should be carried out, such that 90% of COD is removed. The following assumptions may be made:

Kinetic coefficients: $Y = 0.08$ g VSS/g COD, $k_d = 0.022$ d⁻¹, $k = 2.5$ d⁻¹, and $K_s = 800$ mg/L.

Methane production at 0°C = 0.35 L CH₄/g COD
Height of reactor volume = 5 m
Design organic loading rate = 10 kg COD/m³ .d
The average solid concentration in the reactor = 40 kg VSS/m³
0.65 g biodegradable VSS/g VSS and 1.42 g COD/g biodegradable VSS.
VSS/TSS ratio in effluent = 0.85

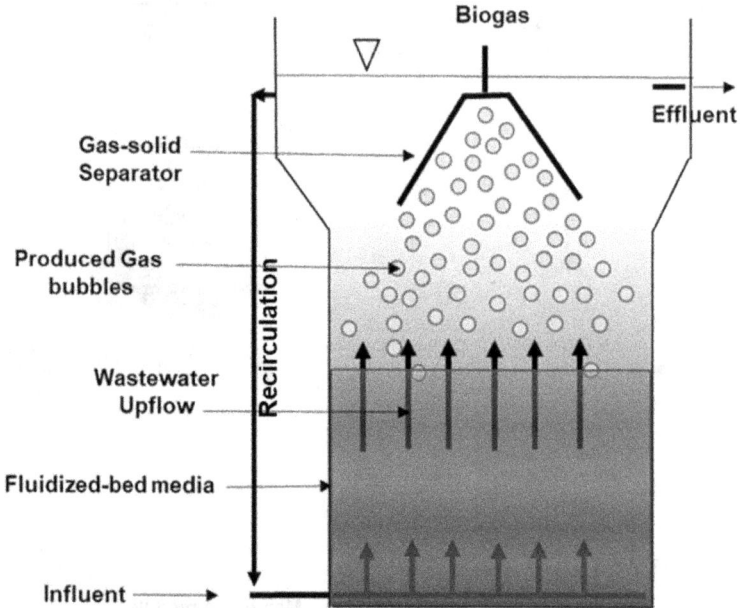

FIGURE 6.16 Schematic of anaerobic fluidized bed reactor.

Solution

Step 1: Determination of volume of the reactor
Volume of the reactor from design loading rate

$$\text{Volume} = \frac{Q \times S_0}{\text{OLR}} = \frac{4000 \text{ m}^3/\text{d} \times 1 \text{ kg COD/m}^3}{10 \text{ kg COD/m}^3.\text{d}} = 400 \text{ m}^3$$

Step 2: Determination of HRT

The HRT is volume / flow rate = 400 m³/ (4000 m³/d) = 0.1 d = 2.4 h

Step 3: Determination of SRT
The effluent COD concentration
   is $1000 - 0.9 \times 1000 = 100$ mg/L or 100 g/m³.

As per Equation 6.18, $\dfrac{1}{\theta_c} = \dfrac{\mu_m \times S}{K_s + S} - K_d$

$\mu_m = Y \times k = 0.08 \times 2.5 \text{ d}^{-1} = 0.2$ VSS/g COD d⁻¹

$$\frac{1}{\theta_c} = \frac{0.2 \text{ d}^{-1} \times 100 \text{ g/m}^3}{800 \text{ g/m}^3 + 100 \text{ g/m}^3} - 0.022 \text{ d}^{-1}$$

$$\theta_c = 45.45 \text{ d}$$

Considering a factor of safety of 1.3, the design θ_c may be taken to be 59.05–60 d.

Step 4: Determination of total solid concentration
As per Equation 6.25

$$P_x = Y_{obs} \times Q \times (S_0 - S)(10^{-3} \text{ kg/g})$$

where $Y_{obs} = Y/(1 + k_d \theta_c)$

$$Y_{obs} = Y/(1 + k_d \theta_c) = 0.08/(1 + 0.03 \text{ d}^{-1} \times 60 \text{ d}) = 0.028$$

$$P_x = 0.028 \times 4000 \times (1000 - 100)(10^{-3} \text{ kg/g}) = 100.8 \text{ kg VSS/d}$$

The amount of TSS produced is $144/0.85 = 118.58$ kg TSS/d
Therefore, TSS concentration on the effluent 118.58 kg TSS/s/400 m³/d = 29.64 m g/L
  VSS concentration in the effluent is $0.85 \times 29.64 = 25.19$ mg/L

$$\text{The total VSS concentration in the reactor} = \frac{\text{SRT} \times \text{Flow} \times \text{VSS}}{\text{Volume}}$$

$$= \frac{32 \times 4000 \times 25.19}{400} = 8061 \text{ mg VSS/L.}$$

Step 5: Determination of area and maximum upflow velocity
The surface area of the reactor is 400 m³/5 m = 80 m².

$$\text{Upflow velocity is} \frac{4000 \text{ m}^3/\text{d}}{80 \text{ m}^2 \times 24 \text{ h/d}} = 2.08 \text{ m/h}$$

Example 7
Design an anaerobic fluidized bed reactor to treat wastewater having COD concentration of 1,000 mg/L and flow rate of 180 m³/d.
  GAC media: diameter = 0.6 mm, specific gravity = 1.4, and porosity ratio = 0.48.
  The fluidized bed ratio = 1.6.
  The design volumetric organic loading = 10 kg COD/m³·d
  Operating temperature = 25 °C.
  kinematic viscosity (v) of wastewater = 0.893×10^{-6} m²/s
  Particle diameter, $d = 0.6$ mm = 6×10^{-4} m, and $G_s = 1.4$.
  Porosity ratio of the unfluidized bed = 0.40

Solution

Step 1: Determination of volume of reactor
Volume of the reactor from design loading rate:

$$\text{Volume} = \frac{Q \times S_0}{\text{OLR}} = \frac{180 \text{ m}^3/\text{d} \times 1 \text{ kg COD/m}^3}{10 \text{ kg COD/m}^3.\text{d}} = 18 \text{ m}^3$$

Step 2: Determination of HRT
The HRT is volume/ flow rate = 18 m³/ (180 m³/d) = 0.1 d = 2.4 h

Step 3: Determination of settling velocity
A circular tank is provided with the heigh (H) being 1.5 times the diameter (D)

Hence, $3.14/4 \, D^2 h = 3.14/4 \, D^2 \times 1.5 \, D = 18 \text{ m}^3$

$$D = 2.84\text{--}3 \text{ m}$$

$$H = 4.5 \text{ m}$$

Heigh of unfluidized bed reactor (H_{ufb}) = 4.5/ fluidized bed ratio = 4.5/1.5 = 3 m
Surface area of the reactor = $3.14/4 \, D^2 = 3.14/4 \times 3^2 = 7.06 \text{ m}^2$
A gas solid separator having a diameter size of 0.5 m greater than that of the reactor is provided. So area of gas solid separator = $3.14/4 \, D^2 = 3.14/4 \times 3.5^2 = 9.61 \text{ m}^2$
Total area for settling is 7.06 + 9.61 = 16.67 m².
Diameter of settling zone = 4.61 m
Under laminar conditions Stokes's law can be valid. Hence, the V_s can be calculated by using Equation 5.13.

$$V_s = \frac{g}{18} \frac{(G_s - 1) \times d^2}{v}$$

$$V_s = \frac{9.81}{18} \frac{(1.4 - 1) \times (6 \times 10^{-4})^2}{0.893 \times 10^{-6}} = 0.088 \text{ m/s}$$

Check for Reynold's number

$$R_e = \frac{V_s d}{v} = \frac{0.088 \times 6 \times 10^{-4}}{0.893 \times 10^{-6}} = 59.1 > 1.0$$

Since $1 > R_e > 10^3$, the settling velocity is in the transition zone.

$$C_D = \frac{24}{R_e} + \frac{3}{\sqrt{R_e}} + 0.34$$

$$C_D = \frac{24}{59.1} + \frac{3}{\sqrt{59.1}} + 0.34 = 1.14$$

$$V_s = \sqrt{\frac{4\,g\,(G_s - 1)d}{3\,C_D}} = \sqrt{\frac{4 \times 9.81 \times (6 \times 10^{-4})^2}{3 \times 1.14}} = 0.052 \text{ m/s}$$

After consecutive iterations, the value of V_s was found to be stable at 0.043 m/s.

Step 4: Determination of upflow velocity
Upflow velocity in fluidized bed reactor is given by the Equation 6.54 (Qasim and Zhu, 2017).
   The porosity of the unfluidized bed is 0.40.
   Equation 6.54 may be used to calculate the porosity of the fluidized bed reactor:

$$\text{Fluidized bed ratio} = \frac{1 - \text{porosity of unfluidized bed}}{1 - \text{porosity of fluidized bed}}$$

$$\text{Porosity of fluidized bed} = 1 - \frac{1 - \text{porosity of unfluidized bed}}{\text{Fluidized bed ratio}}$$

$$\text{Porosity of fluidized bed} = 1 - \frac{1 - 0.40}{1.6} = 0.0625$$

Upflow velocity = settling velocity × (porosity) $^{4.5}$ (6.54)

Upflow velocity = 0.043 m/s × (0.625) $^{4.5}$ = 0.0051 m/s

6.4   SECONDARY SETTLING

The secondary settling tank is an important component in the biological treatment processes that separate the biomass produced during the biological oxidation of organic matter from the liquid portion. The different biological processes require a secondary clarifier to remove the solids. On the other hand, in the suspended growth process, like in oxidation ponds and lagoons, biomass removal is accomplished within the reactor without a secondary clarifier. The design and operation of the secondary clarifier are different in the suspended and attached growth process because of the significant difference in the characteristics of biological solids. Hence, this section has presented a detailed discussion of secondary clarifiers for suspended and attached growth systems.

In suspended growth processes, a secondary clarifier has two main important functions: (i) removal of the solids from treated effluents using gravity settling, that is, clarification and (ii) thickening of settled sludge to reduce the volume before recirculating it to the aeration tank. The type of settling of activated sludge in the secondary clarifier is classified based on the depth in the clarifier. For example, in the upper portion, that is, clear water level, a type 1 (discrete) setting occurs. As the particles sink, they begin to flocculate, and type II settling occurs. Type III settling (hindered) and type IV (compression) settling occur near the bottom of the clarifier (Tchobanoglous et al., 2014).

Two principal parameters, such as surface overflow rate and solids loading rate, are used to design a secondary clarifier. The parameters mentioned above depend on the design and operation of suspended growth processes, and these parameters determine the characteristics of mixed liquor floc.

> *Surface overflow rate (SOR):* SOR is the ratio of influent flow rate to the surface area of the clarifier. The typical SOR in the ASP range from 15 to 35 m^3/m^2.d. Solids having settling velocities greater than the SOR will be removed by gravity settling (CPHEEO, 2013). In contrast, solids with settling velocity less than the SOR will be removed with effluent from the clarifier.

$$SOR = \frac{Q}{A} \qquad (6.55)$$

where
SOR = surface overflow rate (m^3/m^2.d)
Q = Influent flowrate (m^3/d)
A = surface area of the clarifier (m^2)

> *Solids loading rate (SLR):* SLR is the quantity of solids per day that can be removed per square foot of surface area by a clarifier. As discussed before, particles with settling velocities greater than the SOR will settle down due to gravity, forming a thick suspension referred to as a sludge blanket in the bottom of the clarifier. The typical range of SLR in secondary clarifier of ASP is 70 to 140 kg/m^2.d (CPHEEO, 2013).

$$SLR = \frac{(Q + Q_R) \times X}{A} \qquad (6.56)$$

where
SLR = Solids loading rate (Kg/m^2.d)
Q = Influent flowrate to the secondary system (m^3/d)

Q_R = Recirculation flowrate (m³/d)
X = MLSS concentration entering to the secondary clarifier (mg/L)
A = Surface area of the clarifier (m²)

From Equations 6.55 and 6.56, the relation between SOR and SLR can be written as follows:

$$SLR = SOR \times (1 + R) \times X \tag{6.57}$$

where R represents recirculation ratio $= Q_R/Q$.

In the secondary clarifier of suspended growth processes, the thickening characteristics of activated sludge mixed liquor are assessed using the SVI test. Lower SVI represents the more efficient thickening and more efficient performance.

In attached growth systems, the design of a secondary clarifier is similar to that of the primary clarifier. No thickening or hindered settling occurs; hence design criteria are based upon particle size and density. Sludge thickening and recirculation of sludge are not required, hence the underflow is negligible compared to overflow. The quantity of sludge produced during the attached growth system is significantly lower than that of the suspended growth system because of the endogenous nature of biomass near the media.

6.5 POST-AERATION

Post-aeration is often applied to increase the DO concentration of effluent wastewater for satisfying the discharge standards. The post-aeration can be accomplished by either cascade, diffused, or mechanical aerators. Based on the availability of elevation head and the quantity of treated wastewater, the aforementioned techniques can be applied. For example, if sufficient elevation head is available, cascade aeration is the best alternative to improve the DO concentration of wastewater. Otherwise, surface or diffuse aerators are employed.

Cascade aerators are one of the oldest and common practices used to improve DO levels. Cascade aerator consists of a series of steps through which the water flows over. The height of cascade can be calculated by Equation 6.58 (Mackenzie, 2010).

$$H = \frac{R_{\text{deficit}} - 1}{0.289 \times b \times (1 + 0.046\,T)} \tag{6.58}$$

where
H = Height through which wastewater must fall (m)
R_{deficit} = Deficit ratio $= \dfrac{C_s - C_o}{C_s - C}$

C_o = DO concentration of the influent to the cascade (mg/L)
C = DO required (mg/L)
C_s = DO saturation concentration of the wastewater at temperature (T, °C) (mg/L)

b = Weir geometry parameter [for a broad-crested weir b = 1.0; for steps b = 1.1; for step weir b = 1.3]

T = Wastewater temperature (°C)

Example 8

Before the effluent from an oxidation pond can be released, the DO concentration must be raised from 1 to 5 mg/L. This oxidation pond has water temperatures of around 25°C during the summer. Determine the step cascade height required to attain a 5.0 mg/L DO concentration.

Solution

The DO saturation concentration at 25 °C is 8.38 mg/L.

Step 1: Calculation of the deficit ratio

$$R_{deficit} = \frac{8.38 - 1}{8.38 - 5} = 2.18$$

Step 2: Calculation of the height of the cascade

$$H = \frac{2.18 - 1}{(0.289)(1.1)(1 + 0.046(25))} = \frac{1.18}{0.6835} = 1.72 \text{ m}$$

6.6   CHAPTER SUMMARY

- Microorganisms are the driving forces in the secondary treatment process. Their growth depends on appropriate environmental conditions (primarily pH and temperature), macronutrients (carbon, nitrogen, and phosphorous), and micronutrients (trace elements and vitamins).
- There the four phases in biomass growth: lag phase, exponential growth phase, stationary-growth phase, and endogenous/death phase.
- Biological treatment processes may be aerobic or anaerobic based on the presence or absence of oxygen in the treatment system. They can be further classified into suspended growth process, attached growth process, and hybrid growth process.
- In wastewater treatment, the organic matter is usually represented by chemical oxygen demand and the newly formed biomass is represented by volatile suspended solids.
- Activated sludge process, sequencing batch reactor, and membrane bioreactor are the most common suspended growth process. These reactors can be operated both under aerobic and anaerobic conditions.
- Trickling filters, fluidized bed reactors, and other packed bed filters are examples of attached growth processes. Similar to suspended growth processes, attached growth processes can be operated both under aerobic and anaerobic conditions.

- Rotating biological contactors, integrated fixed-film activated sludge process, and moving bed biofilm reactors are examples of hybrid biological processes.
- Aerobic processes have advantages over anaerobic processes, such as no production of foul smell and easier operation. However, they are cost-intensive processes and produce a significant amount of sludge.
- Anaerobic processes have the additional advantages of biogas generation, lesser sludge production, and higher reaction rate.

6.7   CONCLUDING REMARKS

The different processes of biological treatment generate a substantial amount of sludge. Furthermore, many of the recalcitrant organic pollutants and microorganisms still remain in the effluent of the secondary clarifiers. In this context, the tertiary treatment processes and the different ways to handle the sludge produced have been discussed in the subsequent chapters.

REFERENCES

APHA, 2017. Standard methods for the examination of water and wastewater standard methods for the examination of water and wastewater. *Public Health* 51, 1–1546. https://doi.org/10.2105/AJPH.51.6.940-a

Bassin, J.P., Dezotti, M., 2018. *Moving Bed Biofilm Reactor*. Springer, Cham. https://doi.org/10.1007/978-3-319-58835-3

CPHEEO, 2013. *Manual on Sewerage and Sewage Treatment Systems: Part A Enigneering*. Central Public Health and Environmental Engineering Organisation.

Dutta, A., Sarkar, S., 2015. Sequencing batch reactor for wastewater treatment: Recent advances. *Curr. Pollut. Reports* 1, 177–190. https://doi.org/10.1007/s40726-015-0016-y

Gukelberger, E., Gabriele, B., Hoinkis, J., Figoli, A., 2018. MBR and integration with renewable energy toward suitable autonomous wastewater treatment, in: *Current Trends and Future Developments on (Bio-) Membranes: Renewable Energy Integrated with Membrane Operations*. https://doi.org/10.1016/B978-0-12-813545-7.00014-3

Iorhemen, O.T., Hamza, R.A., Tay, J.H., 2016. Membrane bioreactor (MBR) technology for wastewater treatment and reclamation: Membrane fouling. *Membranes (Basel)*. 33, 13–16. https://doi.org/10.3390/membranes6020033

Kim, J.K., Oh, B.R., Chun, Y.N., Kim, S.W., 2006. Effects of temperature and hydraulic retention time on anaerobic digestion of food waste. *J. Biosci. Bioeng.* 102, 328–332. https://doi.org/10.1263/JBB.102.328

Mackenzie, D.L., 2010. *Water and Wastewater Engineering: Design Principles and Practice*. McGraw-Hill Education.

Mara, D., 2003. *Domestic wastewater treatment in developing countries*. EARTHSCAN.

Ødegaard, H., 2014. Compact wastewater treatment with MBBR. Int. DSD Conf. Sustain. Stormwater Wastewater Manag.

Odegaard, H., Rusten, B., Westrum, T., 1994. A new moving bed biofilm reactor – applications and results, in: *Water Science and Technology*. Pergamon Press Inc, pp. 157–165. https://doi.org/10.2166/wst.1994.0757

Peavy, H.S., Rowe, D.R., Tchobanoglous, G., 1985. *Environmental Engineering*. McGraw-Hill.

Qasim, S.R., Zhu, G., 2017. *Wastewater Treatment and Reuse, Theory and Design Examples*, Volume 1 – Google Books. CRC Press.

Saidulu, D., Majumder, A., Gupta, A.K., 2021. A systematic review of moving bed biofilm reactor, membrane bioreactor, and moving bed membrane bioreactor for wastewater treatment: Comparison of research trends, removal mechanisms, and performance. *J. Environ. Chem. Eng.* 9(5), 106112. https://doi.org/10.1016/J.JECE.2021.106112

Tchobanoglous, G., Burton, F.L., Stensel, H.D., 2004. *Wastewater Engineering: Treatment and Reuse (Book)*, 4th Edition, Metcalf & Eddy, Inc. McGraw-Hill Education.

Tchobanoglous, G., Burton, F.L., Stensel, H.D., 2014. *Wastewater Engineering: Treatment and Resource Recovery*, Metcalf & Eddy, Inc. McGraw-Hill Education.

Tchobanoglous, G., Stensel, H.D., Tsuchihashi, R., Burton, F., 2013. *AECOM – Wastewater Engineering: Treatment and Resource*. McGraw-Hill.

7 Overview of Conventional Wastewater Treatment Processes
Tertiary Treatment

CHAPTER OBJECTIVES

The chapter seeks to provide an understanding of the necessity of tertiary treatment of wastewater and discusses the most commonly used tertiary treatment methods, such as chemical precipitation, filtration, adsorption, and disinfection.

7.1 INTRODUCTION

The secondary treatment of wastewater results in significant removal of biological oxygen demand (BOD), chemical oxygen demand (COD), and suspended solids. Furthermore, the biological treatment also removes phosphorous, heavy metals, and nitrogen in small amounts. Post-secondary treatment is tertiary treatment, which is applied to improve the treated effluent quality further. This treatment, known as advanced wastewater treatment, is an additional treatment process used to minimize organics, turbidity, nitrogen, phosphorus, metals, and pathogens left after secondary treatment. If nitrogen and phosphorus are present in the effluent of secondary treatment processes, its discharge can cause eutrophication of the receiving water body. The dissolved, suspended, and colloidal particles must be removed before the treated effluent can be considered for reuse. Thus, tertiary treatment is required to meet the discharge/reuse standards to prevent contamination of land and water. The different tertiary treatments used have been discussed in the following sections.

7.2 CHEMICAL PRECIPITATION

Chemical precipitation is the process of adding chemicals to water/wastewater to enable the dissolved and suspended particles to change their physical state and form precipitates that can be removed by sedimentation. Chemical precipitation involves the addition of counter-ions to limit the solubility of ionic components in water, which facilitates their removal. It is mostly utilized for phosphorus and heavy metals removal.

DOI: 10.1201/9781003364450-7

7.2.1 CHEMICAL PRECIPITATION OF PHOSPHORUS

Removal of phosphorus is required to control the eutrophication in receiving waters. Orthophosphate, polyphosphate, and organically bound phosphorus are the common forms of occurrence of phosphorus. Organically bound phosphorus comes from human bodies and food waste, and when they decompose, orthophosphates are released. One-half of the phosphorus in wastewater is due to the presence of polyphosphate, which is present in synthetic detergents. In an aqueous solution, all polyphosphates hydrolyze and are converted back to ortho form (PO_4^{3-}) from which they were produced. Monohydrogen phosphate (HPO_4^{2-}) is the most common form of phosphorus found in wastewater.

In biological treatment methods, microbes utilize phosphorus for cell formation, energy transmission, and energy storage. Generally, the conventional secondary treatment removes less than 3 mg/L of phosphorus (Peavy et al., 1985). However, currently, phosphate removal is accomplished via chemical precipitation where different chemical agents, such as salts of iron (ferric sulfate ($Fe_2(SO_4)_3$) and ferrous sulfate ($FeSO_4$)), calcium salts (lime $Ca(OH)_2$), aluminum salts (alum ($Al_2(SO_4)_3.18\ H_2O$), and sodium aluminate ($NaAlO_2$)) are added (Mackenzie, 2010).

Alum for phosphate precipitation: Phosphorus removal up to 80–90% can be achieved with an alum dosage between 50 and 200 mg/L. Doses are usually determined through bench-scale testing and sometimes through full-scale experiments especially when polymers are applied. Alum forms a precipitate of aluminum phosphates ($AlPO_4$) and is frequently employed for phosphate precipitation. The orthophosphates combine with trivalent aluminum to form a precipitate at a slightly acidic pH.

Iron-based chemicals for phosphorus removal: For phosphorus removal, ferric chloride or sulfate, as well as ferrous sulfate, often known as copperas, are also commonly used. Ferric phosphate is formed when ferric ions interact and combine with the phosphate ions. Lime is usually added to aid the coagulation process by elevating the pH, as ferric ions take a long time to coagulate in natural alkalinity. It is to be noted that alum and ferric chloride have an effective range of working pH, i.e. 5.5 and 7.0 (Mackenzie, 2010).

7.2.2 CHEMICAL PRECIPITATION OF HEAVY METALS

Metals such as arsenic, barium, cadmium, copper, mercury, and nickel can be precipitated as sulfides or hydroxides. Usually, in sewage treatment plants (STP) addition of lime is done to maintain the pH level for minimum solubility for metal hydroxides. In stirred reaction tanks, hydroxide precipitation is started by adding a suitable hydroxide to the wastewater to generate insoluble heavy metal hydroxide precipitates. Most heavy metals exhibit amphoteric behavior, with their hydroxides attaining their lowest solubility at a specified pH for each metal. Thus, target metals can be precipitated as metal hydroxides by maintaining the pH level. The pH can simply be adjusted to the minimum solubility for the target metal (Duan et al., 2020; He et al., 2022).

Metals such as arsenic are co-precipitated during the precipitation of phosphorus. The addition of coagulants for phosphorus removal also aids the removal of various inorganic ions, mainly heavy metals, via co-precipitation. Metals present in dissolved forms may be adsorbed on the hydroxide complex, while particulate and colloidal forms may be incorporated into the flocculated matter. In the case of combined treatment of domestic and industrial wastewater, chemicals are required to be added to the primary settling facilities, especially if on-site pre-treatment strategies are insufficient. The toxicity of precipitated heavy metals restricts anaerobic digestion and sludge stabilization (Hedrich and Johnson, 2014; Yadav et al., 2021).

7.3 FILTRATION

7.3.1 FILTRATION TECHNOLOGIES

The suspended and colloidal particles carried over from secondary unit processes are removed using granular media filtration. Granular filtration can be divided into surface filtration and depth filtration. The filtration units can be operated under gravity flow conditions or closed reactors pressure filtration units. Filtration is generally required when the total suspended solids in the water matrix are ≤10 mg/L (Mackenzie, 2010).

Filters can be classified as continuous and semi-continuous based on their mode of operation. In continuous filters, filtration and backwashing occur simultaneously. Large wastewater treatment plants (WWTPs) most frequently use conventional downflow filters, deep-bed downflow filters, deep-bed upflow continuous-backwash filter, and traveling-bridge filters. Figure 7.1a–d depicts the configurations of the aforementioned filters.

Conventional downflow filter: Wastewater containing suspended materials is pumped into the filter bed from the top. The filter material in single-medium filters consists of either sand or anthracite, whereas dual-media filters have a combination of activated carbon and sand, anthracite over sand, resin beads and sand, and others. Multimedia filters mainly comprise a layer of anthracite above a sand layer over a layer of garnet (Figure 7.1a)

Deep-bed downflow filter: This filter is similar to that of a conventional filter but has a deeper filter bed and greater size of filtering medium, usually anthracite, as represented in Figure 7.1b. These filters can be run for a more extended period of time due to greater depth and larger medium size. Backwashing operations help decide the filter media size and use air scour and water.

Deep bed upflow continuous backwash filter: The wastewater is applied from the bottom portion of the filter upwards via riser tubes placed in series. An inlet distribution hood with open bottom distributes the wastewater evenly into the filter bed (Figure 7.1c). The upward flow creates turbulence, which scours the contaminants by the sand particles. When the wastewater reaches the top, polluted waste slurry escapes through a central reject compartment. The clean filtrate is discharged from the filter after it exits the sand bed and overflows a weir. The higher settling velocity of the sand prevents its escape from the filter. It is washed and distributed again onto the top of the sand bed.

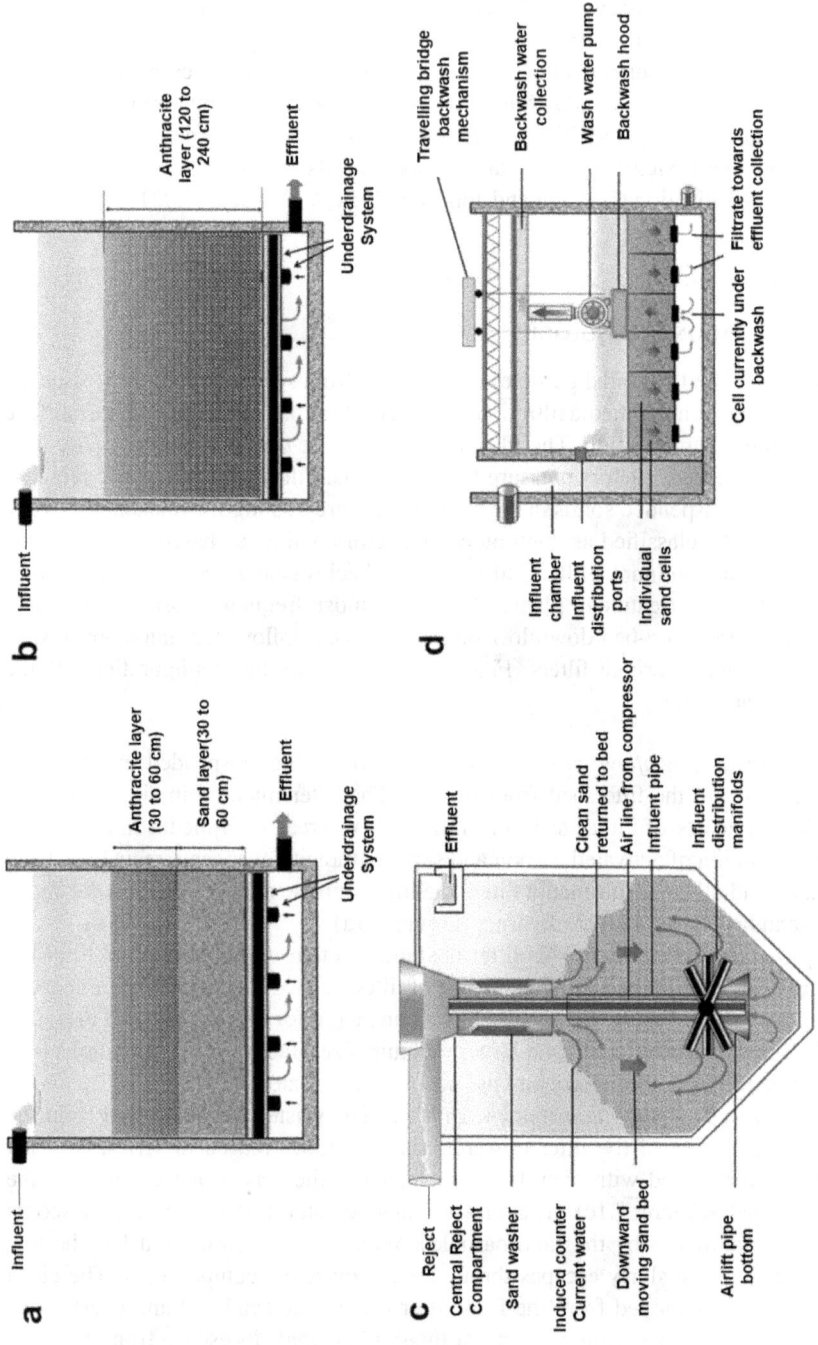

FIGURE 7.1 (a) Conventional downflow filter; (b) deep bed downflow filter; (c) deep bed upflow continuous backwash filter; and (d) traveling-bridge filter.

Traveling bridge filter: It is a granular filter with continuous downflow, low-head operation, and automatic backwashing. The filter bed is separated horizontally into long filter cells. Gravity flow forces the wastewater to flow through the filter medium and exit into the clear well plenum through a porous plate and poly-ethylene underdrains. Backwashing is carried out separately for each cell by operating the overhead traveling bridge. Other cells continue to operate during the backwashing period. A typical traveling bridge filter is shown in Figure 7.1d (Mackenzie, 2010).

7.3.2 DESIGN GUIDELINES

The filters are primarily used as a polishing unit to remove those contaminants that the preceding treatment steps could not remove. The influent of the filtration system should be free from an excessive load of suspended solids. Otherwise, it would easily choke the filter media. Hence, pre-treatment is an essential criterion for filtration. Also, the filter type, head loss, and filtration rate should be carefully considered for designing filters.

Pre-treatment: Generally, provisions for adding organic and inorganic coagulants upstream of the sedimentation tank placed before the filter is considered a good design practice. Polyelectrolyte doses range from 0.5 to 1.5 mg/L for the influent of the sedimentation tank and 0.05 to 0.15 mg/L for the filter influent. Pre-treatment upstream of the filter using coagulation, flocculation, and sedi-mentation is necessary to achieve TSS lower than 3 mg/L if the average TSS in influent to the filter is more than 20 mg/L (Mackenzie, 2010).

Filter type: Dual-media or multimedia, downflow filters are most commonly used these days. The problem of head loss buildup in stratified bed single-medium filters has restricted their applications for wastewater treatment.

Terminal head loss and filtration rate: The range of rate of filtration is from 5 to 20 m/h, and 2.4 to 3 m of terminal head loss is considered (Mackenzie, 2010).

7.3.3 MEMBRANE FILTRATION

A membrane is a porous intervening barrier that is used to separate dissolved materials (solutes), colloids, or fine particulate from solutions. Filtration involves the separation of colloids and particulate matter from water. Membranes act as a selective barrier that allows specific components to pass through and restricts other components present in the liquid by retaining them. Membrane technology can treat a wide range of wastes, including leachate from sanitary landfills comprising constituents of organic and inorganic nature, miscible oil wastes generated in metal industries, solvents mixed with water, and other water–oil mixtures. Based on parameters such as membrane pore size, membrane material, mechanism of separation, and nature of driving force, the membrane processes can be categorized as reverse osmosis (RO), nanofiltration (NF), electrodialysis, microfiltration (MF), and ultrafiltration (UF) (Abdulgader et al., 2013; Fane et al., 2015). Figure 7.2 represents different membranes classified

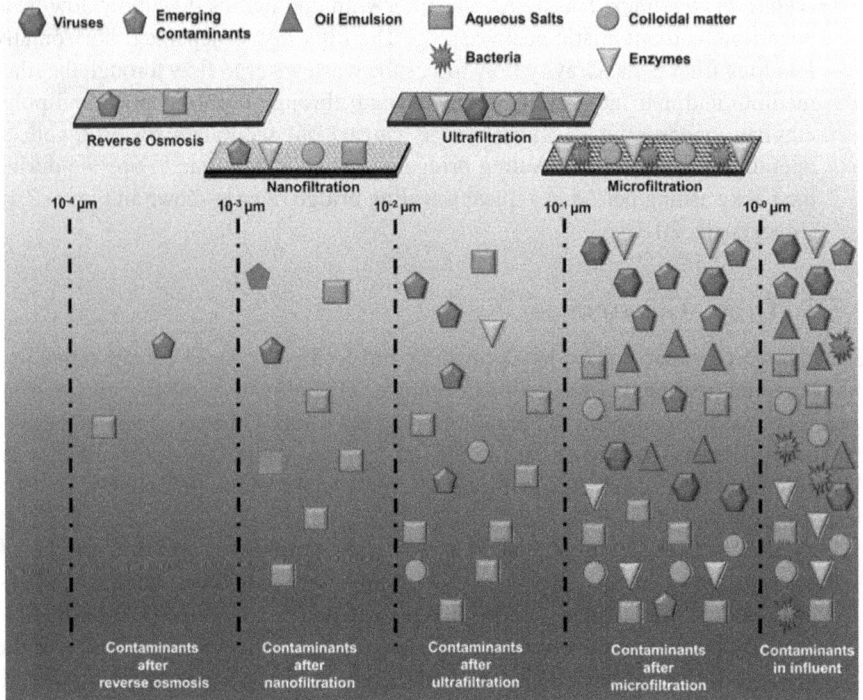

FIGURE 7.2   Classification of membranes based on operating range of pore size and size range of constituents removed.

according to their pore sizes and the constituents they remove. The different types of membrane filtration techniques based on the size of the pores are discussed in the following section. MF can remove bacteria, plant pigment, cell fragments, and debris. UF can remove gelatin, viruses, colloidal matter, oil emulsions, and others. NF can remove sugars, amino acids, colorants, and others. However, RO is required to remove aqueous salts, which the other membrane filtration techniques are unable to do.

7.3.3.1   Types of Membrane Filtration

Membrane filtration can be carried out using different types of membranes with different pore sizes operated at different pressures. The types of membrane filtration are as follows:

Microfiltration (MF): MF membranes are highly porous membranes with pore sizes ranging from 0.1 to 1 µm. MF membranes only retain solid particles over the pores and let the dissolved particles pass through the pores. Pressure-driven MF membranes generally operate in a pressure range of 30–200 kPa (Tchobanoglous et al., 2013). The feed stream flows in a direction that is perpendicular to the surface of the membrane so that cake doesn't build up and the membrane doesn't get clogged up. Cross-flow MF often can't work at its fullest capacity because the membrane gets clogged up. This happens when there are suspended solids in the feed stream (Charcosset, 2012; Lovins et al., 2002; Pontius et al., 2011). Over time, as more and more particles stay on and inside the membrane, the amount of permeate that can pass through the membrane decreases. Cells, cell debris, and other unwanted particles can build up on the surface of a membrane. This is called external fouling or cake formation, and it can usually be cleaned up (Fu et al., 2017). This buildup can be made up of cells, pieces of cells, or other things. On the other hand, internal fouling, also known as the deposition and adsorption of tiny particles or macromolecules within the internal pore structure of the membrane, is usually irreversible. Because of a loss in effective pore area or pore numbers, filtration fluxes may be lower in heavily clogged membranes than in UF. This could happen if the effective pore area gets smaller. This can happen when the membrane is covered with a lot of debris or suspended matter (Charcosset, 2012).

Ultrafiltration (UF): UF membranes can be either asymmetric or composite, having pore sizes from 0.005 to 0.05 μm. Bacteria and protein macromolecules are mainly retained on UF membranes while they let most of the organic molecules and mineral salts pass through. Operating pressure is usually kept in the lower range, that is, 68–350 kPa, as UF membranes do not retain low molecular solutes, and osmotic back pressure can also be ignored (Tchobanoglous et al., 2013). The UF method is a type of membrane separation technology that can be used to filter out particle pollutants that may be present in solutions. This type of separation is typically utilized in industry as well as in research settings for the purposes of purifying and concentrating macromolecular solutions, particularly protein solutions. This is due to the fact that this type of separation is able to separate macromolecular solutions with high levels of specificity. Similar to microfiltration, ultrafiltration operates according to the concept of size exclusion, often known as particle capture (Fane and Fane, 2005; Lovins et al., 2002; Peter-Varbanets et al., 2009). The molecular weight cut-off of the membrane is what is used to define the membrane. This molecular weight cut-off is what is used to generate the ultrafiltration membrane. Depending on the conditions present, ultrafiltration membranes can function either in a cross-flow or a dead-end mode of operation. The ultrafiltration membranes are the only kind of molecules that allow small molecules, such as leachate molecules, inorganic salts, and micromolecular organics, to pass through them. These membranes are impermeable to macromolecules like suspended solids, colloids, proteins, and bacteria, therefore, they cannot be passed through them (Monnot et al., 2016; Youcai, 2018). Ultrafiltration is always used as a pre-treatment method for leachate that has a lower organic concentration, and this is done before nanofiltration is ever used as a treatment method (Youcai, 2018).

Nanofiltration (NF): NF membrane process is driven by pressure, and in terms of its capacity of rejection for ionic and molecular species, it lies between UF and RO processes. The pore size of NF membranes is around 0.001 μm. Particulates larger than the pore sizes, mainly organic solutes and multivalent ions, such as calcium and magnesium, are retained. It removes both organic and inorganic constituents and even bacteria and viruses, which lowers the requirement for disinfection. It is also known as loose RO or low-pressure RO. The usual range of operating pressure used in NF is 700–1400 kPa (Tchobanoglous et al., 2013). NF is less expensive when compared to RO as it shows higher flow rates and has lower operating pressure. This pressure-driven membrane method sits between UF and RO in terms of its ability to remove molecular or ionic species (Nataraj et al., 2009; Shen and Schäfer, 2014). NF membranes, organic membranes, and ceramic membranes each have dense or porous membranes. In nanofiltration membranes, the open space, small pores, or nanovoids available for transport can be much larger than in conventional membranes. The size of these nanovoids, which can range between 0.5 and 1 nm in diameter, can provide a transition between microporous and dense membranes. As is customary, nanofiltration membranes are commonly believed to have a nominal cut-off value between 1,000 and 200 Da (Nagy, 2012; Warsinger et al., 2018). The characteristics that distinguish nanofiltration membranes from other types of membranes are primarily the high rejections for multivalent ions (>99%) average rejections for monovalent ions (approximately =70%), and high rejection (>90%) of organic compounds with a molecular weight greater than the membrane's molecular weight (Nagy, 2012). The manner of mass transfer is greatly dependent on the membrane's structure and the interactions between the membrane and the transported molecules. The separation efficiency may be governed by the sieving effect or by the solution and diffusion properties of the solute molecules. When it comes to transporting charged molecules, the electrical field performs a distinct and crucial role. Three variables are crucial to the operation of a nanofiltration unit: the permeability of the solvent through the membrane (also known as the flux), the rejection of solutes, and the yield or recovery to precisely predict the amount of separation that will occur during nanofiltration (Nagy, 2012).

Reverse osmosis (RO): In recent times, RO has been frequently used as a tertiary treatment option for the treatment of wastewater (Beier et al., 2010; Fane et al., 2015; Holloway et al., 2014; Luo et al., 2014). RO uses fine membranes which retain almost all salts and allows only water to pass through. These membranes can either be asymmetric or composite, having a dense skin surface. The size of the RO membranes is less than 0.001 μm, and the typical operating pressure is around 800–1,900 kPa (Tchobanoglous et al., 2013). After the advanced treatments using granular depth filtration or MF, particularly dissolved constituents, which still remain in the water, can be removed using RO. Removal of 90% of the TDS can be achieved using the RO system. Other constituents, including turbidity, specific organic molecules, viruses, and bacteria, can also be effectively removed (Patel et al., 2019; Pype et al., 2016; Urtiaga et al., 2013). The RO process is the opposite of the osmosis process. In osmosis, a semi-permeable membrane separates two different solutions not having the same solute concentration, and there exists a chemical potential difference across the membranes. The water from the lower concentration side will diffuse through the membrane to the higher concentration side until the difference in pressure balances out the potential chemical difference in a system with a fixed volume. This balancing pressure difference is termed osmotic pressure, and it is a function of solute concentration and temperature (Martin et al., 1999). In RO, a pressure gradient higher than the osmotic pressure is applied across the membrane in the opposite direction, forcing flow from a high concentration toward the lower concentration region. This implies that only water molecules pass through the membrane openings, and other particles are retained on it, thus resulting in water relatively free from ions on the lower concentration side (Mackenzie, 2010).

The different types of RO membranes available are as follows.

Composite polyamide membranes: Composite polyamide (CPA) RO membranes are used for all critical high-purity applications. Spiral-wound CPA RO membranes from CPA offer high salt rejection rates for brackish water applications at an affordable price (Hydranautics, 2022; Nataraj et al., 2009; Sahebi et al., 2020).

Energy-saving polyamide membranes: Energy-saving polyamide membranes (ESPA) can produce high-quality water at a low cost. These are used in the treatment of drinking water, industrial process waters, municipal and industrial wastewater recovery and reuse because of this exceptional performance capability (Hydranautics, 2022).

Low-fouling composite membranes: When treating wastewater and surface water with a high potential for fouling, using low-fouling composite membranes with a neutral surface charge helps limit the amount of fouling that occurs. The differential pressure decreases, resulting in decreasing the accumulation of biological and colloidal fouling within the membrane and an effective lowering of the amount of times the membrane needs to be cleaned (Hydranautics, 2022; Qin et al., 2018; Ye et al., 2020).

Ceramic membranes: The production of membranes typically uses ceramics and polymers as the two primary types of materials. In water treatment and desalination, polymeric membranes now hold the majority of the market share. However, their applications are restricted because they have poor long-term stability, are easily fouled, and have a relatively short lifetime. Ceramic membranes are used most frequently in MF and UF processes, and occasionally in NF and RO as well (Nagy, 2012; Sun et al., 2021). Ceramic membranes have a number of technical advantages over polymeric membranes. These advantages include a narrow and well-defined pore size distribution, higher porosity, better separation, a higher flux, higher thermal, mechanical, and chemical stability, longer membrane lifetimes, higher hydrophilicity, high fluxes at low pressures, and lower fouling. Ceramic membranes also have a higher lifetime than polymeric membranes. (He et al., 2019).

7.3.3.2 Membrane Configurations

The complete unit comprising membrane elements, ports for inlet of feed, outlet for permeate and retentate, and the whole support structure is termed the module. Membrane modules primarily used for the treatment of wastewater are plate and frame, spiral wound membrane, tubular, and hollow fiber modules. Figure 7.3 represents the typical membrane modules described in the following sections.

Plate and frame module: In this module configuration, flat membrane sheets are stacked adjacent to the permeate ports and supported by plates. The feed water is fed to the stacked membranes through which it gets filtered, and the permeate is obtained. Plate provides mechanical support to membranes and also enables the drainage of the permeate. This module is not often used for treating municipal wastewater, but it is better suited to wastewater with high suspended particles and a strong fouling tendency. A packing density of 100–400 m^2/m^3 is provided in the plate and frame module (Wang et al., 2006).

Spirally wound module: This membrane module is extensively used in RO and NF processes. The arrangement provides a high packing density, resulting in a large membrane surface area. A number of membranes, permeate spacers, and feed spacers are looped around a perforated central collection tube in this configuration. These are then inserted into a tube-shaped pressure vessel. To construct a cylindrical element, the membrane is looped around the center tube. The open end of the membrane envelope is wrapped around a tube with holes through which permeate can flow out. Water that has gone through the membrane in service flows through the porous support toward the center tube. Each element has a length of up to 150 cm and a diameter of up to 30 cm. In a single cylindrical pressure vessel, two to six elements are installed. This module is compact with around 700–1,000 m^2/m^3 packing density (Al-Bastaki, 2004; Wang et al., 2006).

Tubular module: This module is made up of a shell which has a tubular shape and forms the outside casing. A perforated or porous stainless steel or fiberglass pipe is contained within this tubular casing, within which a semi-permeable membrane is embedded. Under pressure, the fluid to be treated is pumped into the tube. The permeate from the membrane enters the housing through the perforated pipe and is collected through the outlet (Qasim et al., 2019; Wang et al., 2006).

Hollow fiber module: In this, thousands of hollow fibers are bundled together and placed inside a pressure vessel. The feed may flow from the outward or inward directions of the fiber. Packing density of about 1,000 m^2/m^3 in UF and up to 10,000 m^2/m^3 in RO can be attained with this module, and it is very compact (Cartagena et al., 2013; Qasim et al., 2019).

7.3.3.3 Advantages and Disadvantages of Membrane-Based Processes

Membranes have the ability to decrease pollutants to the required levels specified/ needed for certain reuse applications. In a sewage treatment facility, the membranes can be placed at any time, and they can be designed for the specific volume of required treatment. Phase change does not occur during separation, and it happens at ambient temperature. Furthermore, products do not accumulate inside the membranes like ion exchange resins which require replacement. Also, the addition of chemical additives is not required for separation. Although the membrane filtration technique has produced high-quality effluent over the years, it has a few major drawbacks. The

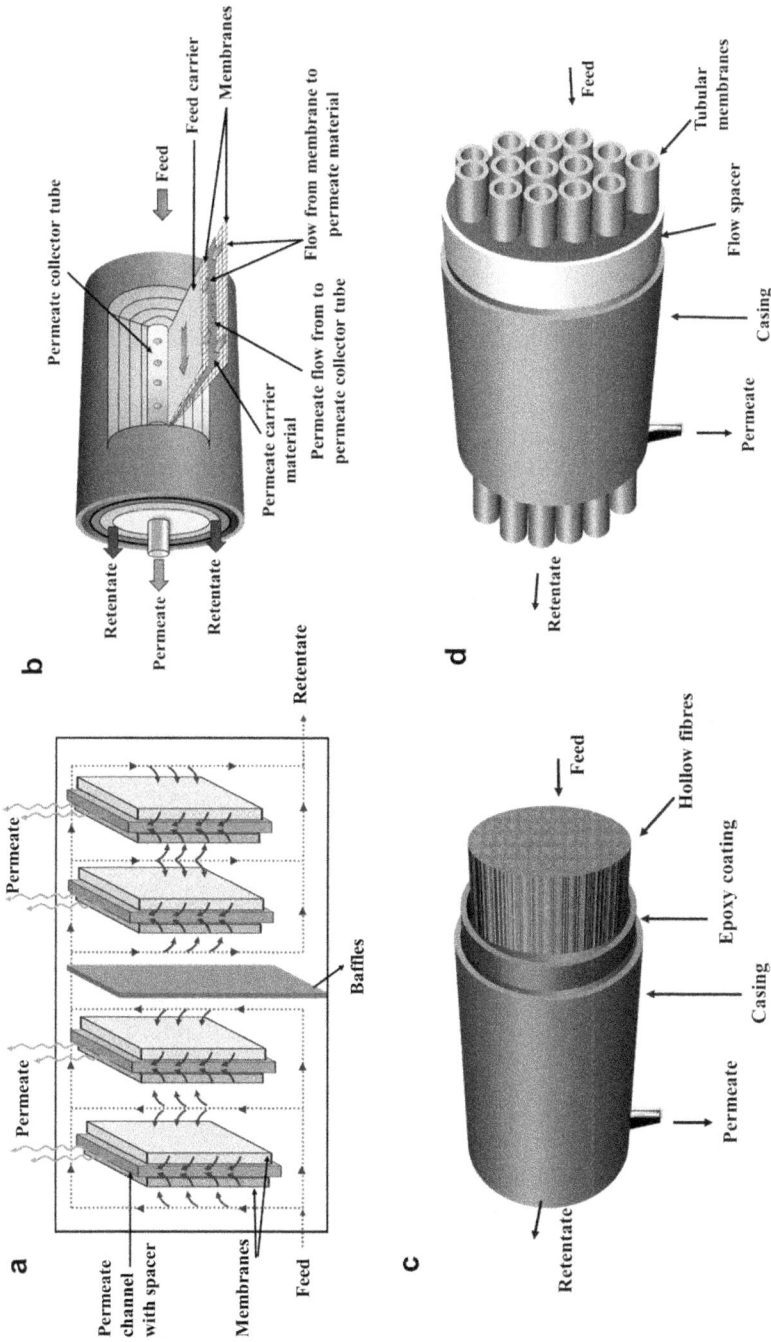

FIGURE 7.3 (a) Plate and frame module; (b) spiral wound membrane module; (c) tubular; and (d) hollow fiber module.

major problems associated with membranes are that they are prone to biofouling, scaling, and degradation (Mackenzie, 2010).

- *Scaling of membrane*: During the process of filtration of wastewater via a membrane, there is an increase in the concentration of various salts that are only marginally soluble. These salts might be divalent or multivalent. When the solubility limit of these salts approaches the degree of supersaturation, they have a tendency to precipitate and create a scale on the membrane surface. This scale reduces the amount of water that can be processed by the RO process (Abdulgader et al., 2013; Shenvi et al., 2015).
- *Fouling of membrane*: Membrane fouling is a process in which particles, colloidal particles, or solute macromolecules are deposited or adsorbed onto the membrane pores or onto a membrane surface by physical and chemical interactions or by mechanical action. This process causes the membrane pores to become smaller or blocked off entirely, which is the result of membrane fouling (Warsinger et al., 2018). Microbial growth on the surface of the membrane can frequently hinder the flow of liquids through the membrane when it is being used for the treatment of wastewater. The fouling of the membrane that takes place as a result of the growth of microorganisms is referred to as biofouling (Shenvi et al., 2015).
- *Degradation of membrane*: Chlorine, chloramines, ozone, hydrogen peroxide, and other chemicals found in wastewater are oxidizers that attack the concentrate layer structure of the membrane. Some layers, such as the thin film composite layer, are more damaged than others. Thermal damage to the membrane occurs when temperature limits are exceeded, rendering it useless. Most RO membranes have a maximum temperature limit of 45°C, after which they begin to degrade (Antony et al., 2014; Shenvi et al., 2015). Hence, research on developing cost-effective RO membranes, resistant to fouling, scaling, and degradation, has been getting significant attention.

Due to the abovementioned drawbacks, appropriate prior treatment is required for the process to remain feasible. Hence, the cost of maintenance and operations is high. Backwash water is required, which is obtained from a portion of treated water. Furthermore, the disposal of concentrate, also referred to as brine, is challenging as it has a high concentration of dissolved solids.

7.4 ADSORPTION

Soluble organic compounds that are resistant to biological breakdown and are usually referred to as refractory organics remain present in the effluent even after the secondary treatments and filtration. To remove such refractory organic compounds, adsorption is a process used widely. It is a mass transfer phenomenon in which the sorption of solute (adsorbate) is adsorbed onto the surface of the pores of the adsorbent by different adsorption mechanisms. Granular activated carbon and powdered activated carbon are the most commonly used adsorbent in wastewater treatment (Pounsamy et al., 2019). Physisorption and chemisorption are the two primary mechanisms by which adsorption takes place. In physisorption, the adsorbate is trapped inside the pores due to physical forces, such as van der Waals, while in chemisorption, the adsorbate is bound to the adsorbent through chemical bonding. Desorption during physisorption is easier to achieve than in chemisorption. The amount of adsorbate that an adsorbent can absorb is determined by a number of factors, including the physical and chemical properties of the adsorbent (pore size, surface morphology, functional groups, and surface area), and the adsorbate (size, polarity, pK_a, molecular weight, and functionality). The experimental conditions (temperature, pH, and ionic strength) also play a pivotal role in the adsorption phenomenon. (Srivastava et al., 2021).

The process takes place in four steps, such as transport of solution in bulk, transport through diffused film, surface and pore transport, and adsorption. The different pore sizes allow the particles to get adsorbed in these pores. Generally, the surface area of micropores is larger as compared to macropores, and mesopores and negligible adsorption are considered to have taken place in these bigger pore sites (Bayantong et al., 2021; Bhaumik et al., 2014).

Since adsorption occurs stepwise, the slowest step is the rate-limiting step of the process. In combination with adsorption, desorption is also a process that takes place simultaneously. During desorption, the adsorbed particles are either released through or from the surface of the adsorbent. The occurrence of desorption depends on the existing state of equilibrium. When the rate of adsorption equals the rate of desorption, equilibrium is achieved, and the adsorbent is said to have reached its capacity. For a particular contaminant, the theoretical adsorption capacity of a given adsorbent can be determined using adsorption isotherms (Lipatova et al., 2018; Yadav et al., 2019).

7.4.1 BATCH AND COLUMN STUDIES

Batch adsorption is an effective method for the removal of a variety of pollutants from real and synthetic wastewater. Batch adsorption studies are mainly carried out for a small quantity of effluent with less contamination. The adsorbent and water to be treated are mixed in the reactor, and after equilibrium is achieved, the adsorbent is removed, and the water is decanted. The batch adsorption process is simple, cheap, and has been frequently used to assess the feasibility of adsorbent–adsorbate system. However, the industrial applications of batch adsorption are limited as large amounts of adsorbent would be required for treating heavily polluted industrial effluent (Ghosal and Gupta, 2018).

Column treatment, also known as continuous pack bed or fixed bed, is a typical adsorption method where the adsorbate is continuously fed through a column filled with the adsorbent at a certain flow rate. The flow of the adsorbate ensures that it is continuously in contact with the adsorbent. This process is generally used for a large quantity of wastewater containing a high pollution load. Various parameters like flow rate, adsorbate concentration, bed height (adsorbent concentration), breakthrough parameters, pH, particle size, and others are evaluated to analyze to performance of the column. A high amount of adsorbate can be adsorbed using column studies, making it feasible for industrial-scale applications. Limitations of column study include exhaustion of adsorbent, channeling through the bed, and uncontrollable feed flow of adsorbent particles. Moreover, pilot-scale establishment and optimization of controlling parameters are also necessary before commercial implementation (Abebe et al., 2020; Singh et al., 2022).

The amount of adsorbate that can be adsorbed is a function of both the temperature and the concentration and characteristics of the adsorbate. Important characteristics of the adsorbate include molecular structure, solubility, molecular weight, polarity, and hydrocarbon saturation. Usually, the quantity of adsorbed material is determined as a function of the concentration at a constant temperature, and the resulting function is called adsorption isotherm. The isotherms are used to analyze and optimize the adsorption mechanism, its pathways, surface characterization of adsorbent, and designing of an adsorption system. Different isotherm models have been applied in adsorption systems, such as the Langmuir, Freundlich, Sips, and the Brunauer, Emmet, and Teller (BET) to model the adsorption phenomenon (Foo and Hameed, 2010; Mozaffari Majd et al., 2022).

If the adsorbent is added to a batch reactor and mass balance is performed, then a general equation for mass balance can be written as Equation 7.1:

$$q_e M = VC_0 - VC_e \qquad (7.1)$$

where q_e = adsorbent phase concentration after equilibrium (mg adsorbate/g adsorbent)
M = mass of adsorbent (g)
V = volume of liquid in the reactor (L)
C_0 = initial solution concentration of adsorbate (mg/L)
C_e = final solution equilibrium concentration of adsorbate after adsorption has occurred (mg/L).

Equation 7.1 can be rewritten as Equation 7.2:

$$q_e = V\left(C_0 - C_e\right)/M \qquad (7.2)$$

The equation. is then used to compute adsorbent phase concentration and develop adsorption isotherms.

Freundlich Isotherm: Freundlich isotherm is one of the most commonly used isotherms to describe the adsorption characteristics of adsorbents used in water and wastewater treatment. The empirical equation which defines Freundlich isotherm is provided in Equation 7.3 (Gupta et al., 2014; Mozaffari Majd et al., 2022).

$$\frac{x}{m} = K_f C_e^{1/n} \qquad (7.3)$$

where $\dfrac{x}{m}$ = mass of adsorbate adsorbed per unit mass of adsorbent (mg adsorbate/g adsorbent)

K_f = Freundlich capacity factor (mg adsorbate/g adsorbent) × (L water/mg adsorbate)$^{1/n}$
C_e = equilibrium concentration of adsorbate in solution after adsorption (mg/L)
$1/n$ = Freundlich intensity parameter.

The nonlinear Freundlich model can be solved by nonlinear regression analysis. The constants can be found by plotting $\log(x/m)$ versus $\log C_e$ and using the linear form of Equation 7.4 (Mozaffari Majd et al., 2022):

$$\log\left(\frac{x}{m}\right) = \log K_f + \frac{1}{n}\log C_e \tag{7.4}$$

Langmuir Isotherm: The Langmuir adsorption isotherm is used to describe the equilibrium between adsorbate and adsorbent system, where the adsorbate adsorption is limited to one molecular layer or before a relative pressure of unity is reached. While developing the Langmuir adsorption isotherm, it was assumed that adsorption is reversible and only a fixed number of adsorption sites having the same energy are available on the surface of the adsorbent. The Langmuir isotherm model can be defined using Equation 7.5 (Ghosal et al., 2018; Mozaffari Majd et al., 2022).

$$\frac{x}{m} = \frac{abC_e}{1+bC_e} \tag{7.5}$$

where
x/m = mass of adsorbate adsorbed per unit gram of adsorbent (mg adsorbate/ g adsorbent)
a and b are empirical constants
C_e = equilibrium concentration of adsorbate in solution after adsorption (mg/L).

The stated assumptions are only valid for a particular system and may not hold true even if the experimental data corresponds to the Langmuir equation, as deviations from the assumptions can have a canceling effect. The constants can be determined using a plot between $C_e/(x/m)$ versus C_e. The linear form of Equation 7.5 can be written as Equation 7.6 (Ghosal et al., 2018; Mozaffari Majd et al., 2022).

$$\frac{C}{(x/m)} = \frac{1}{ab} + \frac{1}{a}C_e \tag{7.6}$$

7.4.2 ADSORPTION CAPACITY

Using the isotherm data, one can figure out how much an adsorbent can hold. By plotting the isotherm data and extrapolating, it is possible to find the adsorption capacity at equilibrium. The breakthrough curves are used to figure out how long an adsorbent will work. For an adsorption study, equilibrium adsorption isotherm data are used to

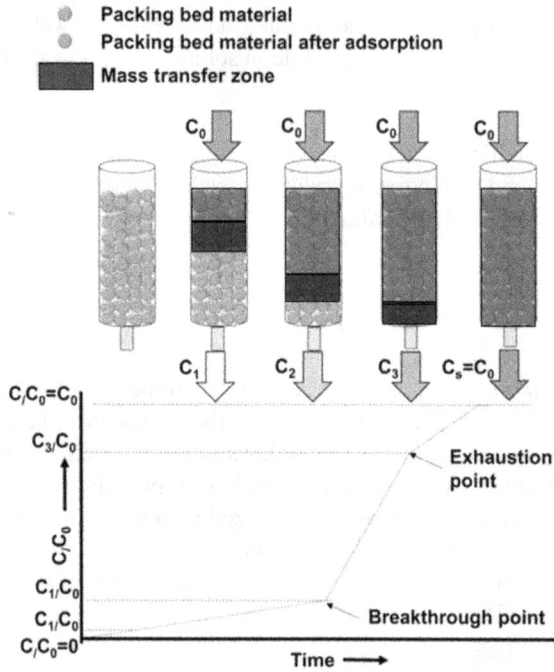

FIGURE 7.4   Breakthrough curve and exhaustion of adsorbent in column studies.

figure out the adsorbent's adsorption capacity. Kinetic models show how fast adsorption happens, and breakthrough curves show how the adsorbent works. A column test can be used to figure out the breakthrough adsorption capacity. A breakthrough curve is a graph of the pollutant-effluent concentration versus time profile in a fixed-bed column. It represents the ratio of the concentration of contaminants in the effluent to the concentration of contaminants in the influent with respect to time (Figure 7.4).

The mass transfer zone (MTZ) can be used to study how breakthrough curves and fixed-bed adsorption are related. MTZ is the area where sorption occurs and is attained near the top or influent end of the column (Reske et al., 2020). At this point, the adsorbate concentration (C) is zero, so the ratio of effluent concentration to initial concentration (C_e/C_0) is also zero (Mozaffari Majd et al., 2022). It is said that during breakthrough operations, the effluent concentration goes down at the beginning of the column study. However, as the adsorption time goes on, the MTZ moves down the column, and when it gets to the bottom, the effluent concentration goes up quickly. Generally, when the effluent concentration reaches 5% of the influent value, a breakthrough is achieved. Also, the bed is said to be exhausted when the effluent (C_s) concentration is 95% of the concentration of the influent. At this point, it is observed that the column adsorption capacity is physically exhausted and the $C_s/C_0 = 0.95$. The best shape for the breakthrough curves is an "S." The shape of the curve is also affected by the rate of hydraulic loading and by whether or not the applied liquid has non-absorbable or biodegradable parts (Mozaffari Majd et al., 2022).

7.4.3 Issues and Concerns Associated with Adsorption

The choice of an adsorbent is the first major challenge in adsorption because apart from easy availability and low cost, the adsorbent should have high chemical and mechanical stability, a high surface area and pore volume, functional groups on the surface, which can interact with the pollutants, and regeneration and reuse potential. (Awad et al., 2020; Dotto and McKay, 2020; Ghosal and Gupta, 2015; Pillai et al., 2020; Yadav et al., 2017). As a result, tremendous stress was placed on the application of many natural and synthetic adsorbents for pollutant abatement. Compared to natural materials or widely applied adsorbents, some novel adsorbents have a high potential for adsorption as those adsorbents are prepared with the elements or properties that favor the targeted solute uptake. Thus, those materials showed very efficient performance in the batch-scale studies (Awad et al., 2020; Dotto and McKay, 2020; Ghosal et al., 2018; Ghosal and Gupta, 2015; Kabir et al., 2020; Pillai et al., 2020; Yadav et al., 2017). However, batch-scale studies involving the adsorption of different pollutants are suitable for the optimization of experimental adsorption conditions and not suitable for handling field-scale consequences, whereas the continuous-mode operation of adsorption is essential in the field (Awad et al., 2020; Dotto and McKay, 2020; Kumari et al., 2021; Sounthararajah et al., 2015; Zubair et al., 2021). Furthermore, many of the adsorbents used in batch studies are only confined to the lab. Adsorbents in powdered form are not suitable for column studies as they tend to cause choking or channeling in the column. Immobilization of the materials in granular beads significantly lowers their removal efficiency (Chai et al., 2013; Dong and Wang, 2016; Ghosal and Gupta, 2015; Kim Phuong, 2014; Mandal and Mayadevi, 2009; Reshmi et al., 2006). The major shortcomings of these adsorbents, which exhibit high removal efficiency in batch studies but fail to perform in column studies, are as follows (Awad et al., 2020; Dotto and McKay, 2020; Ghosal and Gupta, 2018; Kim et al., 2019; Nasir et al., 2019):

- They have low hydraulic conductivity.
- The particle size is very low and cannot be controlled.
- In the case of immobilizing those adsorbents on a suitable base material, the adsorption potential is drastically reduced.
- The adsorbents sometimes have very low desorption capacity, which in turn demands discarding the material after a single cycle.

Fixed-bed column studies are the most widely used continuous-mode adsorption studies conducted both on a small scale and large scale. Continuous operation is mandatory to supply the treated water to the consumers. However, several shortcomings of column-based application in the field were observed as follows (Awad et al., 2020; Bacelo et al., 2020; Dotto and McKay, 2020; Ghosal and Gupta, 2018):

- Maintaining the column in its optimum function requires technical expertise.
- Many drawbacks may arise during operation, such as channeling, choking, exhaustion or even desorption of solute, which cannot be judged without technical knowledge and laboratory testing.

- Regeneration of material is essential, the technical know-how and chemical handling, which is also difficult for common people without proper training.
- Loss of adsorbent during backwash.

7.5 DISINFECTION

Disinfection is a method of destroying disease-causing organisms known as pathogens present in the water. Ideally, disinfections should not only destroy existing germs, microbes, and other pathogens present in the wastewater but also protect water from future possible potential contamination. Hence, it is desirable that the disinfectants have residual power for longer storage of water (Estrada-Arriaga et al., 2016). To know the effectiveness of the disinfection process in wastewater, it is very much important to know the characteristics of an ideal disinfectant. An ideal disinfectant must possess the following characteristics (Mackenzie, 2010; Tchobanoglous et al., 2014).

- Fast rate of removal
- Toxic to only microorganisms and safe for humans and other animals
- Exhibit enough residual power to protect water from future contamination
- Safe to store, transport, handle, and use
- Available in large quantities and cost-friendly
- Not damaging to metals or cause stain clothing
- Efficient in an ambient temperature range

7.5.1 METHODS OF DISINFECTION

7.5.1.1 Physical Methods

Non-Ionizing Radiation: Energy in the form of heat, electromagnetic waves, and sound waves can be transmitted over a wide range without any conductors. This form of transmission is known as radiation. Radiation energy with high intensity can cause severe damage to pathogens in water, making it a useful agent for wastewater disinfection (McDonnell, 2017; Tchobanoglous et al., 2014).

Boiling: Heating water to its boiling point severely damages the cell wall of disease-producing bacteria. The appropriate temperature range is 80–90°C, at which almost 99% of the organisms can be destroyed. Pasteurization of wastewater and sludge is used largely in different countries (Mackenzie, 2010; Tchobanoglous et al., 2014).

Ultraviolet Treatment: Ultraviolet (UV) treatment is one of the most important electromagnetic radiation treatments used for disinfecting. UV rays in the wavelength 250–260 nm have got very high destructive power, and they penetrate into the cell by destroying the protoplasm of the pathogens. The amount of destruction that occurs in water depends on the intensity of UV rays and their time of exposure, and it is provided in Equation 7.7 (Guo et al., 2015; Tchobanoglous et al., 2014).

$$\ln \frac{N_t}{N_o} = -KIt \qquad (7.7)$$

where
N_0 = Initial number of organisms
N_t = Number of organisms remaining at time t
K = Rate constant
I = Intensity of UV rays
t = Time of exposure

> *Ionizing Radiation*: A type of radiation in which subatomic particles (alpha or beta) and electromagnetic waves (gamma or X-rays) having sufficient energy is used to ionize atoms or molecules by removing electrons from them is termed ionizing radiation. The use of high-energy electron beam devices for disinfecting or sterilizing wastewater is associated with high costs. Hence very few commercial uses have been reported yet. For example, gamma rays emitted from radioisotope (usually cobalt-60 or cesium-137) are commonly used for sterilization of disposable medical equipment, such as syringes, cannulas, and similar substances. However, ionizing radiation can be a devastating health threat if handled improperly (McDonnell, 2017; Vadrucci et al., 2020).

7.5.1.2   Chemical Methods

In STPs, chemicals like free chlorine, chlorine dioxide, sodium hypochlorite, chloramines, ozone, and peracetic acid are often used to kill germs (PAA). The most common way to clean water in STPs is with chlorine, which is used because it is cheap and works well (Yao et al., 2020).

- *Chlorination*: Among the different available disinfectant chemicals, chlorine is the most widely used disinfectant throughout the world due to its easy availability, cost-effectiveness, ease of handling, and high residual power. Unlike most physical disinfection methods, chlorine has a high residual effect. Chlorination is most effective against bacteria, while least effective against protozoa. Chlorine has varying disinfective potential against different viruses. Viruses, such as adenovirus and Hepatitis A, are more resistant to chlorination as compared to Coxsackie B virus and SARS-CoV (Majumder et al., 2021). The required dose of chlorine for inactivating different pathogens is provided in Figure 7.5.

Forms of Chlorine
Chlorine is available in powder or solid form, gaseous form, and liquid form. Chlorine gas typically contains pure chlorine and is the most stable form chlorine. Chlorine powder and liquid do not contain pure chlorine. They are mixed with other substances (e.g., calcium, sodium, or water). The strength of chlorine in chlorine powder or chlorine liquid is referred to as the concentration of chlorine in that substance.

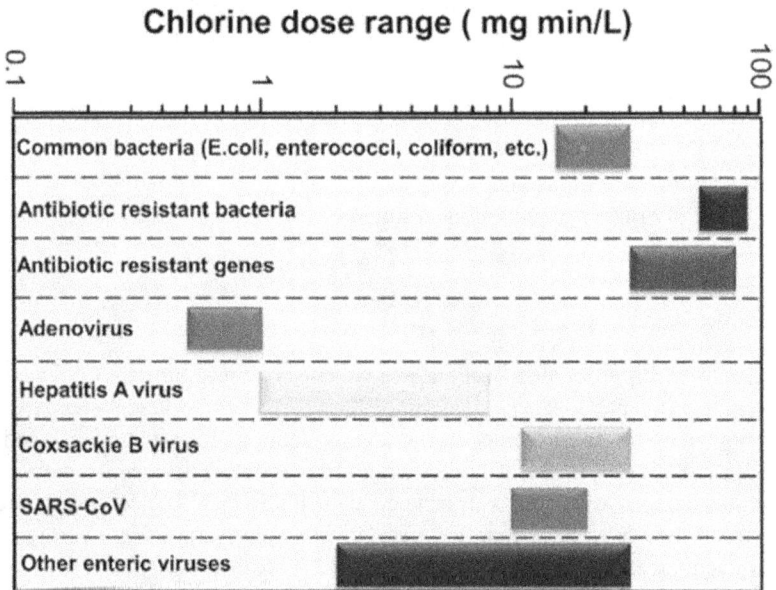

FIGURE 7.5 Chlorine dose required for efficient inactivation of different pathogens (Reprinted with permission from Majumder et al., 2021. Copyright 2023 Elsevier).

Powdered chlorine is typically available in the form of white powder, granules, or tablets. They usually contain around 30–70% active chlorine. They may lose strength over a period of time. However, they are more stable as compared to liquid chlorine, Typical example of powdered form of chlorine is (approximately 35% active chlorine) calcium hypochlorite ($Ca(ClO)_2$), also known as bleaching powder.

Liquid chlorine usually contains around 1–15 % active chlorine. They are less stable as compared to powdered chlorine. Examples of liquid chlorine are sodium hypochlorite (NaClO) (10–15 % active chlorine) and domestic bleach (5–10 % active chlorine) (Mackenzie, 2010; Qasim, 2017; Tchobanoglous et al., 2014).

Chemistry of Chlorine Compounds
In water, chlorine first hydrolyzes to form hypochlorous acid (HOCl) and then it ionizes to form hypochlorite ions (OCL^-). Similarly, HOCL is produced when $Ca(ClO)_2$ and NaClO are added to water. HOCl and OCL^- are collectively known as free chlorine and are responsible for breaking down natural organic matter and killing microorganisms. The reactions involving chlorination using Cl_2 (gas), $Ca(ClO)_2$, and NaClO are provided in Equations 7.8 to 7.11 (Mackenzie, 2010; Qasim, 2017; Tchobanoglous et al., 2014).

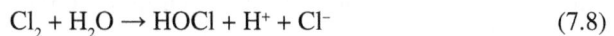

$$Cl_2 + H_2O \rightarrow HOCl + H^+ + Cl^- \tag{7.8}$$

$$NaOCl + H_2O \rightarrow NaOH + HOCl \tag{7.9}$$

$$Ca(OCl)_2 + 2H_2O \rightarrow Ca(OH)_2 + 2HOCl \qquad (7.10)$$

$$HOCl \xrightarrow{\quad pH<7.5 \quad} H^+ + OCl \rightarrow Ionization \qquad (7.11)$$

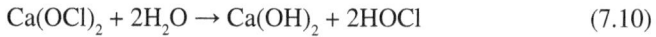

Being radical-free, HOCl is a more powerful oxidant compared to OCL⁻, which is produced in excess when the pH value of water is below 7.5. Hence it is general practice to keep pH value below 7.5 for the effective removal of organisms. The high percentage of HOCl in water mainly depends on two factors, the hydrogen ion concentration of water and the ionization constant. The addition of $Ca(ClO)_2$ and NaClO leads to the formation of NaOH and $Ca(OH)_2$, which are responsible for making the water alkaline. Hence, Cl_2 gas is more effective when it comes to chlorination (Mackenzie, 2010; Qasim, 2017; Tchobanoglous et al., 2014).

Chlorine Dose, Demand, and Residual

During wastewater treatment, chlorine not only reacts with the pathogens present in the wastewater but also with the inorganic and organic materials. The inorganic and organic substances present in the wastewater significantly contribute to the total chlorine demand. In other words, chlorine demand is the difference between the amount of chlorine applied to the wastewater and the amount of residual chlorine after a given contact time. Inorganic compounds, such as hydrogen sulfide, ferrous iron, manganese, nitrite, and ammonia are the major contributors to the total chlorine demand. Ammonia reacts with chlorine to form monochloramine, dichloramine, and trichloramine, which also acts as disinfection agents. Similarly, some organic compounds react with chlorine to form chlororganic compounds. However, these compounds have a much lesser disinfection capacity (Mackenzie, 2010; Qasim, 2017; Tchobanoglous et al., 2014).

The amount of chlorine consumed in reacting with the inorganic and organic matter and microorganisms is known as chlorine demand of water. After meeting the chlorine demand of water, if any excess chlorine remains in water, it appears as residual chlorine (Equation 7.12). The residual chlorine may be in the form of chloramines, chlororganic compounds, and free chlorine (Mackenzie, 2010; Qasim, 2017; Tchobanoglous et al., 2014).

$$Chlorine\ Demand = Chlorine\ Dose - Residual\ Chlorine \qquad (7.12)$$

Chlorine dosage can be determined using laboratory testing or actual field results from a known plant operation. The results are suitable for determining base feed rates, but real-time corrections to account for changing conditions are required. Since field conditions are less controlled than laboratory tests, the actual dosage will be higher than those determined in the laboratory.

Additional chlorine is applied to the wastewater in the form of free chlorine to achieve breakpoint chlorination. Breakpoint chlorination refers to the amount of chlorine required to meet the inorganic, ammonia, and organic demands of the wastewater. The excess chlorine added beyond the breakpoint chlorination exists as free chlorine. The breakpoint chlorination curve has been shown in Figure 7.6. It is to be noted that the term breakpoint chlorination is the addition of enough chlorine

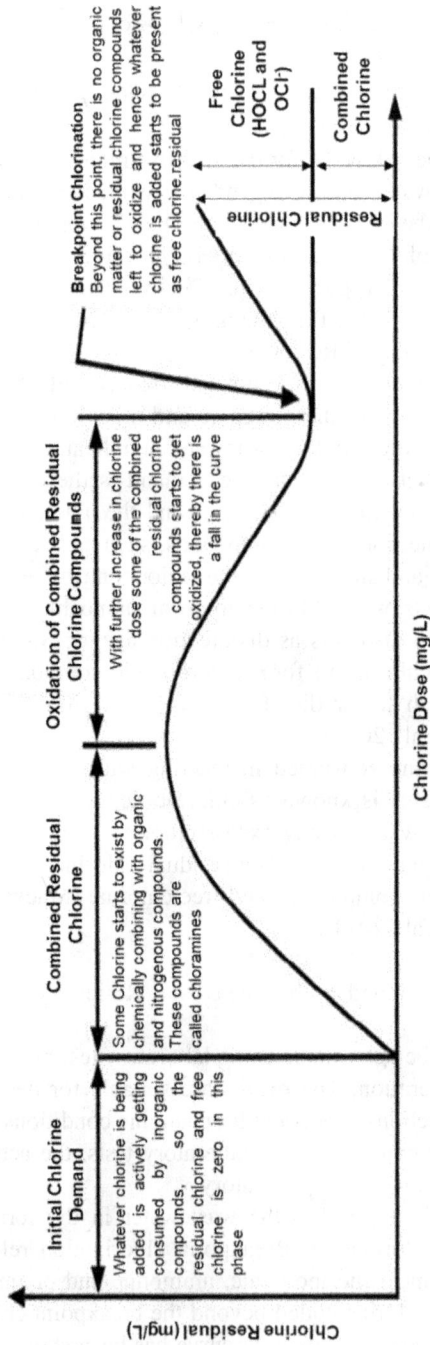

Breakpoint Chlorination
Beyond this point, there is no organic matter or residual chlorine compounds left to oxidize and hence whatever chlorine is added starts to be present as free chlorine.residual

Oxidation of Combined Residual Chlorine Compounds

With further increase in chlorine dose some of the combined residual chlorine compounds starts to get oxidized, thereby there is a fall in the curve

Combined Residual Chlorine

Some Chlorine starts to exist by chemically combining with organic and nitrogenous compounds. These compounds are called chloramines

Initial Chlorine Demand

Whatever chlorine is being added is actively getting consumed by inorganic compounds, so the residual chlorine and free chlorine is zero in this phase

Free Chlorine (HOCL and OCl⁻)

Combined Chlorine

Residual Chlorine

Chlorine Residual (mg/L)

Chlorine Dose (mg/L)

FIGURE 7.6 Breakpoint chlorination curve.

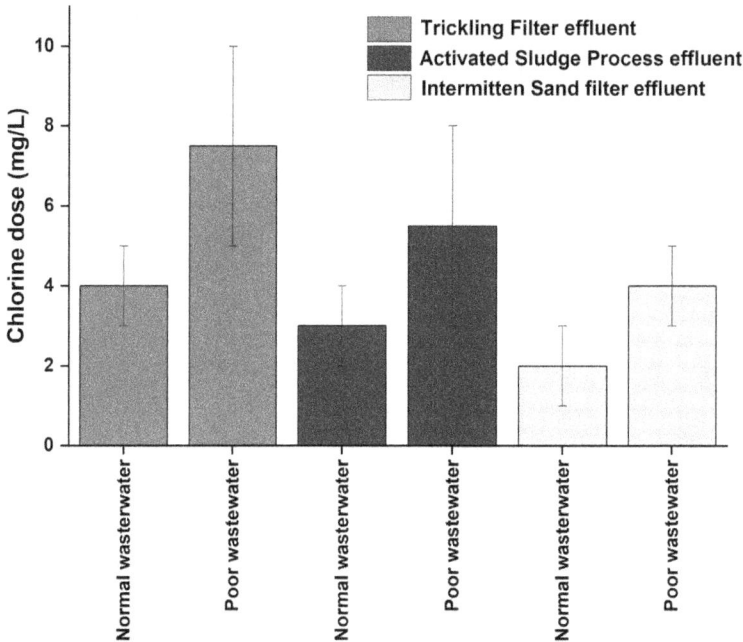

FIGURE 7.7 Chlorine dose (mg/L) to maintain 0.2 mg/L of residual chlorine after 15 min contact time.

Source: Peavy et al., 1985.

to wastewater so that it reacts with all oxidizable matter present. Thereafter, excess chlorine will act as residual chlorine.

The main reason for adding enough chlorine to water is to fulfill the chlorine demand of water and to leave some amount of chlorine for future disinfection. Around 85–90% of the residual chlorine beyond breakpoint chlorination is free chlorine (HOCl or OCl⁻). The concentration of HOCl and OCl⁻ will depend on the pH of the wastewater. Usually, a free chlorine concentration of 5 mg/L should be maintained for the inactivation of coliform (Qasim and Zhu, 2017). In wastewater treatment, if the wastewater is to be discharged directly or to be reused, residual chlorine is desirable. In order to maintain residual chlorine of 0.2 mg/L in different wastewaters, various concentrations of chlorine dose need to be added. The different chlorine doses are depicted in Figure 7.7.

7.5.2 Kinetics of Disinfection

When a single unit of microorganisms is exposed to a single unit of disinfectant, the rate of destruction of microorganisms follows a first-order reaction, which is referred to as Chick's law (Mackenzie, 2010; Qasim, 2017; Tchobanoglous et al., 2014):

$$N_f = N_0 e^{-kt} \qquad (7.12)$$

where
N_0 = initial number of microorganisms
N_t = number of organisms remain after time t
k = disinfectant constant
t = contact time

This relationship means that at constant disinfectant concentrations, pH, temperature, and ionic strength, all species are equally susceptible. The disinfection effect is also highly influenced by the presence of other matter in the wastewater, such as organic matter (Mackenzie, 2010; Qasim, 2017; Tchobanoglous et al., 2014).

The relationship between the dose of disinfectant C and contact time t is given below:

$$C^n \times t = m \tag{7.13}$$

where
N = dilution factor
m = constant

The different factors affecting chlorination are as follows:

- Chlorine is more effective as a disinfectant at lower pH.
- Higher temperature promotes disinfection.
- Longer detention or contact time leads to a higher degree of disinfection.

7.5.3 DISINFECTION BYPRODUCTS FORMED DURING CHLORINATION

Even though chlorination is needed to make sure that recycles treated water is safe, there are still things to figure out about how disinfection byproducts (DBPs) are made and how toxic they are. Recycled water must be treated with chlorine in order to be safe (Priya et al., 2018). During the chlorination process, it has often been found that regulated and emerging DBPs are present in reclaimed water at levels higher than those found in drinking water (Palansooriya et al., 2020). This happens because more chlorine is used to treat reclaimed water. This shows that these DBPs are more dangerous to human health than thought before. After chlorination, there was a rise in the production of DBPs, which caused luminous bacteria and *Daphnia magna* to have more acute toxicity, anti-estrogenic activity, and cytotoxicity (Du et al., 2017). Both genotoxicity and estrogenic activity were found to be lower in water that had been chlorinated. This was because dangerous substances were destroyed. A big thing to think about was how much the quality of the water affected the changes in toxicity that happened after the water was chlorinated. Because of how it reacts with chlorine to make chloramine, ammonium has a tendency to lessen the effects of changes in toxicity. On the other hand, because it makes hypobromous acid, bromide tends to make the effects of changes in toxicity worse (Du et al., 2017). After ozonation and coagulation are used to clean the water first, the disinfection byproduct formation potential (DBPFP) and the toxicity formation potential (TFP) keep going up. This is done along with the removal of DOC, so the drop in DOC was kept to a minimum to show a drop in DBPFP and TFP (Du et al., 2017). It is more important to get rid of the critical percentage of precursors, such as the hydrophobic acids and the hydrophilic neutrals. During the chlorination process, there is a chance that the amount of toxicity will go up as the chlorine dose goes up and the contact time goes up. To prevent too much toxicity, a low dose of chlorine and a short contact time is needed. Also, the use of reductive reagents in the process of quenching residual chlorine can lead to a big drop in the production of compounds that could be harmful (Du et al., 2017).

Determining the amount and quality of natural organic matter and other DBP precursors, figuring out how the different natural organic matter fractions form and whether or not they can be treated, and figuring out how these things affect how DBP is made are essential to control the formation of DBPs (Gora et al., 2018). Use of membrane technologies, improved coagulation, and advanced oxidation processes can all be used to get rid of NOM (Yao et al., 2020). Even though most AOP technologies aren't cost-effective and still need to be improved and scaled up, membrane technologies and enhanced coagulation are particularly useful. Cost is one of the most important factors to consider when choosing a way to treat water. Getting rid of precursors before disinfectants can react with them is an effective and cost-effective way to stop DBPs from forming in sewage treatment plants. It is important to find a balance between a chlorine residual that is high enough to control the water's biostability and a chlorine residual that is low enough to stop DBPs from forming. In general, figuring out what kind of organic matter is in a certain water source and keeping track of it should make it easier for STPs to come up with ways to get rid of DBPs or stop them from forming in the first place (Yao et al., 2020). STPs can come up with protocols that include characterization techniques that track surrogate parameters and how often they change. This is done to track the formation of organic matter and DBPs at STPs (Yao et al., 2020).

7.6 CHAPTER SUMMARY

- Chemical precipitation techniques can be used to remove phosphorous, arsenic, and other heavy metals from the water matrix.
- Filtration is a polishing unit and can be operated under gravity flow conditions or closed reactors pressure filtration units. Filtration is generally required when the total suspended solids in the water matrix are ≤ 10 mg/L.
- The most commonly used filters are conventional downflow filters, deep-bed downflow filters, deep-bed upflow continuous-backwash filtersand, and traveling-bridge filters.
- The different types of membrane filtrations are microfiltration, ultrafiltration, nanofiltration, and reverse osmosis.
- Few of the major problems associated with membrane-based processes are fouling of membrane, scaling of membrane, and degradation of membrane.
- Adsorption is efficient in removing most of the contaminants, but it produces a significant amount of toxic sludge.
- Disinfection is carried out to kill the unwanted pathogens in the effluent and can be carried out using physical and chemical methods. Among physical methods, UV treatment is most commonly used, while chlorination is the most commonly used chemical disinfection treatment.

7.7 CONCLUDING REMARKS

The tertiary treatment is provided to make the treated wastewater suitable for discharge or reuse in various applications. It produces an effluent of high quality. However, most of the steps are cost-intensive. Although the effluent is of good quality after tertiary treatment, a significant amount of sludge is generated from the earlier treatment processes. The management of the sludge produced has been discussed in the following chapter.

REFERENCES

Abdulgader, H. Al, Kochkodan, V., Hilal, N., 2013. Hybrid ion exchange – Pressure driven membrane processes in water treatment: A review. *Sep. Purif. Technol.* 116, 253–264. https://doi.org/10.1016/j.seppur.2013.05.052

Abebe, B., Murthy, H.C.A., Amare, E., 2020. Enhancing the photocatalytic efficiency of ZnO: Defects, heterojunction, and optimization. *Environ. Nanotechnology, Monit. Manag.* 4, 100336. https://doi.org/10.1016/j.enmm.2020.100336

Al-Bastaki, N., 2004. Removal of methyl orange dye and Na_2SO_4 salt from synthetic waste water using reverse osmosis. *Chem. Eng. Process. Process Intensif.* 43, 1561–1567. https://doi.org/10.1016/j.cep.2004.03.001

Antony, A., Blackbeard, J., Angles, M., Leslie, G., 2014. Non-microbial indicators for monitoring virus removal by ultrafiltration membranes. *J. Memb. Sci.* 454, 193–199. https://doi.org/10.1016/j.memsci.2013.11.052

Awad, A.M., Jalab, R., Benamor, A., Nasser, M.S., Ba-Abbad, M.M., El-Naas, M., Mohammad, A.W., 2020. Adsorption of organic pollutants by nanomaterial-based adsorbents: An overview. *J. Mol. Liq.* 301, 112335. https://doi.org/10.1016/j.molliq.2019.112335

Bacelo, H., Pintor, A.M.A., Santos, S.C.R., Boaventura, R.A.R., Botelho, C.M.S., 2020. Performance and prospects of different adsorbents for phosphorus uptake and recovery from water. *Chem. Eng. J.* 381, 122566. https://doi.org/10.1016/j.cej.2019.122566

Bayantong, A.R.B., Shih, Y.J., Ong, D.C., Abarca, R.R.M., Dong, C. Di, de Luna, M.D.G., 2021. Adsorptive removal of dye in wastewater by metal ferrite-enabled graphene oxide nanocomposites. *Chemosphere* 274, 129518. https://doi.org/10.1016/j.chemosphere.2020.129518

Beier, S., Köster, S., Veltmann, K., Schröder, H.F., Pinnekamp, J., 2010. Treatment of hospital wastewater effluent by nanofiltration and reverse osmosis. *Water Sci. Technol.* 61, 1691–1698. https://doi.org/10.2166/wst.2010.119

Bhaumik, M., Choi, H.J., McCrindle, R.I., Maity, A., 2014. Composite nanofibers prepared from metallic iron nanoparticles and polyaniline: High performance for water treatment applications. *J. Colloid Interface Sci.* 425, 75–82. https://doi.org/10.1016/j.jcis.2014.03.031

Cartagena, P., El Kaddouri, M., Cases, V., Trapote, A., Prats, D., 2013. Reduction of emerging micropollutants, organic matter, nutrients and salinity from real wastewater by combined MBR-NF/RO treatment. *Sep. Purif. Technol.* 110, 132–143. https://doi.org/10.1016/j.seppur.2013.03.024

Chai, L., Wang, Y., Zhao, N., Yang, W., You, X., 2013. Sulfate-doped Fe_3O_4/Al_2O_3 nanoparticles as a novel adsorbent for fluoride removal from drinking water. *Water Res.* 47, 4040–4049. https://doi.org/10.1016/j.watres.2013.02.057

Charcosset, C., 2012. Microfiltration. *Membr. Process. Biotechnol. Pharm.* 101–141. https://doi.org/10.1016/B978-0-444-56334-7.00003-4

Dong, S., Wang, Y., 2016. Characterization and adsorption properties of a lanthanum-loaded magnetic cationic hydrogel composite for fluoride removal. *Water Res.* 88, 852–860. https://doi.org/10.1016/j.watres.2015.11.013

Dotto, G.L., McKay, G., 2020. Current scenario and challenges in adsorption for water treatment. *J. Environ. Chem. Eng.* 8, 103988. https://doi.org/10.1016/j.jece.2020.103988

Du, Y., Lv, X.T., Wu, Q.Y., Zhang, D.Y., Zhou, Y.T., Peng, L., Hu, H.Y., 2017. Formation and control of disinfection byproducts and toxicity during reclaimed water chlorination: A review. *J. Environ. Sci. (China).* 58, 51–63. https://doi.org/10.1016/j.jes.2017.01.013

Duan, C., Ma, T., Wang, J., Zhou, Y., 2020. Removal of heavy metals from aqueous solution using carbon-based adsorbents: A review. *J. Water Process Eng.* 37, 101339. https://doi.org/10.1016/j.jwpe.2020.101339

Estrada-Arriaga, E.B., Cortés-Muñoz, J.E., González-Herrera, A., Calderón-Mólgora, C.G., de Lourdes Rivera-Huerta, M., Ramírez-Camperos, E., Montellano-Palacios, L., Gelover-Santiago, S.L., Pérez-Castrejón, S., Cardoso-Vigueros, L., Martín-Domínguez, A., García-Sánchez, L., 2016. Assessment of full-scale biological nutrient removal systems upgraded with physico-chemical processes for the removal of emerging pollutants present in wastewaters from Mexico. *Sci. Total Environ.* 571, 1172–1182. https://doi.org/10.1016/j.scitotenv.2016.07.118

Fane, A.G., Fane, S.A., 2005. The role of membrane technology in sustainable decentralized wastewater systems. *Water Sci. Technol.* 51, 317–325. https://doi.org/10.2166/wst.2005.0381

Fane, A.G., Wang, R., Hu, M.X., 2015. Synthetic membranes for water purification: Status and future. *Angew. Chemie – Int. Ed.* 54, 3368–3386. https://doi.org/10.1002/anie.201409783

Foo, K.Y., Hameed, B.H., 2010. Insights into the modeling of adsorption isotherm systems. *Chem. Eng. J.* 156, 2–10. https://doi.org/10.1016/j.cej.2009.09.013

Fu, C., Yue, X., Shi, X., Ng, K.K., Ng, H.Y., 2017. Membrane fouling between a membrane bioreactor and a moving bed membrane bioreactor: Effects of solids retention time. *Chem. Eng. J.* 309, 397–408. https://doi.org/10.1016/j.cej.2016.10.076

Ghosal, P.S., Gupta, A.K., 2015. An insight into thermodynamics of adsorptive removal of fluoride by calcined Ca-Al-(NO$_3$) layered double hydroxide. *RSC Adv.* 5, 105889–105900. https://doi.org/10.1039/c5ra20538g

Ghosal, P.S., Gupta, A.K., 2018. Application of Sequential Batch Type Reactor (SBTR) for Using in Adsorptive Removal of Pollutant from Drinking Water for Continuous Operation. 201831035015 A.

Ghosal, P.S., Kattil, K.V., Yadav, M.K., Gupta, A.K., 2018. Adsorptive removal of arsenic by novel iron/olivine composite: Insights into preparation and adsorption process by response surface methodology and artificial neural network. *J. Environ. Manage.* 209, 176–187. https://doi.org/10.1016/j.jenvman.2017.12.040

Gora, S., Sokolowski, A., Hatat-Fraile, M., Liang, R., Zhou, Y.N., Andrews, S., 2018. Solar photocatalysis with modified TiO$_2$ photocatalysts: Effects on NOM and disinfection byproduct formation potential. *Environ. Sci. Water Res. Technol.* 4, 1361–1376. https://doi.org/10.1039/C8EW00161H

Guo, M.T., Yuan, Q. Bin, Yang, J., 2015. Distinguishing effects of ultraviolet exposure and chlorination on the horizontal transfer of antibiotic resistance genes in municipal wastewater. *Environ. Sci. Technol.* 49, 5771–5778. https://doi.org/10.1021/acs.est.5b00644

Gupta, V.K., Pathania, D., Kothiyal, N.C., Sharma, G., 2014. Polyaniline zirconium (IV) silicophosphate nanocomposite for remediation of methylene blue dye from waste water. *J. Mol. Liq.* 190, 139–145. https://doi.org/10.1016/j.molliq.2013.10.027

He, N., Hu, L., Jiang, C., Li, M., 2022. Remediation of chromium, zinc, arsenic, lead and antimony contaminated acidic mine soil based on Phanerochaete chrysosporium induced phosphate precipitation. *Sci. Total Environ.* 850, 157995. https://doi.org/10.1016/J.SCITOTENV.2022.157995

He, Z., Lyu, Z., Gu, Q., Zhang, L., Wang, J., 2019. Ceramic-based membranes for water and wastewater treatment. *Colloids Surfaces A Physicochem. Eng. Asp.* 578, 123513. https://doi.org/10.1016/j.colsurfa.2019.05.074

Hedrich, S., Johnson, D.B., 2014. Remediation and selective recovery of metals from acidic mine waters using novel modular bioreactors. *Environ. Sci. Technol.* 48, 12206–12212. https://doi.org/10.1021/es5030367

Holloway, R.W., Regnery, J., Nghiem, L.D., Cath, T.Y., 2014. Removal of trace organic chemicals and performance of a novel hybrid ultrafiltration-osmotic membrane bioreactor. *Environ. Sci. Technol.* 48, 10859–10868. https://doi.org/10.1021/es501051b

Hydranautics, 2022. *Reverse Osmosis (RO) Hydranautics – A Nitto Group Company* [WWW Document]. https://membranes.com/solutions/products/ro/ (accessed 5.18.22).

Kabir, H., Gupta, A.K., Tripathy, S., 2020. Fluoride and human health: Systematic appraisal of sources, exposures, metabolism, and toxicity. *Crit. Rev. Environ. Sci. Technol.* 50, 1116–1193. https://doi.org/10.1080/10643389.2019.1647028

Kim Phuong, N.T., 2014. Layered double hydroxide-alginate/polyvinyl alcohol beads: Fabrication and phosphate removal from aqueous solution. *Environ. Technol. (United Kingdom)* 35, 21–24. https://doi.org/10.1080/09593330.2014.924564

Kim, J.H., Kang, J.K., Lee, S.C., Kim, S.B., 2019. Immobilization of layered double hydroxide in poly(vinylidene fluoride)/poly(vinyl alcohol) polymer matrices to synthesize bead-type adsorbents for phosphate removal from natural water. *Appl. Clay Sci.* 170, 1–12. https://doi.org/10.1016/j.clay.2019.01.004

Kumari, U., Mishra, A., Siddiqi, H., Meikap, B.C., 2021. Effective defluoridation of industrial wastewater by using acid modified alumina in fixed-bed adsorption column: Experimental and breakthrough curves analysis. *J. Clean. Prod.* 279, 123645. https://doi.org/10.1016/j.jclepro.2020.123645

Lipatova, I.M., Makarova, L.I., Yusova, A.A., 2018. Adsorption removal of anionic dyes from aqueous solutions by chitosan nanoparticles deposited on the fibrous carrier. *Chemosphere* 212, 1155–1162. https://doi.org/10.1016/j.chemosphere.2018.08.158

Lovins, W.A., Taylor, J.S., Hong, S.K., 2002. Micro-organism rejection by membrane systems. *Environ. Eng. Sci.* 19, 453–465. https://doi.org/10.1089/109287502320963436

Luo, Y., Guo, W., Ngo, H.H., Nghiem, L.D., Hai, F.I., Zhang, J., Liang, S., Wang, X.C., 2014. A review on the occurrence of micropollutants in the aquatic environment and their fate and removal during wastewater treatment. *Sci. Total Environ.* 473, 619–641. https://doi.org/10.1016/j.scitotenv.2013.12.065

Mackenzie, D.L., 2010. *Water and Wastewater Engineering: Design Principles and Practice.* McGraw-Hill Education.

Majumder, A., Gupta, A.K., Ghosal, P.S., Varma, M., 2021. A review on hospital wastewater treatment: A special emphasis on occurrence and removal of pharmaceutically active compounds, resistant microorganisms, and SARS-CoV-2. *J. Environ. Chem. Eng.* 9, 104812. https://doi.org/10.1016/j.jece.2020.104812

Mandal, S., Mayadevi, S., 2009. Defluoridation of water using as-synthesized Zn/Al/Cl anionic clay adsorbent: Equilibrium and regeneration studies. *J. Hazard. Mater.* 167, 873–878. https://doi.org/10.1016/j.jhazmat.2009.01.069

Martin, D.D., Ciulla, R.A., Roberts, M.F., 1999. Osmoadaptation in archaea. *Appl. Environ. Microbiol.* 65(5), 1815–1825. https://doi.org/10.1128/aem.65.5.1815-1825.1999

McDonnell, G.E., 2017. *Antisepsis, Disinfection, and Sterilization: Types, Action, and Resistance.* Wiley. https://doi.org/10.1002/9781555819682

Monnot, M., Laborie, S., Cabassud, C., 2016. Granular activated carbon filtration plus ultrafiltration as a pretreatment to seawater desalination lines: Impact on water quality and UF fouling. *Desalination* 383, 1–11. https://doi.org/10.1016/j.desal.2015.12.010

Mozaffari Majd, M., Kordzadeh-Kermani, V., Ghalandari, V., Askari, A., Sillanpää, M., 2022. Adsorption isotherm models: A comprehensive and systematic review (2010–2020). *Sci. Total Environ.* 812, 151334. https://doi.org/10.1016/J.SCITOTENV.2021.151334

Nagy, E., 2012. Nanofiltration. *Basic Equations Mass Transp. through a Membr. Layer* 249–266. https://doi.org/10.1016/B978-0-12-416025-5.00010-7

Nasir, A.M., Goh, P.S., Abdullah, M.S., Ng, B.C., Ismail, A.F., 2019. Adsorptive nanocomposite membranes for heavy metal remediation: Recent progresses and challenges. *Chemosphere* 232, 96–112. https://doi.org/10.1016/j.chemosphere.2019.05.174

Nataraj, S.K., Hosamani, K.M., Aminabhavi, T.M., 2009. Nanofiltration and reverse osmosis thin film composite membrane module for the removal of dye and salts from the simulated mixtures. *Desalination* 249, 12–17. https://doi.org/10.1016/j.desal.2009.06.008

Palansooriya, K.N., Yang, Y., Tsang, Y.F., Sarkar, B., Hou, D., Cao, X., Meers, E., Rinklebe, J., Kim, K.H., Ok, Y.S., 2020. Occurrence of contaminants in drinking water sources and the potential of biochar for water quality improvement: A review. *Crit. Rev. Environ. Sci. Technol.* 50(6), 549–611. https://doi.org/10.1080/10643389.2019.1629803

Patel, M., Kumar, R., Kishor, K., Mlsna, T., Pittman, C.U., Mohan, D., 2019. Pharmaceuticals of emerging concern in aquatic systems: Chemistry, occurrence, effects, and removal methods. *Chem. Rev.* 119, 3510–3673. https://doi.org/10.1021/acs.chemrev.8b00299

Peavy, H.S., Rowe, D.R., Tchobanoglous, G., 1985. *Environmental Engineering*. McGraw Hill.

Peter-Varbanets, M., Zurbrügg, C., Swartz, C., Pronk, W., 2009. Decentralized systems for potable water and the potential of membrane technology. *Water Res.* 43(2), 245–265. https://doi.org/10.1016/j.watres.2008.10.030

Pillai, P., Dharaskar, S., Pandian, S., Panchal, H., 2020. Overview of fluoride removal from water using separation techniques. *Environ. Technol. Innov.* 21, 101246. https://doi.org/10.1016/j.eti.2020.101246

Pontius, F.W., Crimaldi, J.P., Amy, G.L., 2011. Virus passage through compromised low-pressure membranes: A particle tracking model. *J. Memb. Sci.* 379, 249–259. https://doi.org/10.1016/j.memsci.2011.05.066

Pounsamy, M., Somasundaram, S., Palanivel, S., Balasubramani, R., Chang, S.W., Nguyen, D.D., Ganesan, S., 2019. A novel protease-immobilized carbon catalyst for the effective fragmentation of proteins in high-TDS wastewater generated in tanneries: Spectral and electrochemical studies. *Environ. Res.* 172, 408–419. https://doi.org/10.1016/j.envres.2019.01.062

Priya, T., Tarafdar, A., Gupta, B., Mishra, B.K., 2018. Effect of bioflocculants on the coagulation activity of alum for removal of trihalomethane precursors from low turbid water. *J. Environ. Sci.* 70, 1–10. https://doi.org/10.1016/J.JES.2017.09.019

Pype, M.L., Lawrence, M.G., Keller, J., Gernjak, W., 2016. Reverse osmosis integrity monitoring in water reuse: The challenge to verify virus removal – A review. *Water Res.* 98, 384–395. https://doi.org/10.1016/j.watres.2016.04.040

Qasim, M., Badrelzaman, M., Darwish, N.N., Darwish, N.A., Hilal, N., 2019. Reverse osmosis desalination: A state-of-the-art review. *Desalination* 459, 59–104. https://doi.org/10.1016/j.desal.2019.02.008

Qasim, S.R., 2017. *Wastewater Treatment Plants: Planning, Design, and Operation*. Elsevier.

Qasim, S.R., Zhu, G., 2017. *Wastewater Treatment and Reuse, Theory and Design Examples*, Volume 1 – Google Books. CRC Press.

Qin, L., Zhang, Y., Xu, Z., Zhang, G., 2018. Advanced membrane bioreactors systems: New materials and hybrid process design. *Bioresour. Technol.* 269, 476–488. https://doi.org/10.1016/j.biortech.2018.08.062

Reshmi, R., Sanjay, G., Sugunan, S., 2006. Enhanced activity and stability of α-amylase immobilized on alumina. *Catal. Commun.* 7, 460–465. https://doi.org/10.1016/j.cat com.2006.01.001

Reske, G.D., da Rosa, B.C., Visioli, L.J., Dotto, G.L., De Castilhos, F., 2020. Intensification of Ni(II) adsorption in a fixed bed column through subcritical conditions. *Chem. Eng. Process. – Process Intensif.* 149, 107863. https://doi.org/10.1016/j.cep.2020.107863

Sahebi, S., Sheikhi, M., Ramavandi, B., Ahmadi, M., Zhao, S., Adeleye, A.S., Shabani, Z., Mohammadi, T., 2020. Sustainable management of saline oily wastewater via forward osmosis using aquaporin membrane. *Process Saf. Environ. Prot.* 138, 199–207. https://doi.org/10.1016/j.psep.2020.03.013

Shen, J., Schäfer, A., 2014. Removal of fluoride and uranium by nanofiltration and reverse osmosis: A review. *Chemosphere.* 117, 679–691. https://doi.org/10.1016/j.chemosph ere.2014.09.090

Shenvi, S.S., Isloor, A.M., Ismail, A.F., 2015. A review on RO membrane technology: Developments and challenges. *Desalination.* 368, 10–26. https://doi.org/10.1016/j.desal.2014.12.042

Singh, H., Raj, S., Rathour, R.K.S., Bhattacharya, J., 2022. Bimetallic Fe/Al-MOF for the adsorptive removal of multiple dyes: optimization and modeling of batch and hybrid adsorbent-river sand column study and its application in textile industry wastewater. *Environ. Sci. Pollut. Res.* 29(37), 56249–56264. https://doi.org/10.1007/s11 356-022-19686-x

Sounthararajah, D.P., Loganathan, P., Kandasamy, J., Vigneswaran, S., 2015. Adsorptive removal of heavy metals from water using sodium titanate nanofibres loaded onto GAC in fixed-bed columns. *J. Hazard. Mater.* 287, 306–316. https://doi.org/10.1016/j.jhaz mat.2015.01.067

Srivastava, A., Gupta, B., Majumder, A., Gupta, A.K., Nimbhorkar, S.K., 2021. A comprehensive review on the synthesis, performance, modifications, and regeneration of activated carbon for the adsorptive removal of various water pollutants. *J. Environ. Chem. Eng.* 9(5), 106177. https://doi.org/10.1016/J.JECE.2021.106177

Sun, H., Liu, H., Zhang, M., Liu, Y., 2021. A novel single-stage ceramic membrane moving bed biofilm reactor coupled with reverse osmosis for reclamation of municipal wastewater to NEWater-like product water. *Chemosphere* 268, 128836. https://doi.org/10.1016/j.chem osphere.2020.128836

Tchobanoglous, G., Burton, F.L., Stensel, H.D., 2014. *Wastewater Engineering: Treatment and Resource Recovery*, Metcalf & Eddy, Inc. McGraw-Hill Education.

Tchobanoglous, G., Stensel, H.D., Tsuchihashi, R., Burton, F., 2013. *AECOM – Wastewater Engineering: Treatment and Resource.* Mc-Graw Hill.

Urtiaga, A.M., Pérez, G., Ibáñez, R., Ortiz, I., 2013. Removal of pharmaceuticals from a WWTP secondary effluent by ultrafiltration/reverse osmosis followed by electrochemical oxidation of the RO concentrate. *Desalination* 331, 26–34. https://doi.org/10.1016/ j.desal.2013.10.010

Vadrucci, M., De Bellis, G., Mazzuca, C., Mercuri, F., Borgognoni, F., Schifano, E., Uccelletti, D., Cicero, C., 2020. Effects of the Ionizing Radiation Disinfection Treatment on Historical Leather. *Front. Mater.* 7, 21. https://doi.org/10.3389/ fmats.2020.00021

Wang, L.K., Hung, Y.-T., Shammas, N.K., 2006. *Advanced Physicochemical Treatment Processes.* Springer. https://doi.org/10.1007/978-1-59745-029-4

Warsinger, D.M., Chakraborty, S., Tow, E.W., Plumlee, M.H., Bellona, C., Loutatidou, S., Karimi, L., Mikelonis, A.M., Achilli, A., Ghassemi, A., Padhye, L.P., Snyder, S.A., Curcio, S., Vecitis, C.D., Arafat, H.A., Lienhard, J.H., 2018. A review of polymeric membranes and processes for potable water reuse. *Prog. Polym. Sci.* 81, 209–237. https://doi.org/10.1016/j.progpolymsci.2018.01.004

Yadav, M.K., Gupta, A.K., Ghosal, P.S., Mukherjee, A., 2017. pH mediated facile preparation of hydrotalcite based adsorbent for enhanced arsenite and arsenate removal: Insights on physicochemical properties and adsorption mechanism. *J. Mol. Liq.* 240, 240–252. https://doi.org/10.1016/j.molliq.2017.05.082

Yadav, M.K., Gupta, A.K., Ghosal, P.S., Mukherjee, A., 2019. Modeling and analysis of adsorptive removal of arsenite by Mg–Fe–(CO$_3$) layer double hydroxide with its application in real-life groundwater. *J. Environ. Sci. Heal. – Part A Toxic/Hazardous Subst. Environ. Eng.* 54, 1318–1336. https://doi.org/10.1080/10934529.2019.1646604

Yadav, M.K., Saidulu, D., Gupta, A.K., Ghosal, P.S., Mukherjee, A., 2021. Status and management of arsenic pollution in groundwater: A comprehensive appraisal of recent global scenario, human health impacts, sustainable field-scale treatment technologies. *J. Environ. Chem. Eng.* 9, 105203. https://doi.org/10.1016/j.jece.2021.105203

Yao, B., Luo, Z., Xiong, W., Song, B., Zeng, Z., Zhou, Y., 2020. Disinfection techniques of human norovirus in municipal wastewater: Challenges and future perspectives. *Curr. Opin. Environ. Sci. Heal.* 17, 29–34. https://doi.org/10.1016/j.coesh.2020.08.003

Ye, W., Liu, R., Chen, X., Chen, Q., Lin, J., Lin, X., Van der Bruggen, B., Zhao, S., 2020. Loose nanofiltration-based electrodialysis for highly efficient textile wastewater treatment. *J. Memb. Sci.* 608, 118182. https://doi.org/10.1016/j.memsci.2020.118182

Youcai, Z., 2018. Physical and chemical treatment processes for leachate. *Pollut. Control Technol. Leachate from Munic. Solid Waste* 31–183. https://doi.org/10.1016/B978-0-12-815813-5.00002-4

Zubair, M., Ihsanullah, I., Abdul Aziz, H., Azmier Ahmad, M., Al-Harthi, M.A., 2021. Sustainable wastewater treatment by biochar/layered double hydroxide composites: Progress, challenges, and outlook. *Bioresour. Technol.* 319, 124128. https://doi.org/10.1016/j.biortech.2020.124128

8 Sludge Management in Wastewater Treatment Plants

CHAPTER OBJECTIVES

This generation of sludge from different processes in a sewage treatment plant has been discussed in the chapter. The quantity of the sludge generated and its characteristics have also been covered. The different processes involved in sludge treatment, such as preliminary treatment, sludge thickening, stabilization, conditioning, dewatering, processing, reduction, disinfection, and finally, disposal, have been delineated.

8.1 INTRODUCTION

One of the major problems associated with wastewater treatment is the production of sludge and biosolids. Sludge refers to the solids formed during various stages of treatment in the STP. Typically, they contain around 3% solids, and the remaining portion is liquids. Biosolids are treated sludge with higher concentrations of solids ("US EPA," 2021).

The amount of sludge produced in conventional sewage treatment plants (STPs) depends on the strength of the wastewater, degree of treatment, and kind of chemicals used in the treatment. On average, small conventional STPs (capacity <2.5 million gallons per day (MGD)) generate around 75,000 kg of dry sludge annually. On the other hand, large STPs (capacity > 100 MGD) annually generate around 5×10^7 kg of dry sludge (Wang et al., 2007). The sludge generated at different stages of the treatment plant may differ in characteristics based on the kind of treatment provided. Furthermore, the untreated sludge may become anaerobic and release a bad odor, creating health hazards. The amount of sludge generated and its negative implications indicate the necessity of proper management of sludge. Hence, all STPs must process and dispose of sewage sludge. Treatment of sludge has two main goals: minimizing its volume and stabilizing the organic contents. The stabilized sludge does not have a foul odor and can be easily disposed of (Gallego-Schmid and Tarpani, 2019). The major steps in sludge processing are sludge thickening, stabilization, conditioning, dewatering, processing, reduction, and finally, disposal. In this chapter, the characteristics of different types of sludge produced from STPs have been addressed. Furthermore, the various technologies available for sludge processing have been described. Also the upcoming new technologies for sludge processing have also been covered in this chapter.

8.2   CLASSIFICATION OF SLUDGE AND THEIR PROPERTIES

Typically, in STPs, sludge may be classified as primary, secondary, or tertiary sludge (Mackenzie, 2010). The different solids generated at different stages of an STP and their characteristics are provided in Figure 8.1. Screens and grit chambers are common in most STPs, which are targeted to remove the plastics, larger debris, and readily settleable solids from the raw wastewater.

Primary sedimentation tanks are provided to remove the readily settleable fraction of the solids. These solids recovered from the bottom of the primary sedimentation tank are classified as primary sludge. The solids are rich in organic matter and comprise a small fraction of nitrogen and phosphorous (Mackenzie, 2010; Wang et al., 2007). Typically, the pH of primary sludge is around 6, and it has alkalinity in the range of 500–1,500 mg/L as $CaCO_3$. The total dry solid content usually lies between 1% and 6 %, while the volatile solid content is around 65% of the total solids. The remaining portion of the total solids comprises grease, fats, protein, nitrogen, phosphorous, cellulose, iron, and silica. The bulk specific gravity of primary sludge is around 1.02 (Gurjar and Tyagi, 2017; Tchobanoglous et al., 2014; Wang et al., 2007).

Secondary or biological sludge usually comprises a mixture of microorganisms and inert materials wasted from secondary treatment processes, such as the activated sludge process, rotating biological contactors, sequencing batch reactors, trickling filters, and others. Typically, the amount of sludge produced in aerobic biological processes is more in anaerobic biological processes. The sludge produced in the biological processes comprises 60–85% organic matter. In the absence of air, the sludge turns anaerobic and creates noxious conditions (Mackenzie, 2010). The characteristics of biological sludge also vary based on the kind of treatment. Sludge from the activated sludge process typically has a pH of around 7. The alkalinity of such sludge usually lies in the range of 580–1,100 mg/L as $CaCO_3$ (Gurjar and Tyagi, 2017; Wang et al., 2007). The volatile content of total solids is slightly higher (75 %) than primary sludge. The sludge characteristics of the trickling filter are quite similar to that of activated sludge. However, the volatile content of the sludge is slightly higher compared to primary and activated sludge (Gurjar and Tyagi, 2017; Wang et al., 2007).

Tertiary sludge may be of various types depending on the type of treatment provided. Often different chemicals, such as iron or aluminum salts, are used to precipitate phosphorous. Furthermore, activated carbon or some other adsorbent may be used in tertiary treatment. The exhausted adsorbents are also treated as chemical sludge (Gurjar and Tyagi, 2017; Mackenzie, 2010; Wang et al., 2007). If not disposed of properly, the chemical sludge derived from filter media may also lead to secondary pollution. Hence, the processing of sludge is required. Apart from the sludge's physical and chemical properties, certain parameters determine the processes to be used for sludge processing.

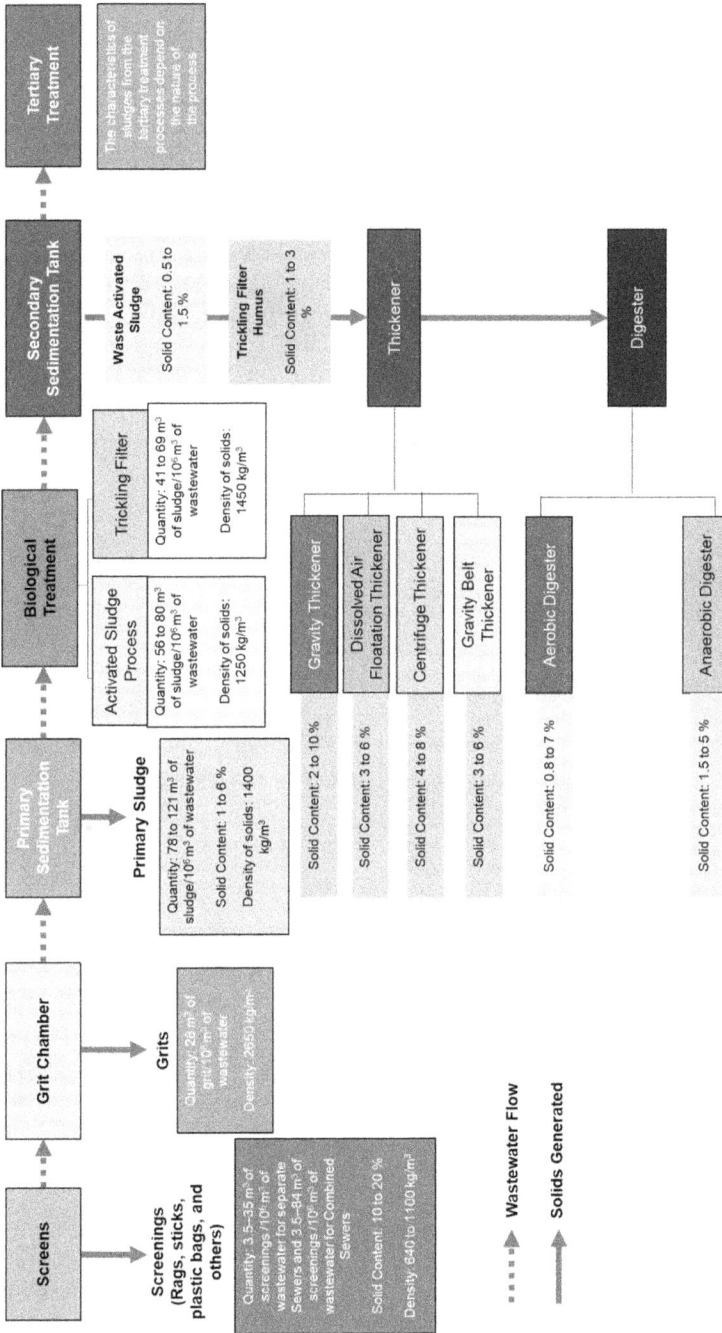

FIGURE 8.1 Sources and characteristics of solid residuals in a municipal wastewater treatment plant.

Source: Mackenzie, 2010; Tchobanoglous et al., 2014.

8.2.1 STABILITY OF THE SLUDGE

The stability of sludge is an indicator of how well the sludge is digested. The stability of the sludge can be defined as per Equation 8.1 (Gurjar and Tyagi, 2017).

$$S_S = 100 \times 1.035 \left[1 - \frac{OUR_{measured}}{OUR_{maximum}} \right] \tag{8.1}$$

where S_s is the sludge stability (%), c is a constant, $OUR_{measured,}$ and $OUR_{maximum}$ are the oxygen uptake rate measured in digested sludge and the maximum oxygen uptake rate of the activated sludge process reactor producing the sludge, respectively. The higher the value of sludge stability, the better is the digestibility of the sludge.

8.2.2 SLUDGE VOLUME INDEX

Sludge volume index (SVI) is defined as volume occupied in ml by 1 g of solids when allowed to settle for 30 min.

SVI can be calculated as per Equation 8.2 (APHA, 2017; Mackenzie, 2010).

$$SVI(mL/g) = \frac{\text{Volume of settled sludge } (mL)}{\text{Mixed Liquor Suspended Solids } \left(\frac{mg}{L} \right)} \times 1000 \text{ mg/g} \tag{8.2}$$

Typically, a good settling sludge has an SVI of 100 mL/g, while sludge having an SVI of more than 150 mL/g is associated with filamentous growth and sludge bulking (Gurjar and Tyagi, 2017).

8.2.3 SLUDGE VOLUME RATIO

Gravity thickening is controlled by the sludge volume ratio (SVR). A sludge blanket formed at the bottom of a gravity thickener helps concentrate the sludge. SVR is used as a determining parameter to find the retention time of the sludge in the thickener. SVR can be defined as the volume of the sludge blanket held in the thickener divided by the volume of the thickened sludge removed daily. SVR values typically fall between 0.5 and 2. A lower SVR value is required during warmer weather (i.e., higher sludge withdrawal rate) because the sludge settles and turns septic more quickly (Gurjar and Tyagi, 2017).

8.3 QUANTIFICATION OF SLUDGE PRODUCTION

The amount of sludge produced can be calculated using mass balance. The amount of sludge generated is equal to the difference between the mass of the solids or dissolved chemicals entering and leaving the system. Hence, if M_{in} is the amount of solids or chemicals entering the system, and M_{out} is the amount of solids or chemicals leaving the system the amount of sludge produced (dS/dt) can be given by Equation 8.3 (Mackenzie, 2010).

$$\frac{dS}{dt} = M_{in} - M_{out} \qquad (8.3)$$

The solids mass balance may be valuable to estimate the average long-term solids loadings on sludge treatment components. This solid mass balance helps to determine variables like running expenses and the amount of sludge that will ultimately be disposed of. It does not, however, specify the amount of solids loading that each piece of equipment must be able to handle. A specific component should be sized to withstand the most demanding loading circumstances that it is likely to experience. Due to storage and plant scheduling issues, steady-state models are typically not used to calculate this loading. As a result, the rate of solids reaching any specific piece of equipment typically fluctuates independently of the amount of solids entering the plant headworks. The steps for the solid mass balance is as follows (Mackenzie, 2010):

The first step involves the identification of each stream and marking what are the inputs and outputs of each of the process. In order to find the solid separation efficiency (S_R) of a particular unit, the ratio of the mass of solids in the effluent of the treatment unit to the mass of solids entering the particular unit should be calculated (Mackenzie, 2010).

$$S_R = \frac{\text{Mass of solids in the effluent}}{\text{Mass of solids in the influent}} \qquad (8.4)$$

Figure 8.2 depicts the incoming and outgoing of solids in a primary sedimentation tank. A represents the mass of solids in the influent of the primary sedimentation tank, B represents the mass of solids in the supernatant of the digester sent back to the primary sedimentation tank, and C represents the amount of solids coming out of the

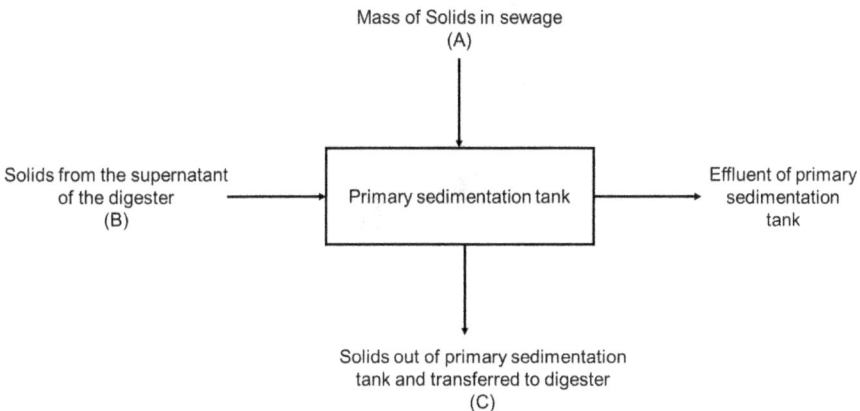

FIGURE 8.2   Flow sheet of solids coming in and ging out of a primary sedimentation tank.

primary sedimentation tank. As per Equation 8.4, the solid separation efficiency of the primary sedimentation tank can be given by Equation 8.5.

$$S_R = \frac{A+B}{C} \tag{8.5}$$

8.4 CONVENTIONAL PROCEDURES FOR THE MANAGEMENT OF WASTEWATER TREATMENT PLANT RESIDUES

Effective sludge management has numerous benefits. The different benefits of effective sludge management have been depicted in Figure 8.3. Sludge treatment often reduces the amount of sludge to be disposed of, thereby reducing significant costs involved in sludge transportation and handling. The recovery of methane is possible during anaerobic digestion. Methane has a significant calorific value, which can be used for different purposes, such as cooking and heat generation. The digested sludge is rich in various nutrients necessary for agriculture. Hence, the digested sludge is often used as a fertilizer. Untreated sludge, if left open, often attracts disease-causing insects and may be a source of various diseases. Proper digestion of sludge and disinfection before its disposal can effectively reduce the negative health risks of sludge (Gurjar and Tyagi, 2017; Muga and Mihelcic, 2008).

There are a few basic processes involved in the management of solid residuals generated from STPs. The schematic describing the various processes is depicted in Figure 8.4.

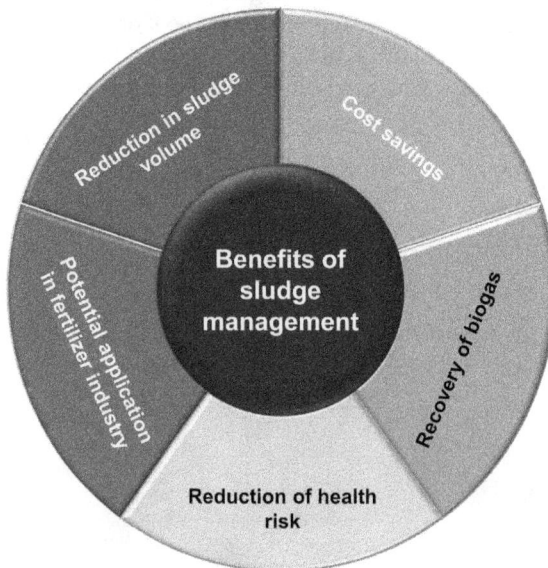

FIGURE 8.3 Benefits of sludge management.

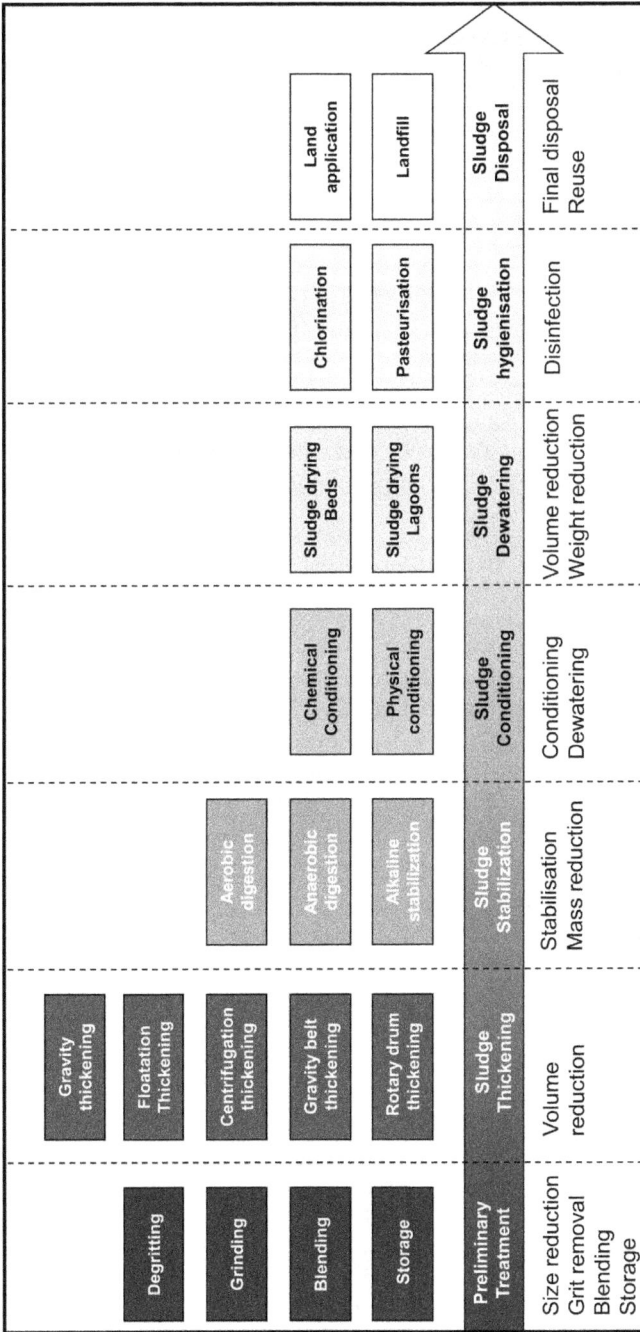

FIGURE 8.4 Different processes in the management of solid residuals from STP.

8.4.1 PRELIMINARY OPERATION

The preliminary operation includes degritting, grinding, blending, and storage of sludge. The first and foremost process is degritting. In this step, solids generated in the screening chamber (screenings) are removed. The screenings are usually disposed of at a sanitary landfill or incinerated along with municipal solid waste (Mackenzie, 2010). The grit generated from the grit chamber is initially separated from the organic sludge. Degritting is usually carried out using classifiers and hydrocyclones. After degritting, the grits are dried and transported to sanitary landfill sites for disposal (Gurjar and Tyagi, 2017; Mackenzie, 2010).

After degritting, the remaining sludge undergoes grinding. The grinding of sludge involves breaking down the sludge into smaller particles. The commonly used methods for sludge grinding are cutting and hammer mill pulverization (Gurjar and Tyagi, 2017; Mackenzie, 2010). After breaking down the sludge, the sludge is blended. The blending of sludge is essential prior to the sludge stabilization, dewatering, and incineration processes. Sludge is blended or mixed to create a homogeneous and uniform sludge mixture (Gurjar and Tyagi, 2017; Mackenzie, 2010). The next step is sludge storage. It is necessary to provide sludge storage to ensure a consistent feed rate for the following sludge handling facilities.

8.4.2 SLUDGE THICKENING

Sludge thickening is the procedure used to increase the solids content of sludge by separating and removing some of the liquid phases from sludge. Physical techniques such as centrifugation, flotation, and gravity settling are frequently used to thicken liquids. Implementing sludge thickening reduces the total cost of treatment. (Gurjar and Tyagi, 2017; Mackenzie, 2010; Muga and Mihelcic, 2008; Pastor et al., 2008; Tchobanoglous et al., 2014; Wang et al., 2007). The different types of sludge thickening processes are described in the subsequent sections.

  Gravity thickening: Gravity thickening is the most common method for dewatering and concentrating sludge prior to digestion. The procedure is simple and inexpensive compared to other thickening methods. Gravity thickening is of two types: plain settling and mechanical thickening. Plain settling results in the formation of scum on the surface and the stratification of sludges near the bottom (Muga and Mihelcic, 2008; Tyagi et al., 2009). Gentle agitation is commonly used to mix the sludge, and the open channels allow water to escape and encourage densification. A slow-rotating circular tank with a sludge collection makes up a typical mechanical thickening. Primary and secondary sludges are frequently combined prior to thickening. Septicity and gasification in sludge are avoided using chlorine. Organic polyelectrolytes (anionic, non-ionic, and cationic) have been used to increase the sludge settling rates (Gurjar and Tyagi, 2017; Mackenzie, 2010; Tchobanoglous et al., 2014; Wang et al., 2007).

  Floatation thickening: The flotation thickening is another thickening technique used in the sludge treatment process. By injecting tiny gas bubbles, the flotation process system artificially creates separation. The gas bubbles cling to the solid particles and create a gas–solid aggregate with a lower bulk density than the liquid. As a result, these aggregates rise to the surface of the fluid. A skimming operation can collect the solid particles once they have moved to the top (Ahmad et al., 2016).

Dissolved air flotation (DAF) is the most commonly used floatation method to thicken sludge. In dissolved air flotation, the air is introduced through a rotating impeller or a porous material to produce air bubbles. This kind of flotation device has a special use in treating wastewater when it separates suspended particles based on surface energy or when wastewater contains surface-active chemicals.

Centrifugation thickening: Sludge thickening can be carried out under the influence of centrifugal forces. Centrifugation raises the centrifugal force, or the artificial gravitational force, which in turn raises the settling rate or settling velocity. The applied centrifugal force can be maintained by varying the speed of rotation and the radius or diameter of the centrifugation chamber. The concentration of solids attained by this process is less. Hence, chemical conditioning agents, such as polymers, are used to raise the concentration of the solids (Bassan et al., 2013; Gurjar and Tyagi, 2017; Muga and Mihelcic, 2008).

Gravity belt thickening: Prior to digestion, a gravity belt thickener thickens the polymer-conditioned sludge by gravity drainage through a filter belt. Diluted sludge is typically injected at the feed end of a horizontal filter belt. Gravity belt thickeners are less expensive, require less energy, and have a lesser environmental impact than other sludge thickening procedures (Gurjar and Tyagi, 2017; Tchobanoglous et al., 2014).

Rotary drum thickening: Rotary drum thickeners raise the sludge solids concentration by stirring the solids in a slowly rotating vessel with porous walls through which the water (or filtrate) gets drained out, and the sludge is thickened. Water drains from the sludge via a porous holding media in a rotary drum thickening, just like in a gravity belt thickener. The cylindrical drum of a rotary drum thickener serves as the porous medium and rotates continuously while the sludge flows through it at a speed between 5 and 20 rpm. The porous wall can be made from various materials, including polymers, steel, and ceramics. A spray system is employed to clean the drum and prevent the pores from clogging (Muga and Mihelcic, 2008; Tchobanoglous et al., 2014).

8.4.3 SLUDGE STABILIZATION OR DIGESTION

The sludge is stabilized during this process. The goal is to lower the biological and chemical reactivity of sludge. Sludge that has been stabilized means that little to no biological or chemical activity is occurring in the sludge zone. The digestion decreases the total mass of the solid and typically kills the pathogens. The procedure makes the sludge simpler to dewater and convert into a consistency that resembles rich, unremarkable potting soil. While aerobic digestion can still be used, anaerobic digestion is typically preferred for sludge stabilization (Gurjar and Tyagi, 2017; Mackenzie, 2010; Tchobanoglous et al., 2014; Wang et al., 2007).

8.4.3.1 Aerobic Digestion

Aerobic digestion is the biochemical oxidative stabilization of wastewater sludge in open or closed tanks. This method can be used to digest sludge produced from secondary and primary treatment. The activated sludge system and the aerobic digester both function in the same ways. In the endogenous phase, the microorganisms run out of food and oxidize the cell tissue to produce CO_2, H_2O, NH_4^+, NO_2^-, and NO_3^-. Diffusers or surface aerators are used to deliver air or oxygen into the system. Aerobic digestion requires other equipment, such as mixers, scum collection baffles, pumps, and pipework for sludge recirculation. Aerobic digesters can have rectangular or circular cross-sections. They are easy to operate and do not produce any foul odor. The supernatant produced in aerobic digesters is not rich in organic content and is usually returned to the head end of the plant. However, the operation cost is more since artificial aeration is required. Additionally, methane gas is not produced, which is a valuable byproduct of the anaerobic digesters (Gurjar and Tyagi, 2017; Mackenzie, 2010; Tchobanoglous et al., 2014; Wang et al., 2007).

8.4.3.1.1 Process Parameters

Aerobic sludge digestion requires aeration. Based on the quantity of sludge, a sufficient amount of oxygen demand is required. Also, the amount of time the sludge is to be retained in the aerobic digestion chamber is also essential. Hence, the retention time and oxygen demand are the two prominent process parameters involved in aerobic sludge digestion (Gurjar and Tyagi, 2017; Mackenzie, 2010; Tchobanoglous et al., 2014; Wang et al., 2007).

> *Retention time*: The amount of volatile solids in the aerobic digester can be reduced by 40% over a period of 10–12 days. The reduction of volatile solids can be further reduced upon increasing the retention time. However, the rate of digestion is significantly reduced. The reduction of solids can be increased up to 70% by increasing the temperature (Gurjar and Tyagi, 2017; Tchobanoglous et al., 2014).
> *Oxygen demand*: The amount of oxygen required to completely oxidize the cell tissue can be calculated using Equation 8.6.

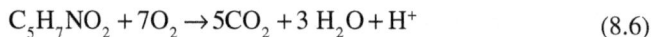

$$C_5H_7NO_2 + 7O_2 \rightarrow 5CO_2 + 3\,H_2O + H^+ \tag{8.6}$$

Approximately, based on Equation 8.6, the total amount of oxygen required is around 2 kg/kg of cells. It has been found that the dissolved oxygen (DO) concentration in the aerobic digester should be maintained at around 1–2 mg/L (Gurjar and Tyagi, 2017; Mackenzie, 2010; Tchobanoglous et al., 2014; Wang et al., 2007).

8.4.3.1.2 Issues in Aerobic Digestion

The different issues in aerobic digestion are listed in Figure 8.5. Acids are often formed during the nitrification step in the aerobic digester. If the feed sludge has low alkalinity, there may be a drop in pH to about 5.5. At around this pH, there may be filamentous growth leading to foaming in the digester. The foaming problem may be

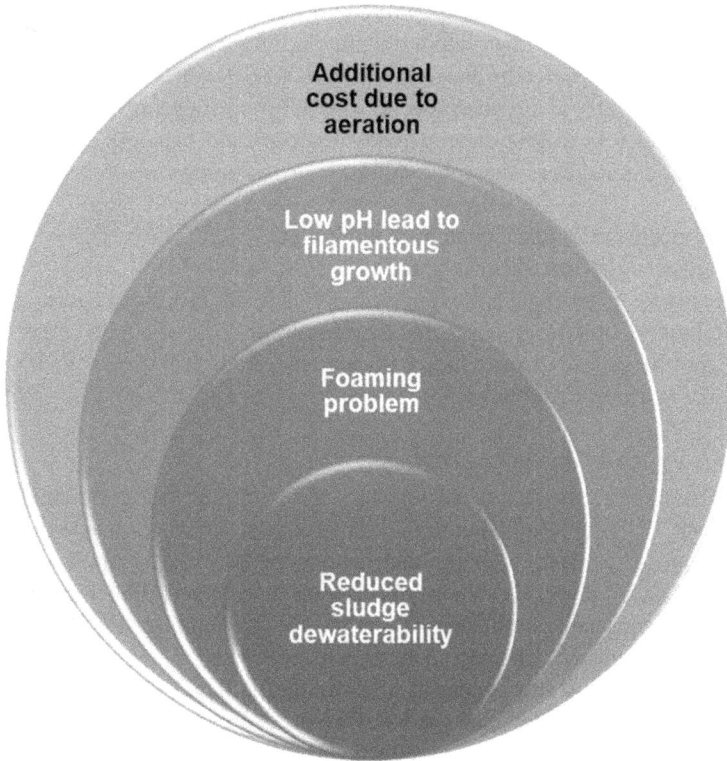

FIGURE 8.5 Limitations of aerobic digestion.

tackled by killing the microorganisms by chlorination, using water sprays, or creating transient anaerobic conditions by switching off the aerator. Due to mechanical aeration and mixing, the structure of the sludge flocs are destroyed, thereby affecting the dewatering properties of the sludge (Gurjar and Tyagi, 2017; Wang et al., 2007).

8.4.3.2 Anaerobic Digestion

Anaerobic digestion is the term used to describe the anaerobic decomposition of organic matter that leads to partial gasification and liquefaction. The process is typically viewed as a two-step biological process that includes waste stabilization and conversion. The principal byproducts are stable organic waste, methane (CH_4), and carbon dioxide (CO_2). Anaerobic digestion of solid waste and/or wastewater sludge has long been used to stabilize organic wastes before disposal (Gurjar and Tyagi, 2017; Wang et al., 2007).

The three stages of anaerobic digestion are solids liquefication (hydrolysis), soluble solids digestion (fermentation), and gas production (methanogenesis). Heterotrophic organisms that produce organic acids use complex organic substrates and the degradation byproducts of those substances. Carbohydrates are initially broken down into simple sugars. The sugar is further broken down into alcohols and aldehydes, which in turn are converted to organic acids. Proteins are initially converted to amino acids, which are further broken down into ammonia and organic acids. The organic acids produced are then converted by anaerobic bacteria to CO_2 and CH_4. A pH in the range of 6.6 and 7.4 usually favors the process. The different stages involved in methane formation in anaerobic digestion are already described in Chapter 6.

The different advantages of anaerobic digestion are as follows:

  i.   The decreased organic content of the sludge.
 ii.   Enhanced dewaterability of sludge.
iii.   Recovery of methane gas, which may be an alternative fuel source.
 iv.   Relatively little residual organic waste is produced.
  v.   Low operation cost since oxygen is not required.
 vi.   Low nutrient requirements.

8.4.3.2.1 Types of Anaerobic Digestion

Low-rate or conventional sludge digesters and high-rate or continuous sludge digesters are the two most common types of sludge digesters used. In the conventional sludge digester, no additional heating or mixing is carried out. All the processes, such as digestion, sludge thickening, and supernatant formation, take place in one single tank simultaneously. The raw sludge enters the digestion tank after the pH is adjusted. The pH adjustment is necessary since methanogenesis is a pH-sensitive process and occurs at a pH range of 6.6–7.4. In the digester, the sludge gets actively digested, and gas is released. A scum layer consisting of sludge particles and other substances is formed when the gas rises to the surface. Naturally, due to bio-flotation, some sludge mixing may occur due to the exchange of solids between the scum and sludge zones. In the conventional sludge digester, due to the lack of mixing, certain areas of the tank can acquire high pH, while certain portions may acquire low pH, thereby reducing the optimum biological activity. A schematic of a conventional anaerobic digester has been provided in Figure 8.6a (Gurjar and Tyagi, 2017; Mackenzie, 2010; Tchobanoglous et al., 2014; Wang et al., 2007). The produced biogas escapes from the top of the digester, where a moisture trap is provided. The moisture trap is provided to remove the moisture from the biogas. Calcium chloride is used for trapping moisture since it is a hygroscopic material and is highly effective when it comes to absorbing moisture from the surrounding air. A flare stack is also provided at the end of the gas outlet to stabilize pressure and manage undesirable gas that cannot be processed.

In the high-rate or continuous sludge digesters, the sludge is continuously added and vigorously mixed, either mechanically or by recirculating some of the digestion gases through a compressor. In order to maintain the highest level of activity in the mesophilic region, the digester is heated. Thus, the key components of high-rate digestion are feeding, mixing, heating, and thickening. The heating generates a mesophilic atmosphere having a temperature of around 35°C. The higher temperature favors anaerobic digestion. Thorough mixing overcomes various drawbacks of a conventional anaerobic digester, such as higher contact time between substrate and inoculum, decreased accumulation of scum, and dilution of toxic byproducts.

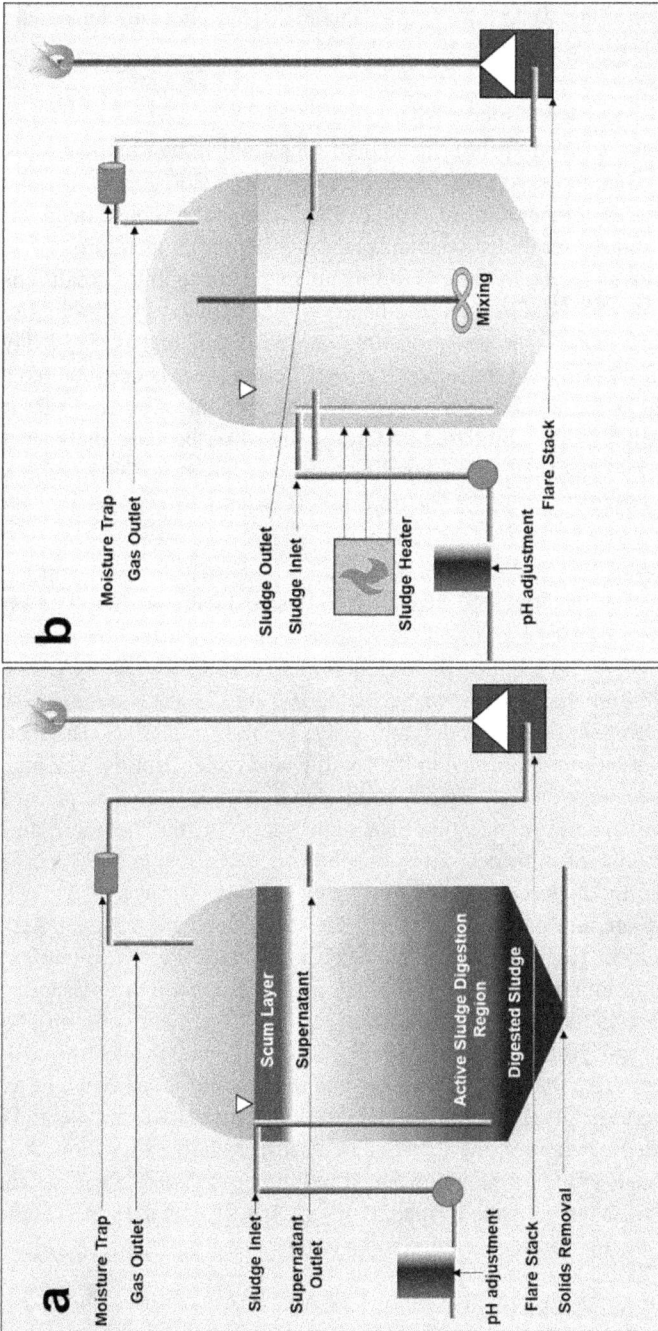

FIGURE 8.6 Schematic of (a) low rate or conventional anaerobic sludge digester and (b) high rate or conventional anaerobic sludge digester.

The other components of the high-rate sludge digester are similar to the conventional sludge digester. A schematic of a high-rate anaerobic digester is provided in Figure 8.6b. (Gurjar and Tyagi, 2017; Mackenzie, 2010; Tchobanoglous et al., 2014; Wang et al., 2007).

8.4.3.2.2 Process Parameters

Temperature plays a pivotal role in anaerobic digestion. Lower temperatures in the range of 15–25°C do not favor a high rate of hydrolysis, acidogenesis, and methanogenesis. At a higher temperature of around 35°C, the sludge constituents are more soluble. As a result, better digestion takes place at a higher temperature.

A 15–20 days of retention time is sufficient under mesophilic conditions (temperature of around 35°C). On the other hand, at thermophilic conditions (temperature of around 50–60°C), the retention time can be lower, that is, 8–12 days. The optimum pH for carrying out the anaerobic digestion process is between 6.6 and 7.4 since methanogens are most active in this pH. The amount of nutrients required in anaerobic digestion is low compared to aerobic digestion because the growth yield of microorganisms under anaerobic conditions is also lower. A COD:N:P:S ratio of 1,600:10:2:1 is desirable for anaerobic digestion (Gurjar and Tyagi, 2017).

8.4.3.2.3 Issues in Anaerobic Digestion

The different issues in anaerobic digestion are listed in Figure 8.7. The optimum performance of the anaerobic digestion system occurs when the pH is maintained between 6.6 and 7.4, and the bicarbonate alkalinity is around 3,000 mg/L as $CaCO_3$. However, conditions may arise when methane production is reduced, thereby increasing CO_2 content (Gurjar and Tyagi, 2017; Wang et al., 2007). The increase in CO_2 content is directly proportional to the volatile acid concentration. The increase in volatile acid concentration will directly bring down the alkalinity and pH of sludge, thereby affecting anaerobic digestion. The supernatant of the anaerobic digester is rich in organic content, nitrogen, and phosphorous. As a result, the supernatant is treated before being disposed of.

An excessive amount of ammonium and phosphate ions are frequently produced during the anaerobic digestion of sludge. Such ions cause the precipitate known as struvite or magnesium ammonium phosphate to develop. Structural issues with the operation and maintenance of anaerobic digester result from the build-up of struvite in pipes (Gurjar and Tyagi, 2017; Wang et al., 2007).

In order to get rid of the deposited sludge and the floating scum, the digester needs to be cleaned periodically. Lack of maintenance will cause scum and grit to build up excessively, reducing the volume that can be used for sludge digestion. Anaerobic digestion also leads to the formation of foul-smelling corrosive H_2S gas, which needs to be separated from the methane formed. Furthermore, the formed supernatant is rich in organic loading and nutrients and requires to be treated.

8.4.3.3 Alkaline Stabilization

Lime stabilization is carried out to increase the pH of stressed anaerobic digesters. Often due to the production of acids during digestion, the pH of the sludge may

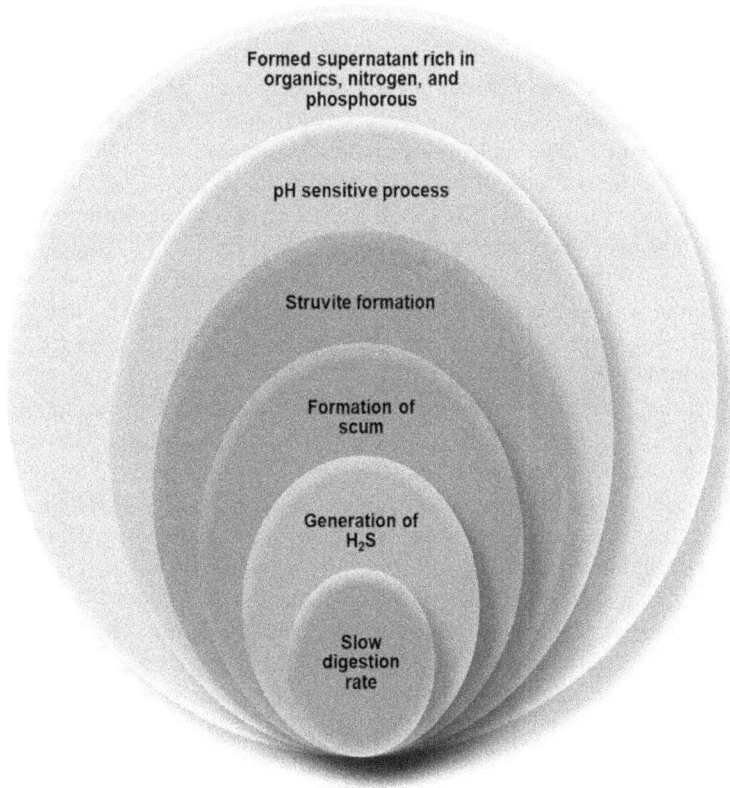

FIGURE 8.7 Limitations of anaerobic digestion.

get reduced. However, methanogenesis occurs best at neutral pH. Hence, in order to maintain such conditions, lime stabilization is carried out. When the treated sludge or biosolids do not meet the desired standards, lime stabilization is carried out to supplement the existing sludge stabilization. Often, lime is added to dewatered stabilized sludge. This process is known as lime post-treatment. In order to establish contact between the sludge and lime particles and to prevent putrescible material pockets, excellent mixing is necessary (Mackenzie, 2010; Wang et al., 2007).

8.4.4 SLUDGE CONDITIONING

The goal of sludge conditioning is to enhance the dewatering properties of sludge or to get the sludge ready for dewatering. It can be achieved by applying inorganic or organic coagulants, such as ferric hydroxide, lime, and alum can be used as coagulants for chemical conditioning purposes. It is also possible to use different organic polymers, such as cationic, ionic, or non-ionic polymers. Sludge conditioning can be carried out primarily by chemical or physical conditioning. The following section discusses the different chemical and physical conditioning processes (Gurjar and Tyagi, 2017; Tunçal and Mujumdar, 2022).

8.4.4.1   Chemical Conditioning

Addition of flocculation agent: Inorganic coagulating agents like ash, lime, or ferric chloride have been the go-to chemicals for sludge conditioning for a long time because they help in the coagulation of the solids and release the absorbed water. Although the combination of ash, ferric chloride, and lime is efficient, the amount of dry solids produced increases by 20–30%. Since organic polymers do not increase the amount of dry solids formed, they are frequently used for sludge conditioning. They are highly efficient and only a small quantity of the polymer is required. Since the polymers have a high molecular weight, they can also be easily separated in the clarifier. However, organic polymers are costly and raise the cost of the process (Mackenzie, 2010; Zhou et al., 2014).

Acid–alkaline treatment: Numerous studies have demonstrated the importance of pH in affecting the flocculation characteristics of the sludge. The stability of flocs in the sludge deteriorates when the pH level falls below 2 due to electrostatic repulsion between the inner surfaces of the sludge. The pH range between 2.6 and 3.6 corresponds to the isoelectric point of sludge, where the best flocculation should theoretically be possible. The isoelectric point is the pH at which the surface charge of the sludge is neutral (Zhou et al., 2014).

Enzyme treatment: The hydrolysis of extracellular polymeric substances and cells in sludge may also be triggered by the addition of enzymes, resulting in the release of bound water from the sludge. Upon using an enzyme called hydrolases for the sludge conditioning, the dewaterability of the sludge improved noticeably. In order to boost the dewatering capacity of sludge, commercial enzyme mixtures are also used. Due to practical challenges and high operating costs, the use of enzymes for sludge conditioning is often still restricted (Zhou et al., 2014).

8.4.4.2   Physical Conditioning

The physical treatment may be heating or freezing. In heat treatment, the typical temperature range is 60–180°C. The heat treatment denatures the proteins in extracellular polymeric substances (EPS). Cell walls of bacteria are also harmed. The rate at which the bound water is removed increases as the structure of the sludge is broken down by the heat hydrolysis of extracellular and intracellular components (Gurjar and Tyagi, 2017). The freezing and thawing process can break down the microbial cells and floc structure, releasing the bound water from the sludge. It was discovered that the density and shape of the flocculent structure significantly change at low freezing speeds. The dewaterability increased by 82% when treated sludge was compared to untreated sludge. In terms of sludge dewaterability, the slow-frozen method performed better than the fast-frozen method (Gurjar and Tyagi, 2017).

8.4.5   Sludge Dewatering

The dewatering systems receive the stabilized sludge. Normally, digested sludge is dewatered before being disposed of or further processed. The drying beds, also

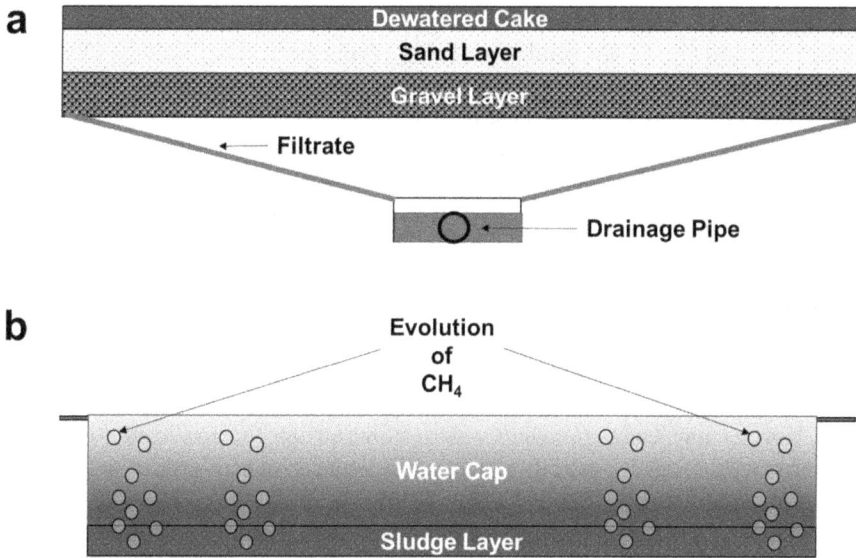

FIGURE 8.8 Schematic of (a) sludge drying bed and (b) sludge drying lagoon.

known as sludge drying beds, are the most basic methods of dewatering. Liquid can actually be removed from these drying beds in two ways: evaporation or draining to the lower strata. However, it has a massive land footprint. Sludge dewatering alternatives include centrifuges and various other mechanical dewatering systems. However, sludge drying beds and sludge drying lagoons are the two most commonly used techniques for municipal sludge dewatering/drying (Figure 8.8).

8.4.5.1 Sludge Drying Beds

Sludge is dewatered in sludge drying beds by draining through the sludge mass and evaporating from the exposed surface. Sand beds are dewatered through two distinct processes: filtration and evaporation. Water drainage is critical during the first 1–3 days because solid concentrations can reach 15–25%. In order to remove more water, evaporation is used. The horizontal contraction of the sludge and the exposure of new sludge regions aid in evaporation. Estimates suggest that 60% of the water can be drained. Secondary sludges can lose up to 85 % of their water due to drainage. The greater the initial water content, the greater the fraction of drainable water. A schematic of a sludge drying bed is depicted in Figure 8.8a.

Sludge drying beds are the most cost-efficient technology for dewatering where land is readily available. The process does not require skilled operators and energy. The minimum amount of chemical is used, and the process is not susceptible to sludge variations. The amount of dry cake solids per volume of sludge produced is more than other mechanical equipment. However, the requirement of a significant amount of land is a drawback of these systems. Furthermore, the drying process is significantly affected during the rainy season. Open sludge drying also generates a foul smell and acts as a breeding ground for insects (Gurjar and Tyagi, 2017).

8.4.5.2 Sludge Drying Lagoons

Sludge drying lagoons are similar to sludge drying beds. They require substantial amounts of land and labor-intensive mechanical removal of the dewatered solids, but it is less complicated to build because drainage of the filtrate is not necessary. In order to stop sludge water from entering groundwater, aquifers, or other environmental water bodies, a lagoon consists of a reservoir with a sealed base. Sludge is dumped into the lagoon, where it is left to settle. Water is lost through evaporation during the drying/dewatering operational cycle, which lasts for several months, while the solids accumulate at the bottom of the lagoon in a thick layer stabilized by anaerobic biological processes. A schematic of a sludge drying lagoon is depicted in Figure 8.8b.

Lagoons require low operational care and expertise and can act as a buffer in the flow stream used to handle sludge. Lagoons can handle shock loading rates. However, the large land requirement is an issue. Apart from the large land requirement, lagoons may lead to problems with vectors and can give rise to odor problems. Furthermore, there is a possibility that nearby groundwater or surface water could become contaminated (Gurjar and Tyagi, 2017).

Municipal sludge is often disinfected to lower the content of pathogenic bacteria in this sludge before going for disposal. This process is called sludge hygienization. Sludge hygienization can be achieved in the following ways (Gurjar and Tyagi, 2017):

- Carrying out pasteurization at 70°C
- Treating the sludge with lime for around 3 h
- Composting at a temperature around 55°C
- Addition of chlorine to disinfect the sludge
- Disinfection may also be carried out by high-energy radiation.

8.5   INNOVATIVE TECHNOLOGIES FOR SLUDGE MANAGEMENT

8.5.1   SLUDGE TREATMENT USING A PARABOLIC SOLAR CONCENTRATOR

One of the traditional ways to use renewable energy is open-air sun drying. However, this kind of system causes issues that render large-scale operations difficult, such as the requirement of a large surface area, long drying time, and process control challenges during drying. In this context, parabolic solar concentrators have proved to be quite efficient in overcoming the drawbacks of conventional solar dryers. In parabolic solar concentrators, the radiation of the sun is focused onto a focal line by means of parabolic reflective surfaces. A sheet of reflective or highly polished material is bent into a parabolic shape to create parabolic solar concentrators. The platform can work with one or two motor tracking systems to keep the aperture plane perpendicular to the incident radiation, thereby increasing the efficiency of direct solar radiation collection (Fendrich et al., 2018). Although parabolic solar concentrators have been used for different applications in water and wastewater treatment, drying of sludge is an area that has been getting attention only recently. Various studies in recent times have shown that parabolic solar concentrators have been highly effective (Ben Othman et al., 2022; Chaanaoui et al., 2021; Di Fraia et al., 2018). The working of a typical parabolic solar concentrator for sludge drying is shown in Figure 8.9.

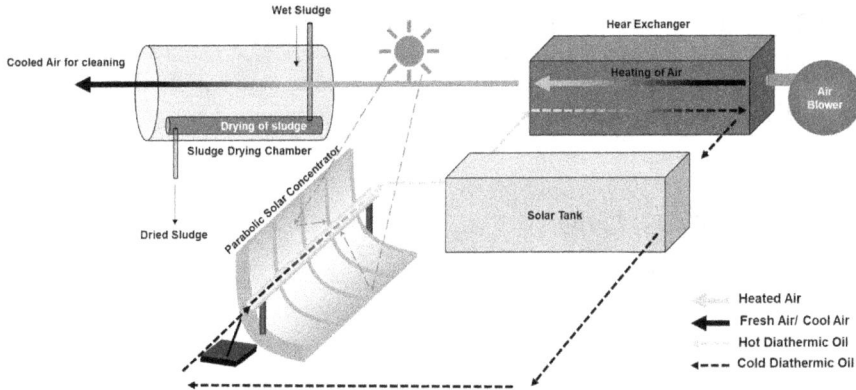

FIGURE 8.9 Schematic of sludge drying using parabolic solar concentrator. (Modified and reprinted with permission from Chaanaoui et al., 2021. Copyright 2023 Elsevier).

Typically, the parabolic solar concentrator is used to heat a diathermic oil. The oil is circulated to the heat exchanger, where cool air is heated. The heated air is passed through a chamber containing wet sludge. As the heated air is passed through the wet sludge, the sludge gets dried. The air coming out of the sludge drying chamber may get contaminated due to exposure to the sludge. Hence, the air coming out is sent for cleaning (Ben Othman et al., 2022; Chaanaoui et al., 2021; Di Fraia et al., 2018).

8.5.2 HYDRODYNAMIC CAVITATION

Due to a local pressure drop during cavitation, small vapor bubbles form in the sludge. When the formed bubbles burst, a lot of energy is released in a condensed space and time, which leads to the emergence of powerful physical forces like oxidative radials, microjets, shock waves, and high temperatures. The extreme conditions produced by the cavitation bubble collapsing offer a special environment for the sludge to disintegrate. Orifice plates or venturi channels are used in the hydrodynamic cavitation method to regulate the sludge velocity at the desired level (Tunçal and Mujumdar, 2022).

8.5.3 MODERN DRYING SYSTEMS

The hot gas circulates around the sewage sludge in convectional drying systems to create direct contact between the sludge and the heat carrier. The drum dryers use a rotating drum to accomplish the drying. The hot gas flow, the guide plates, or the position of an inclined drum all move the sludge. On the other hand, modern sludge drying techniques avoid having the sludge come into direct contact with the heat carrier by uniformly spreading it out over a hot surface. A horizontal stator with a double-walled cylinder and a rotor inside makes up a thin-film contact dryer. Heat transfer is made possible by the double walls of the cylinder. In the drying chamber, the rotor promotes the formation of a thin sludge film. According to reports, thin film dryers experience decreased operational issues. Another common type of contact dryer is the disc dryer, which consists of an interior pipe and a stator. The heat medium circulates through hollow discs that are welded onto a hollow shaft to form the rotor. Infrared radiation and other electromagnetic radiation are used in radiation dryers to transfer heat (Tunçal and Mujumdar, 2022).

8.6 SLUDGE DISPOSAL

The final step in sludge management is the disposal of the treated sludge. Although the moisture content and the volume of the sludge are significantly reduced after undergoing all the above-mentioned steps, the sludge is still a waste product of an STP and must be disposed of. The different sludge disposal alternatives have been mentioned in the following paragraphs.

   Landfilling: Landfilling is the most common disposal technique for municipal sludge. If suitable land is available near the STP, landfilling is the most suitable option for the disposal of solids generated in STP. These solids may include screenings, grit, biosolids, or processed sludge. However, landfilling may not be suitable if biosolids contain heavy metals, pathogens, or toxic organics.

   Land applications: The spreading of biosolids on or just below the surface of the soil is referred to as "land application." The sludge may be biologically broken down into a stable byproduct in the presence of microorganisms (bacteria and fungi) under controlled aerobic conditions. This process is known as composting. Additionally, since the sludge has effectively been pasteurized, it can be utilized as a soil conditioner. Compost made from properly composted sludge has a high fertilizing capacity and hygienic, unobtrusive, and humus-like properties. The compost enhances soil structure, soil aggregation, water-holding capacity, water infiltration, and soil aeration. Additionally, the growth of plants is aided by macronutrients (phosphorus, nitrogen, and potassium) and micronutrients (manganese, copper, iron, and zinc) (Gurjar and Tyagi, 2017; Mackenzie, 2010).

   Incineration: Incineration of sludge is carried out at places where the availability of land for disposing of sludge is minimum. This is primarily because the process is cost-intensive and can have adverse environmental impacts. However, incineration has few advantages over landfilling. In incineration, the pathogens and toxic organics in the sludge are destroyed, and transport costs are drastically reduced. Heavy metals typically end up in the ash, but some of the metals, like mercury, can cause issues. On the other hand, modern flue gas treatment systems are available to remove the most toxic materials from incineration smoke stacks.

8.7 CHAPTER SUMMARY

   • The solids that are formed during the municipal wastewater treatment are referred to as sludge or sewage sludge. On the other hand, biosolids refer to treated sewage sludge. Usually, the water content of sludge is more as compared to biosolids.

- Sludge stability, sludge volume index, and sludge volume ratio are the primary parameters that determine the kind of processing required for sludge treatment.
- The major steps in sludge processing are preliminary treatment, sludge thickening, stabilization, conditioning, dewatering, processing, reduction, disinfection, and finally, disposal.
- In preliminary operations, the solids generated from the preliminary stages of the STP, such as screening and grit chamber, are handled. Furthermore, grinding, degritting, blending, and storing sludge is carried out to provide a homogeneous sludge feed to the subsequent sludge treatment facilities.
- In sludge thickening, the water is separated from the sludge. This is usually achieved by different processes, such as centrifugation, gravity bed filters. rotary drum thickeners, floatation thickeners, and others.
- Sludge stabilization or digestion is carried out to eliminate pathogens, foul odor, and decay organic matter. Biosolids are formed as a result of sludge stabilization. The most common processes of sludge stabilization are alkaline stabilization, aerobic stabilization, and anaerobic stabilization.
- In sludge conditioning, certain chemicals or heat treatment is provided to dewater the sludge further.
- Dewatering of the conditioned sludge is carried out to further reduce the water in order to comply with disposal regulations, facilitate handling, cut costs associated with transportation, stop leachate from disposal sites, and others. Filter presses, drying beds, and centrifugation are some of the separation techniques.
- Hygienization is often carried out to bring down the pathogenic content of the dewatered sludge before its final disposal.
- The final stage of sludge management is to dispose of the dewatered sludge. The dewatered sludge may either be disposed of in a landfill or may be reused in various land applications.

8.8   CONCLUDING REMARKS

Sludge management is essentially the final step in a complete wastewater management system. However, due to the involvement of many processes, from the conveyance of the wastewater to the treatment processes and, finally, sludge management, the entire system is cost-intensive. Hence, energy recovery from wastewater is an area that requires special attention. Furthermore, many of the recalcitrant emerging contaminants that are becoming more prevalent are not completely removed in the conventional STPs. The detection of these compounds is also a challenging task. Hence, in the next few chapters, the occurrence of emerging contaminants, their physicochemical properties, detection, and removal will be discussed.

REFERENCES

Ahmad, T., Ahmad, K., Alam, M., 2016. Sustainable management of water treatment sludge through 3'R' concept. *J. Clean. Prod.* 124, 1–13. https://doi.org/10.1016/j.jclepro.2016.02.073

APHA, 2017. Standard Methods for the Examination of Water and Wastewater. *Public Health* 51, 1–1546. https://doi.org/10.2105/AJPH.51.6.940-a

Bassan, M., Mbéguéré, M., Tchonda, T., Zabsonre, F., Strande, L., 2013. Integrated faecal sludge management scheme for the cities of Burkina Faso. *J. Water Sanit. Hyg. Dev.* 3, 216–221. https://doi.org/10.2166/washdev.2013.156

Ben Othman, F., Eddhibi, F., Bel Hadj Ali, A., Fadhel, A., Bayer, Ö., Tarı, İ., Guizani, A., Balghouthi, M., 2022. Investigation of olive mill sludge treatment using a parabolic trough solar collector. *Sol. Energy* 232, 344–361. https://doi.org/10.1016/j.soler.2022.01.008

Chaanaoui, M., Abderafi, S., Vaudreuil, S., Bounahmidi, T., 2021. Prototype of phosphate sludge rotary dryer coupled to a parabolic trough collector solar loop: Integration and experimental analysis. *Sol. Energy* 216, 365–376. https://doi.org/10.1016/j.soler.2021.01.040

Di Fraia, S., Figaj, R.D., Massarotti, N., Vanoli, L., 2018. An integrated system for sewage sludge drying through solar energy and a combined heat and power unit fuelled by biogas. *Energy Convers. Manag.* 171, 587–603. https://doi.org/10.1016/j.enconman.2018.06.018

Fendrich, M.A., Quaranta, A., Orlandi, M., Bettonte, M., Miotello, A., 2018. Solar concentration for wastewaters remediation: A review of materials and technologies. *Appl. Sci.* 9(1), 118. https://doi.org/10.3390/app9010118

Gallego-Schmid, A., Tarpani, R.R.Z., 2019. Life cycle assessment of wastewater treatment in developing countries: A review. *Water Res.* 153, 63–79. https://doi.org/10.1016/j.watres.2019.01.010

Gurjar, B.R., Tyagi, V.K., 2017. *Sludge Management.* CRC Press. https://doi.org/10.1201/9781315375137

Mackenzie, D.L., 2010. *Water and Wastewater Engineering: Design Principles and Practice.* McGraw-Hill Education.

Muga, H.E., Mihelcic, J.R., 2008. Sustainability of wastewater treatment technologies. *J. Environ. Manage.* 88, 437–447. https://doi.org/10.1016/j.jenvman.2007.03.008

Pastor, L., Marti, N., Bouzas, A., Seco, A., 2008. Sewage sludge management for phosphorus recovery as struvite in EBPR wastewater treatment plants. *Bioresour. Technol.* 99, 4817–4824. https://doi.org/https://doi.org/10.1016/j.biortech.2007.09.054

Tchobanoglous, G., Burton, F.L., Stensel, H.D., 2014. *Wastewater Engineering: Treatment and Resource Recovery*, Metcalf & Eddy, Inc. McGraw-Hill Education.

Tunçal, T., Mujumdar, A.S., 2022. Modern techniques for sludge dewaterability improvement. *Dry. Technol.* 0, 1–13. https://doi.org/10.1080/07373937.2022.2092127

Tyagi, R.D., Surampalli, R.Y., Yan, S., Zhang, T.C., Kao, C.M., Lohani, B.N., 2009. *Sustainable Sludge Management: Production of Value Added Products.* ASCE Library . https://doi.org/10.1061/9780784410516

US EPA [WWW Document], 2021. www.epa.gov/coral-reefs/basic-information-about-coral-reefs; https://portals.iucn.org/library/sites/library/files/documents/CES-001.pdf (accessed 12.9.21).

Wang, L.K., Shammas, N.K., Hung, Y.-T., 2007. *Biosolids Treatment Processes.* Humana Press. https://doi.org/10.1007/978-1-59259-996-7

Zhou, X., Jiang, G., Wang, Q., Yuan, Z., 2014. A review on sludge conditioning by sludge pretreatment with a focus on advanced oxidation. *RSC Adv.* 4(92), 50644–50652. https://doi.org/10.1039/c4ra07235a

9 Emerging Contaminants in the Aqueous Environment

Detection and Quantification, Ecological Impacts, and Legislations

CHAPTER OBJECTIVES

The chapter seeks to provide a broad overview of emerging contaminants in domestic wastewater. The challenges in the detection and quantification have been highlighted and proper methodologies for their analysis have been discussed. The ecological impacts of the emerging contaminants and existing regulations pertaining to them have also been discussed.

9.1 INTRODUCTION

The 21st century has witnessed a boom in the use of various medicines, artificial sweeteners, personal health products, surfactants, perfluoroalkyl, and polyfluoroalkyl (PFAS) substances, and other organic compounds. As a result, there has been increased detection of such compounds, referred to as emerging contaminants (EC), in different aquatic environments (Majumder et al., 2019, 2021; Richardson and Ternes, 2018; Tran and Gin, 2017). Prolonged consumption or use of these products has led to the increase in the levels of ECs and their metabolites in water (Khan et al., 2020; Majumder et al., 2019; Shanmugam et al., 2014). ECs have been reported in hospital wastewater, sewage, industrial wastewater, and effluents of water and wastewater treatment plants (Gupta et al., 2021, 2020; Langford and Thomas, 2009; Majumder et al., 2019; Verlicchi et al., 2010). Usually, ECs do not have any regulatory standards in many countries because either they are present in trace quantities and/or require expensive methods for detection and analysis. Also, the toxic characteristics of these compounds, especially at the concentration in which they are present, are not fully known. However, there is a growing awareness of ECs, and the literature indicates that these compounds can adversely affect human beings and disrupt the environment (Majumder et al., 2019; Parida et al., 2021). Furthermore, many ECs are known to be resistant to degradation processes and are toxic, having the potential to affect various aquatic organisms, microorganisms drastically, and often human beings mostly due to chronic exposure (Barbara Ambrosetti et al., 2015; Białk-Bielińska et al., 2016; Gupta et al., 2021, 2020; Majumder et al., 2019; Matamoros and Bayona, 2006). Hence, it is imperative to detect and quantify the ECs in various aquatic environments so that apply appropriate removal technologies are applied based on their nature and concentration.

This chapter seeks to take a comprehensive outlook at ECs. The sources and transport pathways for various ECs into different aqueous environments is discussed next.

9.2 SOURCES AND PATHWAYS OF EMERGING CONTAMINANTS INTO VARIOUS AQUEOUS ENVIRONMENTS

ECs can enter the aqueous environment by various routes. In sewage, most ECs are typically found at concentrations in the range of pg/L to µg/L (Bayen et al., 2016; Majumder et al., 2019; Parida et al., 2021). Based on their sources, ECs may be broadly classified as pharmaceutically active compounds (Figure 9.1), personal care products,

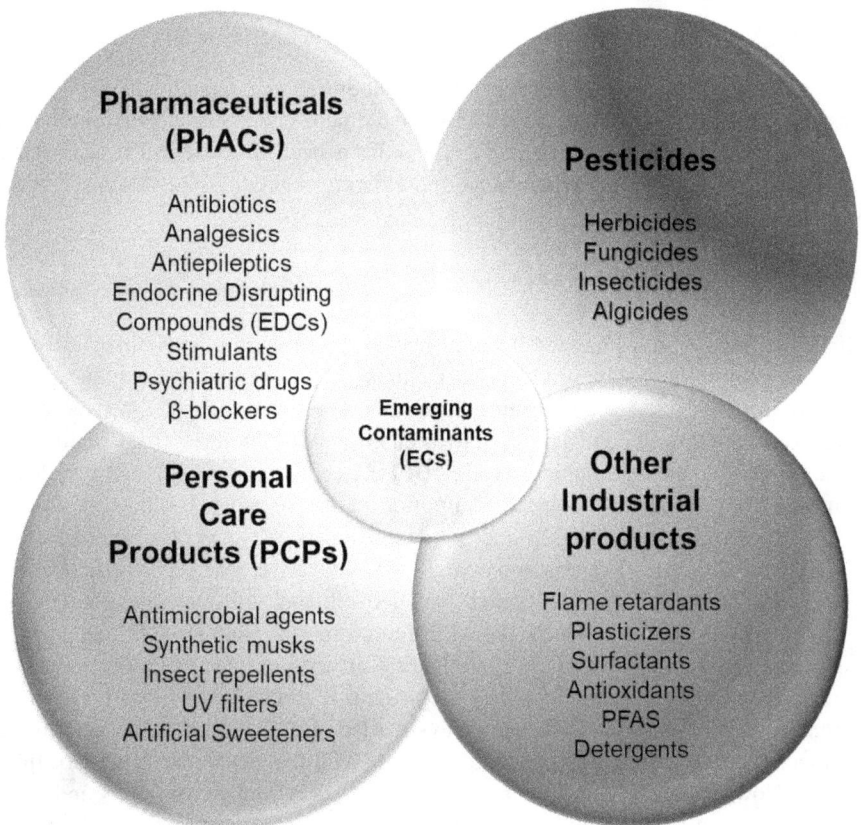

FIGURE 9.1 Classification of emerging contaminants.

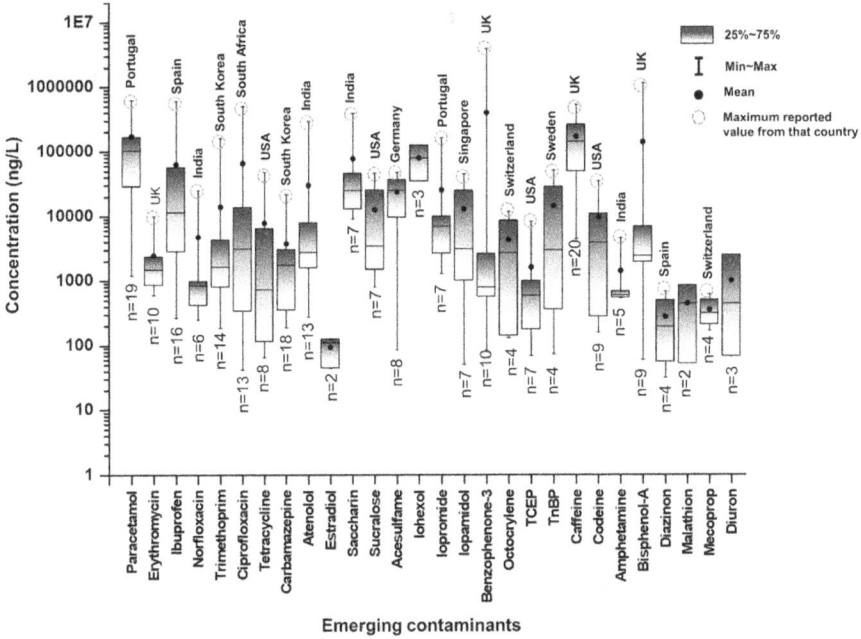

FIGURE 9.2 Box and whisker plots showing variation in the worldwide concentration of selected ECs in the influents of treatment plant (where *n* represents sample size). (Reprinted with permission from Parida et al., 2021. Copyright 2023 Elsevier).

pesticides, and other industrial products. The concentration of few commonly occurring ECs in the influent of wastewater treatment plants in various nations are depicted in Figure 9.2. The country where the highest concentration of a particular contaminant was observed is also depicted in Figure 9.2. Since these compounds are present at such low concentrations in the aqueous environment, their detection is a major challenge. Consequently, research on their removal and toxicological effects is limited.

The pharmaceutically active compounds find their way into sewage primarily through discharges from hospitals and disposal of such products from households. The unmetabolized part of the pharmaceuticals excreted by humans is also found in the sewage. Hospital wastewater streams are often mingled and co-treated with municipal wastewater streams in many countries. As a result, significant amounts of pharmaceutical residues are found in the raw sewage of the sewage treatment plants (STPs). Residues of analgesics, antibiotics, hormones, psychiatric drugs, stimulants, and β-blockers are commonly detected in sewage (Balakrishna et al., 2017; Majumder et al., 2019).

Surfactants and personal care products are used regularly in household activities. Pesticides and insecticides are also common due to their use in gardening and pest control. Hence, they also find their way into the sewage (Tran et al., 2014; Tran and Gin, 2017). PFAS are complex organic chemicals used in everyday products, such as manufacturing non-sticky cookware, clothes resistant to stains, foams, and others. Therefore, PFAS also find their way into municipal sewage due to daily usage of these materials (Al Amin et al., 2020; Domingo and Nadal, 2019).

The recalcitrant nature of the ECs can be attributed to their high polarity and complex organic structures, which prevent them from being easily degraded. For example, the carbon–fluoride bond present in the PFAS makes them one of the most resistant ECs present in the environment (Akhbarizadeh et al., 2020; Xiao, 2017). These chemicals do not degrade naturally in the aqueous environment, and research has shown that the half-lives of PFAS may be around 40–90 years (Kucharzyk et al., 2017). Most of the conventional treatment methods are not effective in completely bringing down ECs concentrations. Furthermore, the ECs may react with other chemicals present in treatment units to form intermediate products, which may have higher toxicity than the parent compounds. In this context, it is essential to understand the analytical procedures necessary for complete characterization of ECs including their metabolites and by-products. Therefore, the analysis methods for ECs are discussed in the subsequent sections.

9.3   ANALYSIS OF EMERGING CONTAMINANTS

9.3.1   Issues in the Analysis of Emerging Contaminants

One of the vital steps in detecting and quantifying organic compounds, such as ECs, is properly handling the samples to be analyzed after collection. Often this part of the research is not given sufficient priority, which leads to inaccurate measurement of the concentration of target compounds. Poor transportation and storage techniques may lead to misinterpretation of data, thereby compromising the accuracy of the work (Omar et al., 2016). Due to the various physicochemical properties of ECs, such as solubility in water, polarity, and mobility, they can be subjected to depletion processes in the sample containers (Mompelat et al., 2013). Although ECs are known to be highly persistent in nature, various studies have indicated a decrease in the concentration of ECs in the aqueous environment (Laws et al., 2011; Mompelat et al., 2013; Radke et al., 2010; Tamtam et al., 2008). In addition, some of the ECs may adsorb to the walls of the sample containers or suspended solids within wastewater. As a result, there is a possibility of losses occurring while transporting and storing EC-containing water samples (Mompelat et al., 2013).

Sewage, hospital wastewater, river, lakes, and others, contain organic and inorganic matter present in the suspended or colloidal form. ECs having a high octanol/water partition coefficient (log K_{ow}) tend to be hydrophobic, and get adsorbed onto the suspended matter present in the aqueous environment (Białk-Bielińska et al., 2016; Majumder et al., 2019; Mompelat et al., 2013). Apart from the hydrophobic or hydrophilic nature of the ECs, physicochemical properties of the container material and suspended matter, adsorption also depends upon the structure and concentration of ECs, composition, temperature, pH of the water matrix, and contact time (Lorphensri et al., 2007; Mompelat et al., 2013, 2009; Omar et al., 2016). Mompelat et al. (2013) reported that 10 out of 30 ECs in a Spanish river were found to be adsorbed onto the suspended solids (Mompelat et al., 2013). Boulard et al. (2020) reported that positively charged ECs were more likely to get adsorbed onto the surface of suspended particulate matter (Boulard et al., 2020).

The presence of suspended solids or organic matter in the samples can interfere with the actual concentration of the ECs in the aqueous medium because of their adsorptive properties. Alternatively, desorption of ECs from the organic matter to the water sample may overestimate the original concentrations. Furthermore, these ECs can also get adsorbed onto the walls of the sample containers (Mompelat et al., 2013; Omar et al., 2016). Various ECs form complexes with different cations present in water (Park et al., 2002). These complexes favor the adsorption of ECs onto the surface of the glassware during sampling and storage (Mompelat et al., 2013). The phenomenon of adsorption and desorption, therefore, becomes crucial during the analysis of EC. Proper selection of sampling containers and the decisions related to filtering of samples are critical for proper characterization of ECs.

Microorganisms in the aqueous sample make ECs susceptible to biodegradation. Recent studies have shown that microbes can use estrogenic ECs as a substrate, which is the primary reason behind their degradation during storage (Baronti et al., 2000; Mompelat et al., 2013; Vanderford et al., 2011). Exposure to visible light irradiation may lead to the transformation of the ECs, leading to the depletion of the parent compound before the analysis. Photo-degradation of ECs depends on the wavelength of light, exposure intensity, duration, and also the nature and content of organic matter. The organic matter (humic acid, nitrates, and others) present may be excited upon exposure to the ambient light leading to the formation of radicals, which may react with the ECs to degrade them (Mompelat et al., 2013, 2009; Omar et al., 2016). Furthermore, hydrolysis leads to declining concentrations of ECs, and this phenomenon is dependent on the stability of the ECs in an aqueous medium (Jiang et al., 2010). The different issues and challenges in analyzing ECs in aqueous environments have been classified into various factors and are depicted in Figure 9.3.

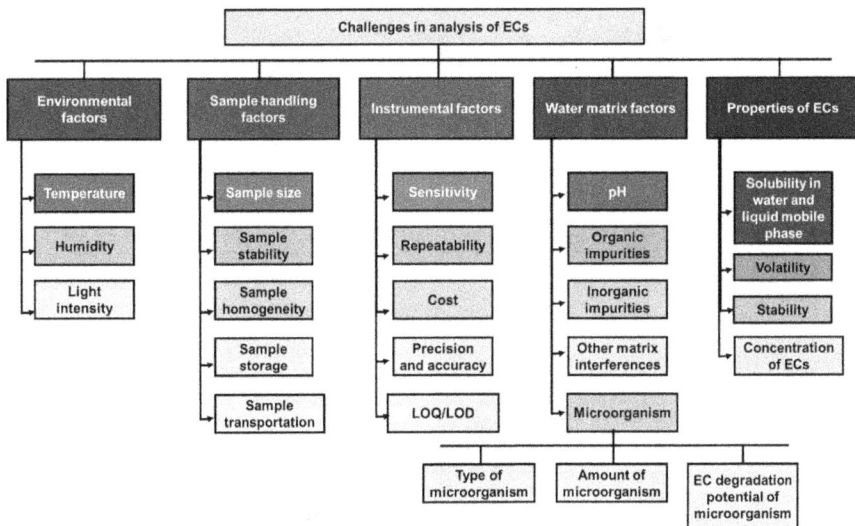

FIGURE 9.3 Different challenges associated with the analysis of ECs.

In order to negate these problems, it is imperative to follow proper protocols for handling of the water and wastewater comprising of the target analytes. These steps essentially include proper collection of the water or wastewater, followed by proper transportation, pre-treatment, and storage. Strictly adhering to these protocols will significantly bring down the error in the analysis of the ECs and help avoid making costly decisions based on erroneous data.

Adhering to proper methods for handling the analytes or the water samples is only a preliminary step in analyzing ECs. Details involving the handling of analytes are provided in the subsequent section and a schematic of the few important steps involved prior to analysis of the ECs are provided in Figure 9.4.

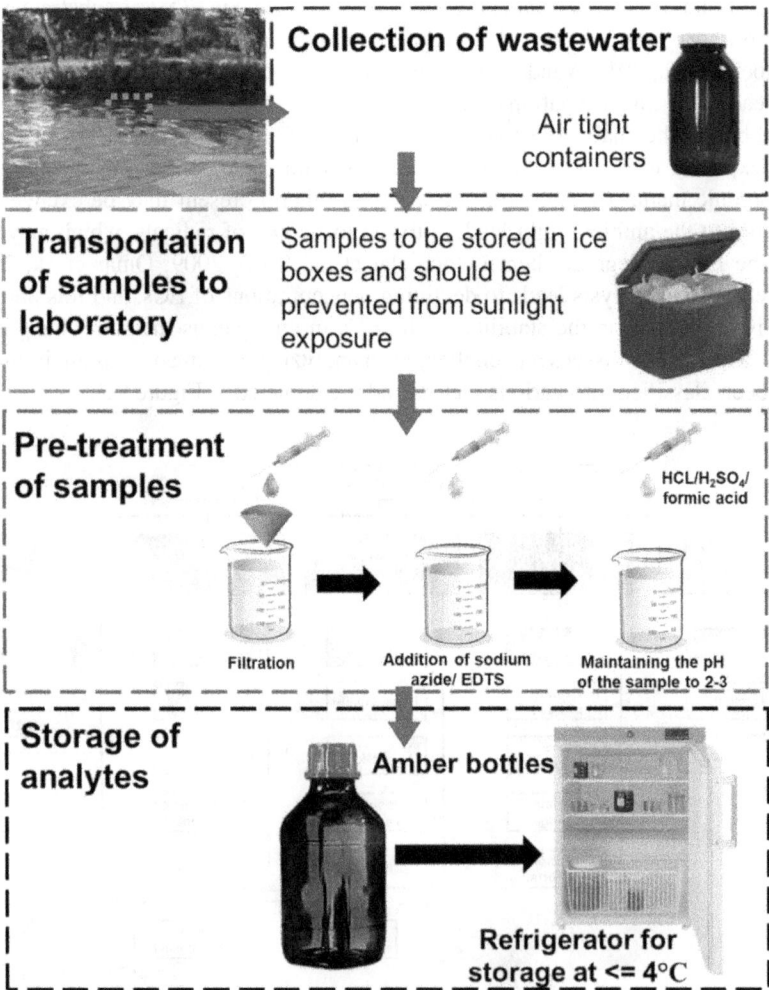

FIGURE 9.4 Necessary steps involved during collection, transportation, and storage of samples for effective analysis of ECs.

9.3.2 TRANSPORTATION AND PRE-TREATMENT OF EC-CONTAINING SAMPLES

Proper sample handling, transportation, and pre-treatment are vital to curtail any physical, chemical, or biological changes to the ECs from the time of sampling to the time of analysis. Refrigeration of the samples, proper selection of sample containers, and prevention of exposure to ambient light can help in minimizing the effects of adsorption, photo-degradation, biodegradation, and others (Omar et al., 2016). The storage containers should be properly cleaned to prevent any chance of cross-contamination. Overestimation of the concentration may be possible if any compounds are already present in the storage container. The analysis of blank samples is necessary since it helps to determine whether any contamination has occurred during sampling, transportation, and storage (Mompelat et al., 2013). As per ISO 5667-3, the samples to be analyzed should be kept in air-tight containers to prevent any leakage and external contamination (ISO-5667-3, 2018; Mompelat et al., 2013). After collection of the samples, they are to be transported immediately to the laboratory for further pre-treatment and storage. During transportation, the samples should be kept in iceboxes, and exposure to light should be prevented to minimize degradation of the ECs (ISO-5667-3, 2018; Mompelat et al., 2013). After transportation, the samples should be properly treated to remove the suspended solids, organic matter, and microorganisms before they are stored. Filtration of the analytes helps to free the analytes of suspended solids that may interfere with the analysis of ECs. Removal of organic matter and suspended solids curtail the effect of adsorption and desorption of ECs, while removal of microorganisms prevents the biodegradation of ECs. A study conducted by Baker and Kasprzyk-Hordern (2011) revealed that ECs were more stable when the analytes were filtered before storage (Baker and Kasprzyk-Hordern, 2011b). Glass fiber filters of various pore sizes are the most commonly used filters for filtering analytes containing ECs (Afsa et al., 2020; Guzel et al., 2019; Lin et al., 2018; Ma et al., 2017; Mompelat et al., 2013; Ying et al., 2009; Zhang et al., 2018). Polytetrafluoroethylene filter papers have also been used recently for the filtration of analytes (Afsa et al., 2020; Biel-Maeso et al., 2018; Ma et al., 2017). Besides that, cellulose acetate membrane, nylon membrane, and other membrane filters are also used to filter EC-containing samples (Mompelat et al., 2013).

9.3.3 PRESERVATION AND STORAGE OF SAMPLES

Preservation of samples until the extraction or sample preparation is essential to ensure the integrity of the analyte (McDowall et al., 2019; Moldoveanu and David, 2015). Various ECs are unstable and can undergo photodegradation, thermal degradation, microbial degradation, oxidation, and reduction during storage. In order to store the samples for a prolonged duration, often preservatives or antioxidants are added, and the pH of the analyte is adjusted to acidic conditions (Moldoveanu and David, 2015; Omar et al., 2016). Often preservatives are added to inhibit bacterial growth, thereby reducing the chances of biodegradation. The addition of preservatives can significantly increase the stability of the ECs. The chemical used for preserving the sample should be carefully chosen so that it does not interfere with the analytical process (McDowall et al., 2019).

Sodium azide is commonly used as a preservative, as it helps inhibit the growth of microorganisms (Mompelat et al., 2013). Acidic conditions also do not favor the growth of microorganisms. Hence, to prevent the biodegradation of ECs, the pH of the analyte is brought down by adding acidifying agents. Hydrochloric acid and sulfuric acid are the most commonly used acidifying agents used for the preservation of ECs (Guzel et al., 2019; Havens et al., 2010; Vanderford et al., 2011, 2003). Formaldehyde (1% v/v) was used as a preservative for storing different hormones (Baronti et al., 2000). Togola and Budzinski (2007) used methanol (5%, v/v) and formaldehyde (5%, v/v) to prevent the degradation of various pharmaceutically active compounds (Togola and Budzinski, 2007). Methanol has been known to prevent the adsorption of ECs onto the surface of the glass containers (Daneshvar et al., 2012).

Researchers have employed various sampling containers, ambient conditions (light and temperature) to store aqueous samples containing ECs and also to assess the stability of these compounds in the provided conditions. Various ECs were found to be stable for up to 7 days when stored at room temperature in amber glass bottles (Managaki et al., 2007; Roig et al., 2012). Vanderford et al. (2011) found that some ECs were stable up to 35 days when stored in amber glass bottles at room temperature and at 4°C (Vanderford et al., 2011). Ciprofloxacin and oxolonic acid were stable up to 62 days and 90 days, respectively, when they were kept in the dark and at room temperature and stored in low-density polyethylene bottles (Turiel et al., 2004). The samples were stable up to 124 days when stored at 4°C (Turiel et al., 2004). Different hormones could be stored up to 14 days in silanized amber glass bottles at 4°C (Havens et al., 2010). Various illicit drugs, stimulants, and antidepressants were analyzed after keeping them in the dark at 4°C for 1–3 days (Boleda et al., 2007; Castiglioni et al., 2006; Huerta-Fontela et al., 2007; Kasprzyk-Hordern et al., 2010). The summary of storage conditions for some of the ECs and their stability is provided in Table 9.1.

9.3.4 SAMPLE BLANKING

The blanking of samples or the analysis of blank samples is necessary to estimate if there is any contamination during sampling, transportation, and storage of the samples. There are various kinds of blanks that can help to trace the sources of artificial contamination.

a. *Instrument blank*: Instrument blank is carried out to check the presence or absence of any contamination after the instrument has analyzed a high concentration of the target analyte.

b. *Method blank*: The method blank is carried out to check whether there has been contamination during sample preparation. The method blank is prepared and analyzed using the same method that is being used for the analysis of the analytes.

c. *Trip blank*: The trip blank is carried out to check whether there are chances of contamination during the transportation and field handling of the samples. This is carried out by analyzing a clean sample that is taken out from the

TABLE 9.1
Summary of Storage Conditions of Various Emerging Contaminants and Their Stability

Containers	Light	Analytes	Stability (days)	References
Temperature: Room Temperature				
Amber glass	Dark	Sulfamethoxazole, Erythromycin, Ofloxacin, Ciprofloxacin, Trimethoprim	10	(Roig et al., 2012)
Amber glass		Trimethoprim, Sulfamethoxazole, Ibuprofen, Naproxen, Diclofenac, Carbamazépine, Phenytoin, Primidone, Fluoxetine, Diazepam, Meprobamate, Gemfibrozil, Iopromide, 17β-Oestradiol, Progesterone, Testosterone, Oestrone, 17α-Ethynyloestradiol, Atenolol	35	(Vanderford et al., 2011)
Low-density polyethylene	Dark	Ciprofloxacin	62	(Turiel et al., 2004)
Low-density polyethylene	Dark	Oxolonic acid	90	(Turiel et al., 2004)
Polyethylene		Ibuprofen, Naproxen, Ketoprofen, Diclofenac		(Farré et al., 2008)
Amber glass		Ketoprofen, Salicylic acid, Diclofenac, Gemfibrozil		(Togola and Budzinski, 2007)
Temperature: 4°C				
Amber glass	Dark	Erythromycin, Trimethoprim, Ciprofloxacin, Ofloxacin, Sulfamethoxazole	10	(Roig et al., 2012)
Amber glass	Ambient light	Trimethoprim, Sulfamethoxazole, Ibuprofen, Naproxen, Diclofenac, Carbamazépine, Phenytoin, Primidone, Fluoxetine, Diazepam, Meprobamate, Gemfibrozil, Iopromide, 17β-Oestradiol, Progesterone, Testosterone, Oestrone, 17α-Ethynyloestradiol, Atenolol	35	(Vanderford et al., 2011)
Low-density polyethylene	Dark	Ciprofloxacin, Oxolonic acid	124	(Turiel et al., 2004)
Amber glass		Ketoprofen, Salicylic acid, Diclofenac, Gemfibrozil	8	(Togola and Budzinski, 2007)
Silanized Amber glass	Dark	17β-Oestradiol, Oestrone, Oestriol, Zearalanol, Zearalenone, Zearalanone, Testosterone, 5α-androstan-17β-ol-3-one, Androsterone, 5α-androstane-3,17-dione, 4-androstene-3,17-dione, Boldenone, Nandrolone, 17β-trenbolone, 17α-trenbolone, Progesterone, 17,20-dihydroxyprogesterone, Melengestrol, Melengestrol acetate	14	(Havens et al., 2010)

(*continued*)

TABLE 9.1 (Continued)
Summary of Storage Conditions of Various Emerging Contaminants and Their Stability

Containers	Light	Analytes	Stability (days)	References
Polyethylene		Ibuprofen, Naproxen, Ketoprofen, Diclofenac	10	(Farré et al., 2008)
Temperature: −18°C				
Low-density polyethylene	Dark	Ciprofloxacin, Oxolonic acid	124	(Turiel et al., 2004)
Temperature: −20°C				
Amber glass		Trimethoprim, Sulfamethoxazole, Ibuprofen, Naproxen, Diclofenac, Carbamazepine, Phenytoin, Primidone, Fluoxetine, Diazepam, Meprobamate, Gemfibrozil, Iopromide, 17β-Oestradiol, Progesterone, Testosterone, Oestrone, 17α-Ethynyloestradiol, Atenolol	35	(Vanderford et al., 2011)
Glass		Acetaminophen, Salbutamol/Albuterol, Atenolol	7	(Bayen et al., 2016)
		Atorvastatin, Caffeine, Carbamazepine, Cimetidine, Citalopram, Cotinine, Diclofenac, Dilantin, Diltiazem, Diphenhydramine, Enalapril		
		Gemfibrozil, Ibuprofen, Naproxen, Norfluoxetine, Paroxetine, Propanolol, Risperidone, Sertraline, Sotalol, Sulfamethazine, Sulfamethoxazole, Triclosan, Trimethoprim, Venlafaxine, Warfarin, Atrazine, Bisphenol A, Estrone, Estradiol, Linuron, Thiabendazole		
Polyethylene		Oxolinic Acid, Trimethoprim Florfenicol, Sulfadiazine	56	(Sørensen and Elbæk, 2004)
Polypropylene		3,4-29 methylenedioxymethamphetamine, cocaine, benzoylecgonine, cotinine, oxycodone,	21	(Chiaia et al., 2008)
		32 hydrocodone, methadone, 3,4- 30 methylenedioxyamphetamine amphetamine, flunitrazepam, Caffeine, creatinine, 31 methamphetamine		

laboratory to the sampling site and then carried back to the laboratory. Usually, volatile organic compounds are absorbed in the samples, and their presence is determined.

d. *Field blank*: Field blank is carried out to check contamination during sampling. The blank, which is free from any analytes, is preserved and shipped back to the laboratory for analysis in the same manner as the other samples. This also involves analyzing for contaminants that may be present in the air, which might have got absorbed during sampling.

e. *Equipment blank*: Equipment blanks are carried out by running an analyte-free sample collected using the same equipment used for the collection of other samples. This assesses the presence of contaminants in the sampling equipment.

As depicted in Figure 9.5, the equipment blank results include total contamination from field sources (field blank and trip blank) and laboratory sources (method blank and instrument blank). Field blanks include contamination during sampling, transportation, and laboratory analysis. Likewise, trip blanks include contaminants during transportation and laboratory analysis. Analyte-free water is exposed to the atmosphere during transport of samples to evaluate any potential contamination when moving the samples from the field to the laboratory. Method blanks involve contamination during sample preparation and instrument contamination.

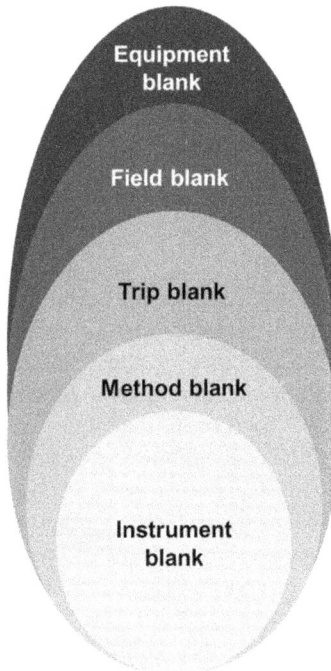

FIGURE 9.5 Different types of blanks.

9.3.5 Sample Preparation and Extraction

The ECs occur in the aqueous environment at concentrations in the range of pg/L to μg/L (Majumder et al., 2019; Oliveira et al., 2015; Tran et al., 2018; Tran and Gin, 2017), while the average limit of detection (LOD) and limit of quantification (LOQ) values for various analytical instruments are usually much higher (Afsa et al., 2020; Białk-Bielińska et al., 2016; Biel-Maeso et al., 2018; Lin et al., 2018; Santos et al., 2009, 2005). The minimum concentration of an analyte that can be detected using a particular instrument is the instrument detection limit for a particular compound. If the instrument detection limit is higher, different methods are implemented to concentrate the sample. After extraction and pre-concentration, the lowest concentration that can be detected by the instrument is called the method detection limit for that particular compound (Pawliszyn, 2012). LOD is the lowest concentration of an analyte that can be detected with statistical significance.

LOQ is the lowest concentration of an analyte that can be measured or determined with established accuracy, precision, and uncertainty.

LOQ values are always higher than the LOD values of the same analyte using a particular analytical method (Pawliszyn, 2012). In many instances, a suitable concentration factor may be required to commensurate with the instrument detection levels. ECs are accompanied by various other organic compounds and ions, which may interfere with the analytical procedures (Kostopoulou and Nikolaou, 2008). Subsequently, high volume extractions lead to a high matrix effect during the analysis of the target compounds. Hence, adequate preparation of samples is required for the efficient quantification of ECs. The extraction processes are implemented to not only free the target analytes from interferences, but also for pre-concentration of the target analytes. The extraction of ECs can be carried out using solid-phase extraction (SPE), solid-phase micro-extraction (SPME), liquid–liquid extraction (LLE), accelerated solvent extraction (ASE), microwave-assisted extraction (MAE), ultrasonic-assisted extraction (UAE), Soxhlet extraction (SE), membrane extraction, and lyophilization, which are described next.

9.3.5.1 Solid-Phase Extraction (SPE)

SPE is one of the most widely used extraction technologies used. In SPE, the target analytes are transferred from the liquid phase, which is usually the water sample, to the solid phase, which is the adsorbent of the SPE cartridge. The adsorbed target analytes (ECs) are extracted into another solvent, which may be used for detection purposes. The purpose of SPE is to bring the concentration of the ECs to detectable or quantifiable limits. If the concentration of a particular EC (X) in an aqueous sample is 1 ng/L and 1 L of the sample is passed through the SPE cartridge, then 1 ng of X is getting accumulated inside the SPE cartridge. Now, if 10 mL of an organic solvent (Y) is used, which can dissolve X, then the concentration of X in Y becomes 1 ng in 10 mL or 100 ng/L. In this way, the concentration of X has been increased by 10^2 times.

Apart from pre-concentrating the analyte, the aqueous sample may have several interfering ions and compounds. These interfering ions and compounds can also be removed using the SPE. If the cartridge material is chosen in such a way that only the

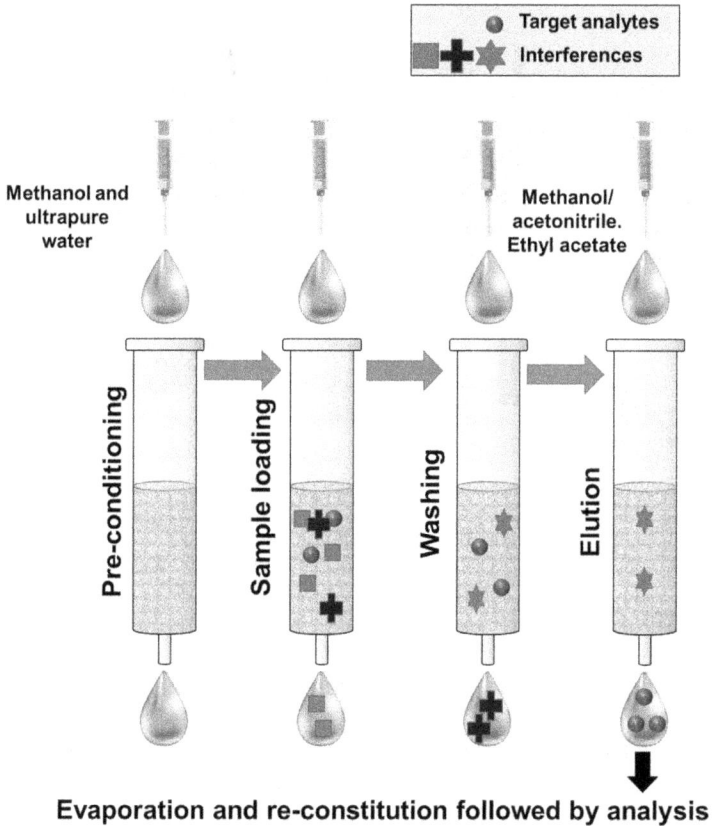

Evaporation and re-constitution followed by analysis

FIGURE 9.6   The major steps involved in solid-phase extraction.

ECs are getting adsorbed and not the interfering agents, then the resultant analyte will be free from the interfering agents. Also, suitable solvents may be used to wash away the interfering agents or simply dissolve the target analytes. The main steps involved in SPE have been depicted in Figure 9.6.

Although SPE has been widely used for extracting ECs, the cost of the cartridges is a significant drawback of this process. The process is also not suitable when the analytes are highly soluble in water. Additionally, if this method is used to directly extract ECs from wastewater, the suspended matter present in the wastewater may clog the cartridge. Hence, the wastewater should be freed from suspended solids by centrifuge or filtration technique.

9.3.5.2   Solid-Phase Microextraction (SPME)

Unlike SPE, SPME is a solvent-free method, where extraction and pre-concentration are carried out in one single step (Guo and Lee, 2012). The SPME requires much less sample volume than SPE, and the LOD values obtained during chromatography are also comparable (Prieto et al., 2009). The solutes get captured in the adsorptive

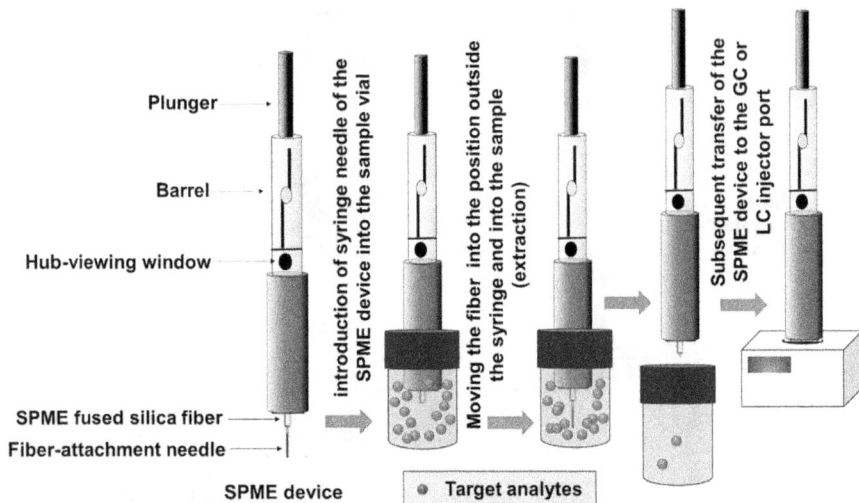

FIGURE 9.7 The major steps involved in solid-phase micro-extraction.

layer on the SPME fiber from where the solutes are transferred into an inlet system, which desorbs the solutes into a liquid or gas for LC and GC, respectively (Guo and Lee, 2012; Prieto et al., 2009). The major steps involved in SPME are depicted in Figure 9.7.

The SPME comes with numerous advantages, such as short analysis time, elimination of solvent, and others. However, SPME is also sensitive to suspended matter in the water. The suspended matter may compete with the target analytes during their sorption on to the adsorptive layer.

9.3.5.3 Liquid–Liquid Extraction (LLE)

In LLE, the target analytes are transferred from one liquid sample to another. Usually, the target analytes or the ECs present in the water sample are transferred to another liquid that will be used for subsequent analysis. In LLE, it is absolutely necessary that the extraction solvent should be immiscible with the aqueous solution containing the target analytes. The extraction is usually carried out in a Soxhlet apparatus. The Soxhlet apparatus is shaken vigorously to transfer the ECs from the aqueous matrix onto the solvent. The aqueous matrix and the solvent are then separated in the Soxhlet apparatus based on their densities. The solvent consisting of the ECs is taken for analysis. The major steps involved in LLE are depicted in Figure 9.8.

Although the LLE process is simple and do not require expensive cartridge or complex apparatus, it requires a large volume of high purity solvents. The use of such solvents is both expensive and hazardous to the environment.

9.3.5.4 Other Extraction Techniques

Soxhlet extraction (SE) has been used to extract ECs from wastewater and wastewater sludge samples (De Sena et al., 2010; Mohapatra et al., 2014). SE is a time-consuming

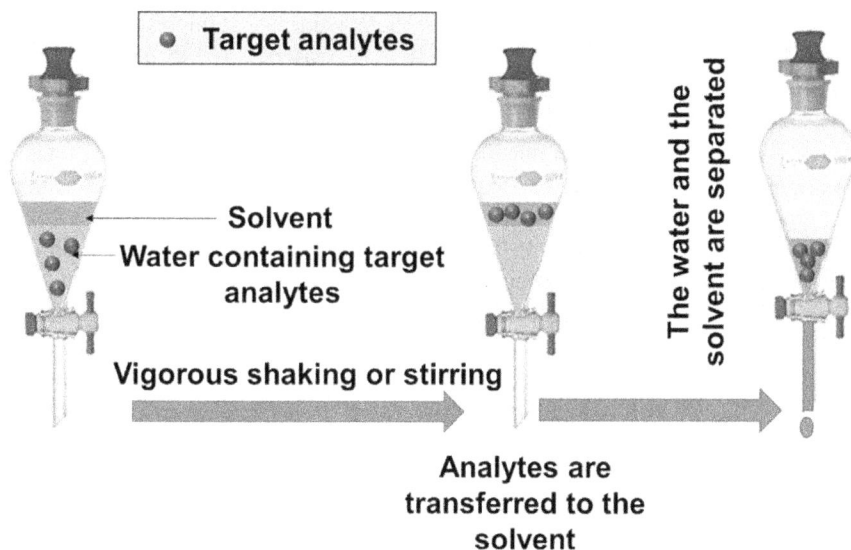

FIGURE 9.8 Schematic for liquid–liquid extraction.

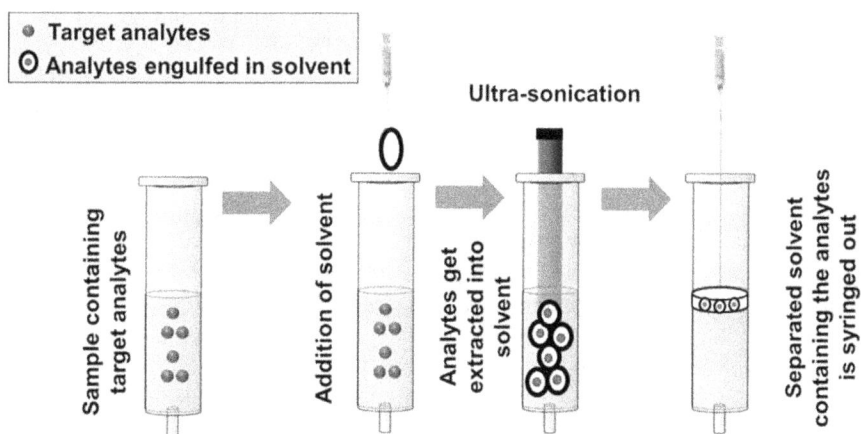

FIGURE 9.9 The major steps involved in ultrasonic-assisted extraction.

process and requires large volumes of organic solvents. Extraction is carried out by boiling, rinsing, and solvent recovery (Mohapatra et al., 2014). Ultrasonic-assisted extraction (UAE) is another extraction process that has been conveniently used for the extraction of ECs. Although this process required less time and less volume of organic solvent than SE, this method's reproducibility is lower (Luque-García and Luque De Castro, 2003). The processes involved in the UAE are depicted in Figure 9.9. Various researchers have used UAE to extract ECs from wastewater and wastewater sludge (Carballa et al., 2007; Martín et al., 2012; Mohapatra et al., 2012). MAE and ASE techniques also reduce the extraction time and solvent requirement. MAE utilizes non-ionizing radiation, microwaves to heat the solvent, and samples for extracting the target analytes (Mohapatra et al., 2014). ASE was reported to be used for the extraction of various ECs (Mohapatra et al., 2014). ASE is usually carried at high temperatures up to 200°C and pressure of around 1,000–2,500 psi to accelerate the process of elution. High temperature increases the analyte solubility and mass transfer rate and decreases the surface tension and viscosity of the solvents, which facilitate the extraction process from complex matrices (Mohapatra et al., 2014; Ramos, 2012).

Among the different extraction techniques used, SPE has numerous advantages compared to other extraction techniques. SPE do not consume large volume of toxic solvents like in the case of LLE. Furthermore, the process is simple as compared to SPME and has high enrichment factors. As a result, SPE is more suitable to extract and pre-concentrate ECs. The next step following extraction of the ECs is their quantification. The different quantification techniques used to analyze ECs and their metabolites have been discussed in the following section.

9.3.6 Quantification of ECs

The quantification of ECs requires the use of sophisticated instruments, such as liquid chromatography (LC) or high-performance liquid chromatography (HPLC), liquid chromatography coupled with mass spectroscopy (LC-MS), gas chromatography (GC), gas chromatography coupled with mass chromatography (GC-MS), spectrophotometry, nuclear magnetic resonance spectroscopy (NMRS), near-infrared spectroscopy (NIS), high-resolution mass spectroscopy (HRMS), and others. These instruments are expensive and not readily available at wastewater treatment plants. Therefore, the samples for analysis are often shipped to specialized laboratories. Commercially available HRMS instruments, such as time-of-flight mass spectrometry (TOF-MS), Orbitrap mass spectrometry, and Fourier transform ion cyclotron resonance mass spectrometry, are known to be efficient in the analysis of ECs, such as pharmaceuticals, PFAS, and others (Al Amin et al., 2020; Meng et al., 2021; Rodriguez et al., 2020). Although researchers have used UV-Vis spectrophotometry for the detection of ECs, they require a high degree of concentration (Ferraro et al., 2003; Khoshayand et al., 2008; Sifuna et al., 2016; Zhou and Jiang, 2012). On the other hand, LC and GC have much lower LOD and LOQ values as compared to UV-Vis spectrophotometer. Hence, LC, LC-MS, GC, and GC-MS are the preferred instruments for the analysis of ECs. The working principle of these instruments have been discussed in the following sections.

9.3.6.1 Liquid Chromatography

At present, reverse phased HPLC is the most commonly used LC for detection. In the HPLC, the ECs are detected based on their polarity. The compound that escapes the column faster is detected earlier and has a low retention time, while the compound that take longer time to escape the column are detected later and has a higher retention time (Meng et al., 2021; Rees, 2020; Siddiqui et al., 2017; Waters corporation, 2020). In reverse phase HPLC, the column is non-polar, and the mobile phase is polar, where the target analytes are eluded. So, when there is a non-polar analyte, there is an interaction between the non-polar stationary phase, which allows the non-polar

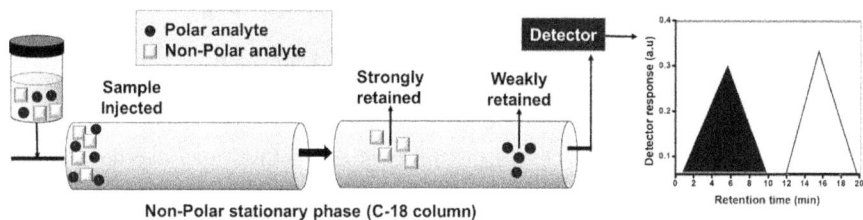

FIGURE 9.10 Working principle of liquid chromatography.

analyte to be retained for a longer time than polar compounds. However, in order to reduce the retention time of the non-polar analytes, a less polar mobile phase may be used, which will decrease the interaction of the non-polar analyte with the stationary phase (Meng et al., 2021; Rees, 2020; Siddiqui et al., 2017; Waters Corporation, 2020). The compounds are most commonly detected using diode array, fluorescence, and refractive index detectors (Meng et al., 2021; Rees, 2020; Siddiqui et al., 2017; Waters corporation, 2020). The working of LC and the principle behind its working has been depicted in Figure 9.10.

9.3.6.2 Gas Chromatography

In GC, the target analytes are separated based on their volatility. The more volatile compounds are weakly retained, while the less volatile compounds are retained for a longer duration in the column. Hence, the volatile compounds have a less retention time as compared to the non-volatile compounds. The target analytes are required to be thermally stable. In order to achieve a low retention time, solvents with low boiling points, such as diethyl ether, dichloromethane, and hexane are usually used (Agüera et al., 2005; Meng et al., 2021; Rees, 2020; Togola and Budzinski, 2007). The retention time varies with polarity of the target analytes and the stationary phase. Similar to LC, the interaction between the polar stationary phase and polar analyte, or non-polar stationary phase and non-polar analyte, increases the retention time. Helium, argon, and nitrogen are the most commonly used gases for the mobile phase. The most commonly used detectors of GC are flame ionization, thermal conductivity, and electron capture (Agüera et al., 2005; Meng et al., 2021; Rees, 2020; Togola and Budzinski, 2007). The working of GC systems and the principle behind their working have been depicted in Figure 9.11.

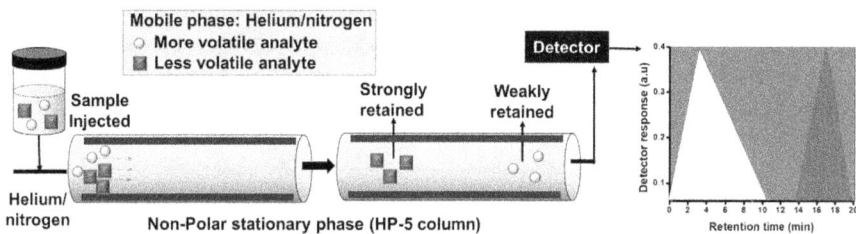

FIGURE 9.11 Working principle of gas chromatography.

9.3.6.3 Mass Spectroscopy

When the sample has more than one organic compound, the retention time of the target analyte may be close to other organic compounds. Such problems in LC or GC arise if the compounds have the same polarity or volatility. Similarly, when there is more than one target analyte, the peaks of the different compounds may overlap and give erroneous readings. Further, if there are unknown compounds in the sample, a GC or LC alone is not sufficient to identify the analytes and quantify them. In order to solve this problem, the GC and LC systems are coupled with a provision for mass spectroscopy (MS). After getting eluded in the column, the analytes are directed toward the MS system, where the analytes are broken down into ionized fragments using a high-energy beam of electrons. Each of the charged fragments has a certain mass by charge (m/z) ratio. The fragments are accelerated through focusing lenses and mass analyzer before finally hitting a detection plate. The m/z ratio and relative abundance are calculated at the detection plate. The mass spectroscopy gives us the m/z ratio of the target analytes, and subsequently, they can be identified and quantified.

The two most common types of mass analyzers are quadrupole mass analyzer and time-of-flight mass analyzers. In the quadrupole mass analyzer, the ions with a specific m/z ratio are able to reach the detector. A radiofrequency electric field is used to guide the ions through the central axis of the quadrupole. Another superimposed direct current field is used to select the required ions and eject them from the quadrupole. The strength of the fields can be adjusted such that ions with the desired range of m/z ratio pass through the quadrupole. The ions that pass through the quadrupole are detected. In case of the time-of-flight analyzer, the ions are identified based on the time required by the ions to travel from the extractor to the detector. The travel path of the ions is called the flight region or flight tube. It works on the principle that molecules with higher mass will travel slower than molecules with lower mass. On this basis, the molar mass of the compounds is determined (Al Amin et al., 2020; Boulard et al., 2020; Kim et al., 2013; Payán et al., 2010).

9.3.6.4 Tandem Mass Spectrometry

In mass spectrometry, the organic molecules are ionized and the m/z ratios are analyzed to gather information about their molecular structure. Additional information about the organic compound can be obtained by fragmenting the intermolecular ion into fragments that we can evaluate by mass spectroscopy. This is called tandem mass spectroscopy or MS–MS. This is usually done to identify the presence of any particular functional groups present in the organic compound. In mass spectrometry, we initially generate the parent ion or precursor ion. The precursor ion corresponds to a molecule that has acquired a positive or negative charge by gaining or losing an electron. After first ionization, in tandem mass spectroscopy, a secondary ionization occurs, where energy is provided to break some of the weak chemical bonds within the molecule to produce fragments. These fragments are known as daughter ions or product ions. The product ion is the second-generation parent ion for subsequent ionizations. This step is essential to identify the degradation pathways of a compound based on which bonds in the compounds are more likely to break (Alygizakis et al., 2016; Grover et al., 2011; Hernández et al., 2015).

With the advent of these new detection technologies, the ECs have started to get detected in most aqueous environments across the world. Consequently, they have become a thing of concern because of their potential to adversely impact the environment. The different ecological impacts of ECs have been discussed next.

9.4   ECOLOGICAL IMPACTS OF EMERGING CONTAMINANTS

The toxicity of ECs is attributed to their chemical structure and the ability of the body to absorb it. The presence of different functional groups, such as phenol, benzene, amide, carboxyl, ketone, among others, are responsible for the toxicity. Most of the ECs have been anthropogenically manufactured for certain beneficial purposes, such as pesticides to control pest, insecticides to kill insects, and antibiotics to control bacteria and infection. Other pharmaceuticals are used to alter the secretion of hormones and the normal activity of human or animals. However, when the ECs find their way into the non-target species, their impact may be negative. Prolonged exposure to such ECs may have severe detrimental effects to all living species. The ecological impacts of a few of the most commonly detected ECs are given in Table 9.2.

The negative impacts of ECs necessitate the requirement of strict regulatory standards pertaining to their discharge and consumption. The following section discusses the different guidelines and standards pertaining to ECs in the aquatic environment.

9.5   GUIDELINES AND STANDARDS PERTAINING TO EMERGING CONTAMINANTS

As noted earlier, the lack of sufficient data on the ecotoxicological effect of ECs, and occurrence data, difficulty in analysis, behavioral patterns, and other aspects has posed difficulty in development of regulatory standards for ECs. However, European Union (EU), United States Environment Protection Agency (USEPA), National Health and Medical Research Council, Australia (NHMRC), and World Health Organization (WHO) have come up with standards for some ECs. The details have been provided in Table 9.3.

Since there are no regulatory guidelines for most of the ECs, the drinking water equivalent limit (DWEL) can be used to estimate what can be the tolerable limit in water for that particular EC. The formula for calculating DWEL is provided in Equation 9.1 (Majumder et al., 2019; Parida et al., 2021):

$$DWEL = \frac{ADI \times BW}{WI \times FOE} \qquad (9.1)$$

DWEL (µg/L) may provide an approximate value or concentration of any contaminant below which it may not be harmful to human beings. The calculation of DWEL takes into consideration the acceptable daily intake (ADI) (µg/kg-day) of the EC, frequency of exposure (FOE), body weight (BW) (kg) of individuals, amount of water intake/day (WI) (L/day), and gastrointestinal absorption rate (AB). ADI is the

TABLE 9.2
The Different ECs and Their Ecological Impact

Class	ECs	Ecological Impact	References
Antibiotics	Azithromycin Tetracycline Erythromycin Trimethoprim Norfloxacin	Inhibit the growth of microorganisms Lead to the formation of antibiotic-resistant bacteria and antibiotic-resistant genes Affects the immune system of humans and animals	(Majumder et al., 2019; Parida et al., 2021)
Analgesics	Diclofenac Naproxen Ketoprofen Paracetamol Ibuprofen	Higher concentrations of analgesics may hepatotoxicity in vertebrates, resulting in liver failure and death. Ibuprofen has been reported to hinder postembryonic development among amphibians. Analgesics is also known to affect the reproduction system of aquatic organisms. Prolonged exposure may lead to gastric ulceration, mucosal damages, dyspepsia, and bowel inflammation and affects the internal organs of animals and humans	(Majumder et al., 2019; Parida et al., 2021; Saidulu et al., 2021)
B-blockers	Atenolol Propranolol	Human embryonic stem cell growth is hampered. It is extremely toxic to aquatic organisms and plants.	(Majumder et al., 2019; Saidulu et al., 2021)
Endocrine-disrupting compounds	17 β-estradiol	Fish and human sexual and reproductive systems are severely harmed, sperm count is reduced in males, and birth defects, abnormal sexual development, and cancer are all possible outcomes. Animals and humans' nervous systems and immune systems may be harmed by endocrine disrupting compounds.	(Majumder et al., 2019)
	Bisphenol-A Estrone-1		
Artificial sweeteners	Saccharin Sucralose Acesulfame	Artificial sweeteners interfere with gut bacteria and digestive enzymes, causing inflammatory bowel disease. Saccharin may cause liver inflammation in mice by altering metabolic functions.	(Parida et al., 2021; Saidulu et al., 2021)
X-ray contrast media	Iohexol	Metabolites of such compounds are toxic to animals and humans.	(Parida et al., 2021; Saidulu et al., 2021)
	Iopromide Iopamidol		
UV filters	Benzophenone-3 Octocrylene	These compounds may cause disruption of the hypothalamic–pituitary–gonadal system in organisms.	(Parida et al., 2021)

TABLE 9.2 (Continued)
The Different ECs and Their Ecological Impact

Class	ECs	Ecological Impact	References
Fire retardants	Tris(2-chloroethyl) Phosphate (TCEP) Tri-n-butyl phosphate (TNBP)	TCEP may cause cancer in humans and can affect animals' reproductive systems and pose a significant risk to fertility. TnBP was found to have a minor inhibitory effect on human plasma cholineste rase. Reports have shown that TnBP hinders the cell multiplication in microorganism.	(Parida et al., 2021)
Stimulant/ Illicit drugs	Caffeine Codeine Amphetamine	Caffeine causes anxiety and panic attacks in humans, and it has been linked to endometrial, hepatocellular, and colorectal cancer. Because of the increase in plasma concentrations, codeine has been shown to have toxic effects on exposed organisms. Amphetamine causes increased arousal or wakefulness, hyperactivity, anorexia, and perseverative movements	(Parida et al., 2021)
Plasticizers	Bisphenol-A	They have estrogenic and hormonal effects in rats and increase the risk of breast cancer in humans	(Parida et al., 2021)
Pesticides	Diazinon Malathion Mecoprop Diuron Mecoprop Triclosan 2,4-D Nonylphenol	Suspected animal or human carcinogen, mutagen, reproductive hazards like teratogenesis or other reproductive impairment, low acute toxicity leads to lower the blood cholinesterase activities in adult volunteers, causes structural changes in spleen and thymus and changes the number of blood lymphocytes and granulocytes in rats, and others.	(Debnath et al., 2019; Parida et al., 2021)
PFAS	Perfluorobutanoic acid Perfluorooctanoic acid Perfluorobutane sulfonate Perfluorohexane sulfonate Perfluorohexanoic acid Perfluorooctane sulfonate	PFAS have been found in the blood and urine of human beings. Over time, the PFAS in the system would increase due to bioaccumulation. There has been toxicological evidence that over prolonged bioaccumulation, the PFAS may affect the reproductive and immune systems of humans and animals.	(Du et al., 2014; US EPA, 2018)

TABLE 9.3

Statutory Guidelines Given by International Regulatory Bodies along with the PNEC Values of a Few ECs

Contaminants	Statutory Guidelines (ng/L)				References
	EU	US EPA	NHMRC	WHO	
Paracetamol			175000		(NHMRC, 2008)
Ibuprofen	11				(Korkaric et al., 2019)
Ciprofloxacin			250000		(NHMRC, 2008)
Erythromycin			17500		(NHMRC, 2008)
Trimethoprim	120000		70000		(NHMRC, 2008; Korkaric et al., 2019)
Norfloxacin			400000		(NHMRC, 2008)
Tetracycline			105000		(NHMRC, 2008)
Carbamazepine	2000		100000		(NHMRC, 2008; Korkaric et al., 2019)
Atenolol	150000				(Korkaric et al., 2019)
17 β -estradiol		175	175		(Carvalho et al., 2016; NHMRC, 2008)
Iohexol			720000		(NHMRC, 2008)
Iopromide			750000		(NHMRC, 2008)
Iopamidol			400000		(NHMRC, 2008)
Benzophenone-3		152000			(Careghini et al., 2015)
TCEP			1000		(NHMRC, 2008)
TnBP	100				(Carvalho et al., 2016)
Caffeine	87000		350		(Carvalho et al., 2016; NHMRC, 2008)
Codeine			50000		(NHMRC, 2008)
Bisphenol-A	240	77000	200000	100	(Careghini et al., 2015; Korkaric et al., 2019; NHMRC, 2008)
Diazinon	100		3000		(Korkaric et al., 2019)
Malathion	100	70000	900	900000	(Carvalho et al., 2016; NHMRC, 2008; WHO Guidelines, 2004)
Mecoprop	3600			10000	(Carvalho et al., 2016; Korkaric et al., 2019)

maximum concentration of a compound that can be ingested over a lifetime without giving rise to any adverse health effects. ADI values may be derived from the NOEAL (No-Observed-Effect Level) of the compounds. The NOAEL values may be found in the safety data sheets of the respective compounds (Parida et al., 2021). FOE is the number of days in a year the individual is getting exposed to a particular contaminant (Majumder et al., 2019; Parida et al., 2021). The DWEL also depends on the ability of the human body to absorb the contaminant and detoxify it.

9.6 CHAPTER SUMMARY

- ECs are present in sewage at very low concentrations (μg/L to pg/L), and have the potential to adversely affect the environment.
- Detection of ECs requires the use of sophisticated instruments, such as LC and GC, and a specific methodology is required to be followed to accurately analyze these compounds.

- Pharmaceutically active compounds, personal care products, surfactants, and PFAS are the commonly occurring ECs in the environment.
- The presence of carbon–fluoride bond in the PFAS makes them one of the most resistant ECs present in the environment. These chemicals do not degrade naturally in the aqueous environment, and research has shown that the half-lives of PFAS might be around 40–90 years.
- The primary factors, which may affect the concentration of the ECs after sampling, include adsorption, photodegradation, biodegradation, hydrolysis, and chemical abiotic transformations.
- Microorganisms in the aqueous sample make the ECs susceptible to biodegradation.
- The blanking of samples is essential to estimate if there is any contamination during sampling, transportation, and storage of the samples.
- Among the various extraction technologies available, the most commonly used techniques for extraction and pre-concentration of ECs are SPE, SPME, LPE, and UAE.
- HPLC, LC-MS, GC, and GC-MS have been commonly used to detect and quantify ECs.
- DWEL provides an approximate concentration of any contaminant below which it may not be harmful to human beings upon prolonged exposure. The calculation of DWEL takes considers the acceptable daily intake of the EC, frequency of exposure, bodyweight of individuals, amount of water intake/day, and gastrointestinal absorption rate.

9.7   CONCLUDING REMARKS

The ECs are ubiquitous in the aqueous environment and pose a significant threat to all the living organisms coming in contact with it. Due to the low concentration of these contaminants, they are difficult to detect and quantify. As a result, it is difficult to make standards pertaining to the ECs. However, ECs are required to be removed from the wastewater stream. Due to the inherent properties of the ECs, they are not easily removed by the conventional treatment plants. In this context, advance treatment technologies for the removal of the ECs are discussed in the following chapter.

REFERENCES

Afsa, S., Hamden, K., Lara Martin, P.A., Mansour, H. Ben, 2020. Occurrence of 40 pharmaceutically active compounds in hospital and urban wastewaters and their contribution to Mahdia coastal seawater contamination. *Environ. Sci. Pollut. Res.* 27, 1941–1955. https://doi.org/10.1007/s11356-019-06866-5

Agüera, A., Perez Estrada, L.A., Ferrer, I., Thurman, E.M., Malato, S., Fernandez-Alba, A.R., 2005. Application of time-of-flight mass spectrometry to the analysis of phototransformation products of diclofenac in water under natural sunlight. *J. Mass Spectrom.* 40, 908–915. https://doi.org/10.1002/jms.867

Akhbarizadeh, R., Dobaradaran, S., Schmidt, T.C., Nabipour, I., Spitz, J., 2020. Worldwide bottled water occurrence of emerging contaminants: A review of the recent scientific literature. *J. Hazard. Mater.* 392, 122271. https://doi.org/10.1016/j.jhazmat.2020.122271

Al Amin, M., Sobhani, Z., Liu, Y., Dharmaraja, R., Chadalavada, S., Naidu, R., Chalker, J.M., Fang, C., 2020. Recent advances in the analysis of per- and polyfluoroalkyl substances (PFAS) – A review. *Environ. Technol. Innov.* 19, 100879. https://doi.org/10.1016/j.eti.2020.100879

Alygizakis, N.A., Gago-Ferrero, P., Borova, V.L., Pavlidou, A., Hatzianestis, I., Thomaidis, N.S., 2016. Occurrence and spatial distribution of 158 pharmaceuticals, drugs of abuse and related metabolites in offshore seawater. *Sci. Total Environ.* 541, 1097–1105. https://doi.org/10.1016/j.scitotenv.2015.09.145

Baker, D.R., Kasprzyk-Hordern, B., 2011a. Critical evaluation of methodology commonly used in sample collection, storage and preparation for the analysis of pharmaceuticals and illicit drugs in surface water and wastewater by solid phase extraction and liquid chromatography-mass spectrometry. *J. Chromatogr. A* 1218, 8036–8059. https://doi.org/10.1016/j.chroma.2011.09.012

Baker, D.R., Kasprzyk-Hordern, B., 2011b. Multi-residue determination of the sorption of illicit drugs and pharmaceuticals to wastewater suspended particulate matter using pressurised liquid extraction, solid phase extraction and liquid chromatography coupled with tandem mass spectrometry. *J. Chromatogr. A* 1218, 7901–7913. https://doi.org/10.1016/j.chroma.2011.08.092

Balakrishna, K., Rath, A., Praveenkumarreddy, Y., Guruge, K.S., Subedi, B., 2017. A review of the occurrence of pharmaceuticals and personal care products in Indian water bodies. *Ecotoxicol. Environ. Saf.* 137, 113–120. https://doi.org/10.1016/J.ECOENV.2016.11.014

Barbara Ambrosetti, Luigi Campanella, Raffaella Palmisano, 2015. Degradation of antibiotics in aqueous solution by photocatalytic process: Comparing the efficiency in the use of ZnO or TiO_2. *J. Environ. Sci. Eng. A* 4, 273–281. https://doi.org/10.17265/2162-5298/2015.06.001

Baronti, C., Curini, R., D'Ascenzo, G., Di Corcia, A., Gentili, A., Samperi, R., 2000. Monitoring natural and synthetic estrogens at activated sludge sewage treatment plants and in a receiving river water. *Environ. Sci. Technol.* 34, 5059–5066. https://doi.org/10.1021/es001359q

Bayen, S., Estrada, E.S., Juhel, G., Kit, L.W., Kelly, B.C., 2016. Pharmaceutically active compounds and endocrine disrupting chemicals in water, sediments and mollusks in mangrove ecosystems from Singapore. *Mar. Pollut. Bull.* 109, 716–722. https://doi.org/10.1016/j.marpolbul.2016.06.105

Białk-Bielińska, A., Kumirska, J., Borecka, M., Caban, M., Paszkiewicz, M., Pazdro, K., Stepnowski, P., 2016. Selected analytical challenges in the determination of pharmaceuticals in drinking/marine waters and soil/sediment samples. *J. Pharm. Biomed. Anal.* 121, 271–296. https://doi.org/10.1016/j.jpba.2016.01.016

Biel-Maeso, M., Baena-Nogueras, R.M., Corada-Fernández, C., Lara-Martín, P.A., 2018. Occurrence, distribution and environmental risk of pharmaceutically active compounds (PhACs) in coastal and ocean waters from the Gulf of Cadiz (SW Spain). *Sci. Total Environ.* 612, 649–659. https://doi.org/10.1016/j.scitotenv.2017.08.279

Boleda, M.R., Galceran, M.T., Ventura, F., 2007. Trace determination of cannabinoids and opiates in wastewater and surface waters by ultraperformance liquid chromatography-tandem mass spectrometry. *J. Chromatogr. A* 1175, 38–48. https://doi.org/10.1016/j.chroma.2007.10.029

Boulard, L., Dierkes, G., Schlüsener, M.P., Wick, A., Koschorreck, J., Ternes, T.A., 2020. Spatial distribution and temporal trends of pharmaceuticals sorbed to suspended particulate matter of German rivers. *Water Res.* 171, 115366. https://doi.org/10.1016/j.watres.2019.115366

Carballa, M., Manterola, G., Larrea, L., Ternes, T., Omil, F., Lema, J.M., 2007. Influence of ozone pre-treatment on sludge anaerobic digestion: Removal of pharmaceutical and personal care products. *Chemosphere* 67, 1144–1152. https://doi.org/10.1016/j.chem osphere.2006.10.004

Careghini, A., Mastorgio, A.F., Saponaro, S., Sezenna, E., 2015. Bisphenol A, nonylphenols, benzophenones, and benzotriazoles in soils, groundwater, surface water, sediments, and food: A review. *Environ. Sci. Pollut. Res.* 22, 5711–5741. https://doi.org/10.1007/s11 356-014-3974-5

Carvalho, R.N., Marinov, D., Loos, R., Napierska, D., Chirico, N., Lettieri, T., 2016. European Commission – Monitoring Data Report [WWW Document]. https://circabc.europa. eu/sd/a/7fe29322-946a-4ead-b3b9-e3b156d0c318/Monitoring-basedExercise Report_ FINAL DRAFT_25nov2016.pdf (accessed 3.30.21).

Castiglioni, S., Zuccato, E., Crisci, E., Chiabrando, C., Fanelli, R., Bagnati, R., 2006. Identification and measurement of illicit drugs and their metabolites in urban wastewater by liquid chromatography-tandem mass spectrometry. *Anal. Chem.* 78, 8421–8429. https://doi.org/10.1021/ac061095b

Chiaia, A.C., Banta-Green, C., Field, J., 2008. Eliminating solid phase extraction with large-volume injection LC/MS/MS: Analysis of illicit and legal drugs and human urine indicators in US wastewaters. *Environ. Sci. Technol.* 42, 8841–8848. https://doi.org/ 10.1021/es802309v

Daneshvar, A., Aboulfadl, K., Viglino, L., Broséus, R., Sauvé, S., Madoux-Humery, A.S., Weyhenmeyer, G.A., Prévost, M., 2012. Evaluating pharmaceuticals and caffeine as indicators of fecal contamination in drinking water sources of the Greater Montreal region. *Chemosphere* 88, 131–139. https://doi.org/10.1016/j.chemosphere.2012.03.016

De Sena, R.F., Tambosi, J.L., Moreira, R.F.P.M., José, H.J., Gebhardt, W., Schröder, H.F., 2010. Evaluation of sample processing methods for the polar contaminant analysis of sewage sludge using liquid chromatography – Mass spectrometry (LC/MS). *Quim. Nova* 33, 1194–1198. https://doi.org/10.1590/s0100-40422010000500034

Debnath, D., Gupta, A.K., Ghosal, P.S., 2019. Recent advances in the development of tailored functional materials for the treatment of pesticides in aqueous media: A review. *J. Ind. Eng. Chem.* 70, 51–69. https://doi.org/10.1016/J.JIEC.2018.10.014

Domingo, J.L., Nadal, M., 2019. Human exposure to per- and polyfluoroalkyl substances (PFAS) through drinking water: A review of the recent scientific literature. *Environ. Res.* 177, 108648. https://doi.org/10.1016/j.envres.2019.108648

Du, Z., Deng, S., Bei, Y., Huang, Q., Wang, B., Huang, J., Yu, G., 2014. Adsorption behavior and mechanism of perfluorinated compounds on various adsorbents: A review. *J. Hazard. Mater.* 274, 443–454. https://doi.org/10.1016/j.jhazmat.2014.04.038

Farré, M., Petrovic, M., Gros, M., Kosjek, T., Martinez, E., Heath, E., Osvald, P., Loos, R., Le Menach, K., Budzinski, H., De Alencastro, F., Müller, J., Knepper, T., Fink, G., Ternes, T.A., Zuccato, E., Kormali, P., Gans, O., Rodil, R., Quintana, J.B., Pastori, F., Gentili, A., Barceló, D., 2008. First interlaboratory exercise on non-steroidal anti-inflammatory drugs analysis in environmental samples. *Talanta* 76, 580–590. https://doi.org/10.1016/ j.talanta.2008.03.055

Ferraro, M.C.F., Castellano, P.M., Kaufman, T.S., 2003. Chemometrics-assisted simultaneous determination of atenolol and chlorthalidone in synthetic binary mixtures and pharmaceutical dosage forms. *Anal. Bioanal. Chem.* 377, 1159–1164. https://doi.org/10.1007/ s00216-003-2185-6

Grover, D.P., Zhou, J.L., Frickers, P.E., Readman, J.W., 2011. Improved removal of estrogenic and pharmaceutical compounds in sewage effluent by full scale granular activated carbon: Impact on receiving river water. *J. Hazard. Mater.* 185, 1005–1011. https://doi.org/10.1016/j.jhazmat.2010.10.005

Guo, L., Lee, H.K., 2012. One step solvent bar microextraction and derivatization followed by gas chromatography-mass spectrometry for the determination of pharmaceutically active compounds in drain water samples. *J. Chromatogr. A* 1235, 26–33. https://doi.org/10.1016/j.chroma.2012.02.068

Gupta, B., Gupta, A.K., Ghosal, P.S., Tiwary, C.S., 2020. Photo-induced degradation of bio-toxic Ciprofloxacin using the porous 3D hybrid architecture of an atomically thin sulfur-doped g-C₃N₄/ZnO nanosheet. *Environ. Res.* 183, 109154. https://doi.org/10.1016/j.envres.2020.109154

Gupta, B., Gupta, A.K., Tiwary, C.S., Ghosal, P.S., 2021. A multivariate modeling and experimental realization of photocatalytic system of engineered S – C₃N₄/ZnO hybrid for ciprofloxacin removal: Influencing factors and degradation pathways. *Environ. Res.* 196, 110390. https://doi.org/10.1016/j.envres.2020.110390

Guzel, E.Y., Cevik, F., Daglioglu, N., 2019. Determination of pharmaceutical active compounds in Ceyhan River, Turkey: Seasonal, spatial variations and environmental risk assessment. *Hum. Ecol. Risk Assess.* 25, 1980–1995. https://doi.org/10.1080/10807039.2018.1479631

Havens, S.M., Hedman, C.J., Hemming, J.D.C., Mieritz, M.G., Shafer, M.M., Schauer, J.J., 2010. Stability, preservation, and quantification of hormones and estrogenic and androgenic activities in surface water runoff. *Environ. Toxicol. Chem.* 29, 2481–2490. https://doi.org/10.1002/etc.307

Hernández, F., Ibáñez, M., Botero-Coy, A.M., Bade, R., Bustos-López, M.C., Rincón, J., Moncayo, A., Bijlsma, L., 2015. LC-QTOF MS screening of more than 1,000 licit and illicit drugs and their metabolites in wastewater and surface waters from the area of Bogotá, Colombia. *Anal. Bioanal. Chem.* 407, 6405–6416. https://doi.org/10.1007/s00216-015-8796-x

Huerta-Fontela, M., Galceran, M.T., Ventura, F., 2007. Ultraperformance liquid chromatography tandem mass spectrometry analysis of stimulatory drugs of abuse in wastewater and surface waters. *Anal. Chem.* 79, 3821–3829. https://doi.org/10.1021/ac062370x

ISO-5667-3, 2018. Water Quality – Sampling – Part 3: Preservation and Handling of Water Samples.

Jiang, M., Wang, L., Ji, R., 2010. Biotic and abiotic degradation of four cephalosporin antibiotics in a lake surface water and sediment. *Chemosphere* 80, 1399–1405. https://doi.org/10.1016/j.chemosphere.2010.05.048

Kasprzyk-Hordern, B., Kondakal, V.V.R., Baker, D.R., 2010. Enantiomeric analysis of drugs of abuse in wastewater by chiral liquid chromatography coupled with tandem mass spectrometry. *J. Chromatogr. A* 1217, 4575–4586. https://doi.org/10.1016/j.chroma.2010.04.073

Khan, N.A., Khan, S.U., Ahmed, S., Farooqi, I.H., Yousefi, M., Mohammadi, A.A., Changani, F., 2020. Recent trends in disposal and treatment technologies of emerging-pollutants – A critical review. *TrAC – Trends Anal. Chem.* 122, 115744. https://doi.org/10.1016/j.trac.2019.115744

Khoshayand, M.R., Abdollahi, H., Shariatpanahi, M., Saadatfard, A., Mohammadi, A., 2008. Simultaneous spectrophotometric determination of paracetamol, ibuprofen and caffeine in pharmaceuticals by chemometric methods. *Spectrochim. Acta – Part A Mol. Biomol. Spectrosc.* 70, 491–499. https://doi.org/10.1016/j.saa.2007.07.033

Kim, H., Hong, Y., Park, J.E., Sharma, V.K., Cho, S. Il, 2013. Sulfonamides and tetracyclines in livestock wastewater. *Chemosphere* 91, 888–894. https://doi.org/10.1016/j.chemosphere.2013.02.027

Korkaric, M., Junghans, M., Pasanen-Kase, R., Werner, I., 2019. Revising environmental quality standards: Lessons learned. *Integr. Environ. Assess. Manag.* 15, 948–960. https://doi.org/10.1002/ieam.4192

Kostopoulou, M., Nikolaou, A., 2008. Analytical problems and the need for sample preparation in the determination of pharmaceuticals and their metabolites in aqueous environmental matrices. *TrAC – Trends Anal. Chem.* 27, 1023–1035. https://doi.org/10.1016/j.trac.2008.09.011

Kucharzyk, K.H., Darlington, R., Benotti, M., Deeb, R., Hawley, E., 2017. Novel treatment technologies for PFAS compounds: A critical review. *J. Environ. Manage.* 204, 757–764. https://doi.org/10.1016/j.jenvman.2017.08.016

Langford, K.H., Thomas, K.V., 2009. Determination of pharmaceutical compounds in hospital effluents and their contribution to wastewater treatment works. *Environ. Int.* 35, 766–770. https://doi.org/10.1016/j.envint.2009.02.007

Laws, B.V., Dickenson, E.R.V., Johnson, T.A., Snyder, S.A., Drewes, J.E., 2011. Attenuation of contaminants of emerging concern during surface-spreading aquifer recharge. *Sci. Total Environ.* 409, 1087–1094. https://doi.org/10.1016/j.scitotenv.2010.11.021

Lin, H., Chen, L., Li, H., Luo, Z., Lu, J., Yang, Z., 2018. Pharmaceutically active compounds in the Xiangjiang River, China: Distribution pattern, source apportionment, and risk assessment. *Sci. Total Environ.* 636, 975–984. https://doi.org/10.1016/j.scitotenv.2018.04.267

Lorphensri, O., Sabatini, D.A., Kibbey, T.C.G., Osathaphan, K., Saiwan, C., 2007. Sorption and transport of acetaminophen, 17α-ethynyl estradiol, nalidixic acid with low organic content aquifer sand. *Water Res.* 41, 2180–2188. https://doi.org/10.1016/j.watres.2007.01.057

Luque-García, J.L., Luque De Castro, M.D., 2003. Ultrasound: A powerful tool for leaching. *TrAC – Trends Anal. Chem.* 22, 41–47. https://doi.org/10.1016/S0165-9936(03)00102-X

Ma, R., Wang, B., Yin, L., Zhang, Y., Deng, S., Huang, J., Wang, Y., Yu, G., 2017. Characterization of pharmaceutically active compounds in Beijing, China: Occurrence pattern, spatiotemporal distribution and its environmental implication. *J. Hazard. Mater.* 323, 147–155. https://doi.org/10.1016/j.jhazmat.2016.05.030

Majumder, A., Gupta, A.K., Ghosal, P.S., Varma, M., 2021. A review on hospital wastewater treatment: A special emphasis on occurrence and removal of pharmaceutically active compounds, resistant microorganisms, and SARS-CoV-2. *J. Environ. Chem. Eng.* 9, 104812. https://doi.org/10.1016/j.jece.2020.104812

Majumder, A., Gupta, B., Gupta, A.K., 2019. Pharmaceutically active compounds in aqueous environment: A status, toxicity and insights of remediation. *Environ. Res.* 176, 108542. https://doi.org/10.1016/j.envres.2019.108542

Managaki, S., Murata, A., Takada, H., Bui, C.T., Chiem, N.H., 2007. Distribution of macrolides, sulfonamides, and trimethoprim in tropical waters: Ubiquitous occurrence of veterinary antibiotics in the Mekong Delta. *Environ. Sci. Technol.* 41, 8004–8010. https://doi.org/10.1021/es0709021

Martín, J., Camacho-Muñoz, D., Santos, J.L., Aparicio, I., Alonso, E., 2012. Occurrence of pharmaceutical compounds in wastewater and sludge from wastewater treatment plants: Removal and ecotoxicological impact of wastewater discharges and sludge disposal. *J. Hazard. Mater.* 239–240, 40–47. https://doi.org/10.1016/j.jhazmat.2012.04.068

Matamoros, V., Bayona, J.M., 2006. Elimination of pharmaceuticals and personal care products in subsurface flow constructed wetlands. *Environ. Sci. Technol.* 40, 5811–5816. https://doi.org/10.1021/es0607741

McDowall, R.D., Pawliszyn, J., Boyaci, E., 2019. Sample handling I sample preservation, in: *Encyclopedia of Analytical Science*. Elsevier, pp. 133–142. https://doi.org/10.1016/B978-0-12-409547-2.00483-2

Meng, Y., Liu, W., Liu, X., Zhang, J., Peng, M., Zhang, T., 2021. A review on analytical methods for pharmaceutical and personal care products and their transformation products. *J. Environ. Sci. (China).* 101, 260–281. https://doi.org/10.1016/j.jes.2020.08.025

Mohapatra, D.P., Brar, S.K., Tyagi, R.D., Picard, P., Surampalli, R.Y., 2014. Analysis and advanced oxidation treatment of a persistent pharmaceutical compound in wastewater and wastewater sludge-carbamazepine. *Sci. Total Environ.* 470–471, 58–75. https://doi.org/10.1016/j.scitotenv.2013.09.034

Mohapatra, D.P., Brar, S.K., Tyagi, R.D., Picard, P., Surampalli, R.Y., 2012. Carbamazepine in municipal wastewater and wastewater sludge: Ultrafast quantification by laser diode thermal desorption-atmospheric pressure chemical ionization coupled with tandem mass spectrometry. *Talanta* 99, 247–255. https://doi.org/10.1016/j.talanta.2012.05.047

Moldoveanu, S., David, V., 2015. Comments on sample preparation in chromatography for different types of materials, in: *Modern Sample Preparation for Chromatography*. Elsevier, pp. 411–446. https://doi.org/10.1016/b978-0-444-54319-6.00012-8

Mompelat, S., Jaffrezic, A., Jardé, E., LeBot, B., 2013. Storage of natural water samples and preservation techniques for pharmaceutical quantification. *Talanta.* 109, 31–45. https://doi.org/10.1016/j.talanta.2013.01.042

Mompelat, S., Le Bot, B., Thomas, O., 2009. Occurrence and fate of pharmaceutical products and by-products, from resource to drinking water. *Environ. Int.* 35, 803–814. https://doi.org/10.1016/j.envint.2008.10.008

NHMRC, 2008. Australian Guidelines for Water Recycling: Augmentation of Drinking Water Supplies. https://doi.org/waterquality.gov.au/sites/default/files/documents/water-recycling-guidelines-augmentation-drinking-22.pdf

Oliveira, T.S., Murphy, M., Mendola, N., Wong, V., Carlson, D., Waring, L., 2015. Characterization of pharmaceuticals and personal care products in hospital effluent and waste water influent/effluent by direct-injection LC-MS-MS. *Sci. Total Environ.* 518–519, 459–478. https://doi.org/10.1016/J.SCITOTENV.2015.02.104

Omar, T.F.T., Ahmad, A., Aris, A.Z., Yusoff, F.M., 2016. Endocrine disrupting compounds (EDCs) in environmental matrices: Review of analytical strategies for pharmaceuticals, estrogenic hormones, and alkylphenol compounds. *TrAC – Trends Anal. Chem.* 85, 241–259. https://doi.org/10.1016/j.trac.2016.08.004

Parida, V.K., Saidulu, D., Majumder, A., Srivastava, A., Gupta, B., Gupta, A.K., 2021. Emerging contaminants in wastewater: A critical review on occurrence, existing legislations, risk assessment, and sustainable treatment alternatives. *J. Environ. Chem. Eng.* 9, 105966. https://doi.org/10.1016/J.JECE.2021.105966

Park, H.R., Kim, T.H., Bark, K.M., 2002. Physicochemical properties of quinolone antibiotics in various environments. *Eur. J. Med. Chem.* 37(6), 443–460.https://doi.org/10.1016/S0223-5234(02)01361-2

Pawliszyn, J., 2012. *Comprehensive Sampling and Sample Preparation*. Elsevier. https://doi.org/10.1016/C2009-1-60918-7

Payán, M.R., López, M.Á.B., Fernández-Torres, R., Mochón, M.C., Ariza, J.L.G., 2010. Application of hollow fiber-based liquid-phase microextraction (HF-LPME) for the determination of acidic pharmaceuticals in wastewaters. *Talanta* 82, 854–858. https://doi.org/10.1016/j.talanta.2010.05.022

Prieto, A., Araujo, L., Navalon, A., Vilchez, J., 2009. Comparison of solid-phase extraction and solid-phase microextraction using octadecylsilane phase for the determination of pesticides in water samples. *Curr. Anal. Chem.* 5, 219–224. https://doi.org/10.2174/157 341109788680309

Radke, M., Ulrich, H., Wurm, C., Kunkel, U., 2010. Dynamics and attenuation of acidic pharmaceuticals along a river stretch. *Environ. Sci. Technol.* 44, 2968–2974. https://doi. org/10.1021/es903091z

Ramos, L., 2012. Critical overview of selected contemporary sample preparation techniques. *J. Chromatogr. A* 1221, 84–98. https://doi.org/10.1016/j.chroma.2011.11.011

Rees, V., 2020. How is liquid chromatography used in the pharmaceutical industry? *Eur. Pharm. Rev.* www.europeanpharmaceuticalreview.com/article/116263/how-is-liquid-chromatography-used-in-the-pharmaceutical-industry/ (accessed 11.12.20).

Richardson, S.D., Ternes, T.A., 2018. Water analysis: Emerging contaminants and current issues. *Anal. Chem.* 90(1), 398–428. https://doi.org/10.1021/acs.analchem.7b04577

Rodriguez, K.L., Hwang, J.H., Esfahani, A.R., Sadmani, A.H.M.A., Lee, W.H., 2020. Recent developments of PFAS-detecting sensors and future direction: A review. *Micromachines.* 11(7), 667. https://doi.org/10.3390/mi11070667

Roig, B., Brogat, M., Mompelat, S., Leveque, J., Cadiere, A., Thomas, O., 2012. Inter-laboratory exercise on antibiotic drugs analysis in aqueous samples. *Talanta* 98, 157–165. https://doi.org/10.1016/j.talanta.2012.06.064

Saidulu, D., Gupta, B., Gupta, A.K., Ghosal, P.S., 2021. A review on occurrences, eco-toxic effects, and remediation of emerging contaminants from wastewater: Special emphasis on biological treatment based hybrid systems. *J. Environ. Chem. Eng.* 9, 105282. https:// doi.org/10.1016/j.jece.2021.105282

Santos, J.L., Aparicio, I., Alonso, E., Callejón, M., 2005. Simultaneous determination of pharmaceutically active compounds in wastewater samples by solid phase extraction and high-performance liquid chromatography with diode array and fluorescence detectors. *Anal. Chim. Acta* 550, 116–122. https://doi.org/10.1016/j.aca.2005.06.064

Santos, J.L., Aparicio, I., Callejón, M., Alonso, E., 2009. Occurrence of pharmaceutically active compounds during 1-year period in wastewaters from four wastewater treatment plants in Seville (Spain). *J. Hazard. Mater.* 164, 1509–1516. https://doi.org/10.1016/ j.jhazmat.2008.09.073

Shanmugam, G., Sampath, S., Selvaraj, K.K., Larsson, D.G.J., Ramaswamy, B.R., 2014. Non-steroidal anti-inflammatory drugs in Indian rivers. *Environ. Sci. Pollut. Res.* 21, 921–931. https://doi.org/10.1007/s11356-013-1957-6

Siddiqui, M.R., AlOthman, Z.A., Rahman, N., 2017. Analytical techniques in pharmaceut-ical analysis: A review. *Arab. J. Chem.* 10, S1409–S1421. https://doi.org/10.1016/j.ara bjc.2013.04.016

Sifuna, F.W., Orata, F., Okello, V., Jemutai-Kimosop, S., 2016. Comparative studies in elec-trochemical degradation of sulfamethoxazole and diclofenac in water by using various electrodes and phosphate and sulfate supporting electrolytes. *J. Environ. Sci. Heal. Part A* 51, 954–961. https://doi.org/10.1080/10934529.2016.1191814

Sørensen, L.K., Elbæk, T.H., 2004. Simultaneous determination of trimethoprim, sulfadia-zine, florfenicol and oxolinic acid in surface water by liquid chromatography tandem mass spectrometry. *Chromatographia* 60, 287–291. https://doi.org/10.1365/s10 337-004-0383-9

Tamtam, F., Mercier, F., Le Bot, B., Eurin, J., Tuc Dinh, Q., Clément, M., Chevreuil, M., 2008. Occurrence and fate of antibiotics in the Seine River in various hydrological conditions. *Sci. Total Environ.* 393, 84–95. https://doi.org/10.1016/j.scitotenv.2007.12.009

Togola, A., Budzinski, H., 2007. Analytical development for analysis of pharmaceuticals in water samples by SPE and GC-MS. *Anal. Bioanal. Chem.* 388, 627–635. https://doi.org/ 10.1007/s00216-007-1251-x

Tran, N.H., Gin, K.Y.H., 2017. Occurrence and removal of pharmaceuticals, hormones, personal care products, and endocrine disrupters in a full-scale water reclamation plant. *Sci. Total Environ.* 599–600, 1503–1516. https://doi.org/10.1016/j.scitotenv.2017.05.097

Tran, N.H., Li, J., Hu, J., Ong, S.L., 2014. Occurrence and suitability of pharmaceuticals and personal care products as molecular markers for raw wastewater contamination in surface water and groundwater. *Environ. Sci. Pollut. Res.* 21, 4727–4740. https://doi.org/10.1007/s11356-013-2428-9

Tran, N.H., Reinhard, M., Yew-Hoong Gin, K., 2018. Occurrence and fate of emerging contaminants in municipal wastewater treatment plants from different geographical regions-a review. *Water Res.* 133, 182–207. https://doi.org/10.1016/j.watres.2017.12.029

Turiel, E., Martín-Esteban, A., Bordin, G., Rodríguez, A.R., 2004. Stability of fluoroquinolone antibiotics in river water samples and in octadecyl silica solid-phase extraction cartridges. *Anal. Bioanal. Chem.* 380, 123–128. https://doi.org/10.1007/s00216-004-2730-y

US EPA, 2018. Understanding PFAS in the Environment [WWW Document]. www.epa.gov/sciencematters/understanding-pfas-environment (accessed 11.11.21).

Vanderford, B.J., Mawhinney, D.B., Trenholm, R.A., Zeigler-Holady, J.C., Snyder, S.A., 2011. Assessment of sample preservation techniques for pharmaceuticals, personal care products, and steroids in surface and drinking water. *Anal. Bioanal. Chem.* 399, 2227–2234. https://doi.org/10.1007/s00216-010-4608-5

Vanderford, B.J., Pearson, R.A., Rexing, D.J., Snyder, S.A., 2003. Analysis of endocrine disruptors, pharmaceuticals, and personal care products in water using liquid chromatography/tandem mass spectrometry. *Anal. Chem.* 75, 6265–6274. https://doi.org/10.1021/ac034210g

Verlicchi, P., Galletti, A., Petrovic, M., Barcelo, D., 2010. Hospital effluents as a source of emerging pollutants: An overview of micropollutants and sustainable treatment options. *J. Hydrol.* 389, 416–428. https://doi.org/10.1016/J.JHYDROL.2010.06.005

Waters Corporation, 2020. Waters Corporation: The Science of What's Possible. www.waters.com/nextgen/us/en.html (accessed 11.1.20).

WHO Guidelines, 2004. WHO guidelines. www.who.int/water_sanitation_health/dwq/chemicals/malathion.pdf (accessed 3.30.21).

Xiao, F., 2017. Emerging poly- and perfluoroalkyl substances in the aquatic environment: A review of current literature. *Water Res.* 124, 482–495. https://doi.org/10.1016/j.watres.2017.07.024

Ying, G.G., Kookana, R.S., Kolpin, D.W., 2009. Occurrence and removal of pharmaceutically active compounds in sewage treatment plants with different technologies. *J. Environ. Monit.* 11, 1498–1505. https://doi.org/10.1039/b904548a

Zhang, Y., Wang, B., Cagnetta, G., Duan, L., Yang, J., Deng, S., Huang, J., Wang, Y., Yu, G., 2018. Typical pharmaceuticals in major WWTPs in Beijing, China: Occurrence, load pattern and calculation reliability. *Water Res.* 140, 291–300. https://doi.org/10.1016/j.watres.2018.04.056

Zhou, Z., Jiang, J.Q., 2012. Detection of ibuprofen and ciprofloxacin by solid-phase extraction and UV/Vis spectroscopy. *J. Appl. Spectrosc.* 79, 459–464. https://doi.org/10.1007/s10812-012-9623-1

10 Emerging Contaminants in the Aqueous Environment

Treatment Technologies for Emerging Contaminants and Its Application

CHAPTER OBJECTIVES

The chapter seeks to provide a broad overview of the different remediation techniques emerging contaminants (ECs) in domestic wastewater. The performance of different conventional and advanced wastewater treatment processes are discussed. Various case studies of full-scale treatment plants and their performance in terms of removing emerging contaminants are also discussed.

10.1 INTRODUCTION

Emerging contaminants (ECs) are highly mobile in the aqueous environment and persistent in nature (Majumder et al., 2019; Parida et al., 2021). They derive their persistent nature largely due to their complex molecular structure. Other physico-chemical properties, such as dissociation constants values (pK_a) and octanol-water partition (log k_{ow}) values of the ECs, also give us a fair idea about their stability in the aqueous environment (Table 10.1) (Majumder et al., 2019; Parida et al., 2021). Usually, ECs have low k_{ow} values, which indicates that they are hydrophilic in nature (Majumder et al., 2019; Parida et al., 2021). The hydrophilicity of such ECs prevents them from easily being separated from the aqueous solution. The sorption affinity of the ECs may be estimated from the solid-water distribution coefficient (k_d) values as provided in Table 10.1. The pK_a values are responsible for imparting surface charges to these compounds (Majumder et al., 2019; Parida et al., 2021). Often the charged compounds are difficult to settle down because of the intermolecular repulsion. All these physicochemical properties severely hinder the performance of conventional primary treatment processes from removing them. It has been observed that the microorganism-based treatment systems have proven to be effective for the removal of some ECs based on degradation and phytoremediation mechanisms (Ahmed et al., 2017). Some of the enzymes produced by microorganisms are responsible

DOI: 10.1201/9781003364450-10

TABLE 10.1
The Physicochemical Properties of Different Emerging Contaminants

Class	ECs	pK_a	Log k_{ow}	K_d (L/Kg MLSS)	K_{bio} (L/g MLSS-d)	References
Antibiotics	Azithromycin	8.5	4.02	34–2156	0.24–0.75	(Majumder et al., 2019; Parida et al., 2021; Saidulu et al., 2021)
	Tetracycline	3.3	–1.37	995–6386	0.44	
	Erythromycin	8.9	3.06	74–309	0.125–6	
	Trimethoprim	7.12, 17.33	1.26, 0.91	119–427	0.291–5.04	
	Norfloxacin	6.34,8.75	0.46	–	–	
Analgesics	Diclofenac	4.15	4.52	16–723	0.619–1.38	(Majumder et al., 2019; Parida et al., 2021; Saidulu et al., 2021)
	Naproxen	4.15	3.18	13–30	0.005–1.9	
	Ketoprofen	4.45	4.24	16–226	0.24–3.36	
	Paracetamol	9.5, 9.38	0.46	1.5–1160	58–240	
	Ibuprofen	4.91, 4.9	3.5, 3.97	6–102.8	1.31–38.7	
β-blockers	Atenolol	9.6	0.16	5.9–95	0.432–5.28	(Majumder et al., 2019; Parida et al., 2021; Saidulu et al., 2021)
	Propranolol	9.52	–0.45	460–1120	0.098–18.24	
Endocrine-disrupting compounds	17 β – estradiol	10.46 ± 0.03	4.01	533–771	–	(Majumder et al., 2019; Parida et al., 2021; Saidulu et al., 2021)
	Bisphenol-A	9.6	3.32	314–505	0.48–12.2	
	Estrone-1	10.34	3.13	645	< 200	
Artificial sweeteners	Saccharin	1.31	0.91	4.1–678	0.1551–0.556	(Parida et al., 2021; Saidulu et al., 2021)
	Sucralose	4.2	–1	5.1–765	0.0139–0.05	
	Acesulfame	2, 5.67	–1.33	5.1–765	0.0227–0.060	
X-ray contrast media	Iohexol	11.73	–3.05	–	0.0910–2.4	(Parida et al., 2021; Saidulu et al., 2021)
	Iopromide	–	–2.05	14	0.0197–2.5	
	Iopamidol	4.15, 10.7	1.62, –2.42	<10	0.111–2.664	

Category	Compound					References
UV filters	Benzophenone-3	7.1	3.79	1200–1500	–	(Parida et al., 2021; Saidulu et al., 2021)
	Octocrylene	14	6.9	14,000	–	(Parida et al., 2021; Saidulu et al., 2021)
Fire retardants	TCEP	–	1.78	<30–162	–	
	TnBP	–	4	–	–	
Stimulant/Illicit drugs	Caffeine	10.4, –0.92	–0.07, –0.13	<30–140	0.48–156.24	(Parida et al., 2021; Saidulu et al., 2021)
	Codeine	13.78	1.2	13–31	4.2–5.4	
	Amphetamine	10.13	1.76	407.9–2158.2	–	
Plasticizers	Bisphenol-A	9.6	3.32	314–505	0.24–16.56	(Parida et al., 2021; Saidulu et al., 2021)
Pesticides	Diazinon	2.6	3.81	–	–	(Debnath et al., 2019; Parida et al., 2021; Saidulu et al., 2021)
	Malathion	–	2.36	–	–	
	Mecoprop	3.78	3.13	–	–	
	Diuron	13.18	2.68	–	–	
	Mecoprop	3.1–3.78	3.13	–	–	
	Triclosan	7.9	4.76	3610	0.024–14.73	
	2,4-D	2.73	2.81	–	–	
	Nonylphenol	–5.4, 10.31	5.76	–	0.7–13.3	

for the biodegradation of ECs. Algal ponds with microalgae (phytoplankton) based wastewater treatment have proven very effective. However, the ECs are toxic in nature and their ability to inhibit the growth of microorganisms makes them less biodegradable. The degradability of the compound can be estimated from the values of normalized biomass rate constant, k_{bio} (Table 10.1).

Over the years, the performance of different treatment systems (physical, chemical, biological, and hybrid) have been assessed in terms of their ability to remove ECs. The following sections discuss the different treatment systems that have been used to remove ECs (Figure 10.1). A schematic demonstrating the different treatment units used for removing ECs have been presented in Figure 10.1.

10.2   BIOLOGICAL PROCESSES

The two primary mechanisms involved in the removal of ECs in biological processes are biodegradation and sorption. During biodegradation, the desired removal of ECs is observed when the ECs are neither toxic nor inhibit the microbial growth. In biological systems, ECs can be biotransformed by microbial groups via metabolic or cometabolic pathways. The microorganisms use ECs as their primary carbon source in the metabolic pathway. Heterotrophic microorganisms have been found to be more effective at degrading ECs (Saidulu et al., 2021). However, during cometabolic degradation, an additional growth substrate is required for transforming the non-growth substrate. Although cometabolic degradation may occur in both aerobic and anaerobic conditions, better biodegradation of ECs was observed under aerobic conditions (Saidulu et al., 2021). This may be due to the presence of greater microbial biodiversity. The steps involved in the biological degradation of ECs have been provided in Figure 10.2.

The other mechanism involved in the removal of ECs in biological process is adsorption. The different mechanisms by which ECs get adsorbed onto the sludge have been provided in Figure 10.3. On the other hand, sorption largely depends on the physicochemical properties of the ECs and the sludge produced in the biological processes. Hydrophobic contaminants are more readily adsorbed onto the sludge. Pharmaceutically active compounds are more likely to be adsorbed by electrostatic attraction and hydrogen bond interaction between the adsorbent and the adsorbate. On the other hand, endocrine-disrupting compounds mostly rely on the hydrophobic interactions with the sludge. Personal care products get adsorbed onto the sludge particles by inner-sphere complexation and π–π electron donor–acceptor reaction (Saidulu et al., 2021). Researchers have found that pesticides get adsorbed via pore diffusion (Saidulu et al., 2021).

10.2.1   ACTIVATED SLUDGE PROCESS

ECs usually have a low biodegradation constant (K_{bio}), making them persistent in the environment and are difficult to remove completely through biodegradation processes. Many researchers have observed a low removal efficiency in conventional STPs using the activated sludge processes (ASP), suggesting further treatment is necessary for

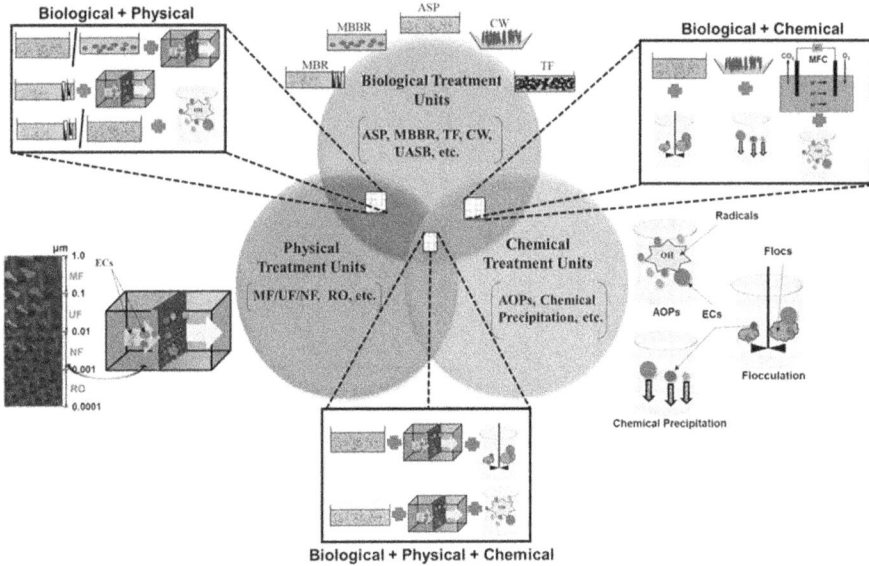

FIGURE 10.1 Different treatment systems used in the removal of emerging contaminants. (Reprinted with permission from Saidulu et al., 2021. Copyright 2023 Elsevier).

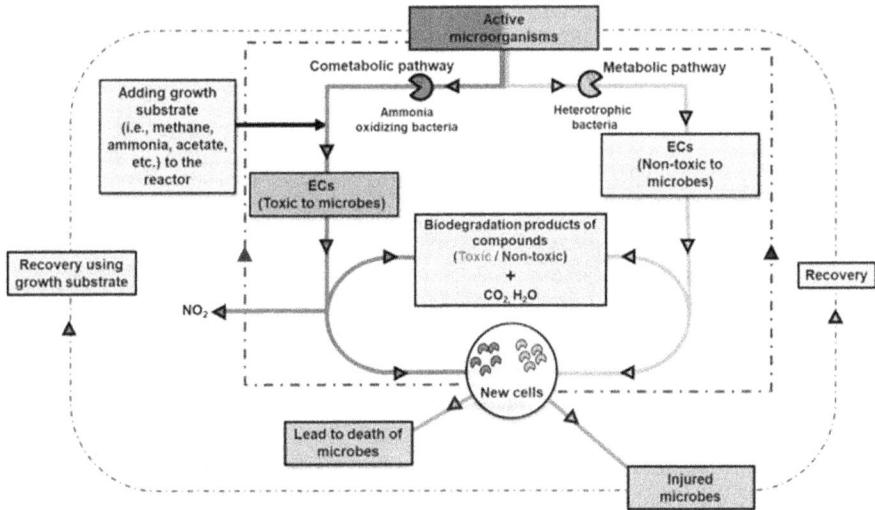

FIGURE 10.2 The steps involved in biological degradation of emerging contaminants. (Reprinted with permission from Saidulu et al., 2021. Copyright 2023 Elsevier).

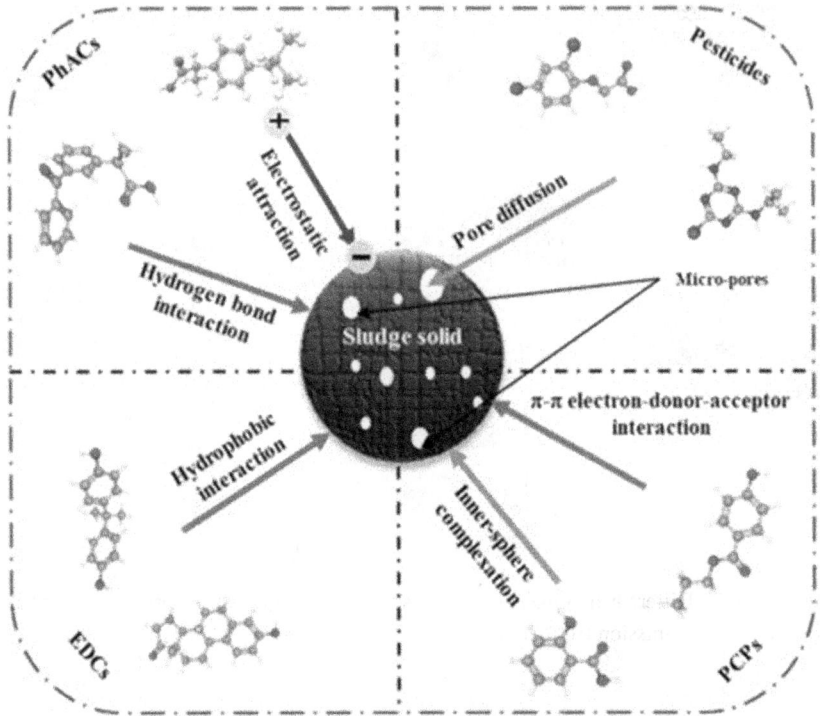

FIGURE 10.3 The mechanisms by which emerging contaminants get adsorbed onto the sludge. (Reprinted with permission from Saidulu et al., 2021. Copyright 2023 Elsevier).

achieving high removal efficiencies. The performance of ASP-based processes in removing ECs has been provided in Figure 10.4. Kasprzyk-Hordern et al. (2009) reported the removal efficiencies of ECs like trimethoprim, erythromycin, codeine, atenolol, and benzophenone-3 to be around 47%, 13%, 48%, 77%, and 63%, respectively, in a conventional STP of UK equipped with ASP. Subedi et al. (2015) reported very low removal efficiencies of 11%, 28%, 34%, and 34% for acesulfame, sucralose, atenolol, and trimethoprim, respectively, using ASP. Based on various studies in the literature, the average removal efficiency of ECs using ASP was found to be only around 70% (Figure 10.4). Researchers have reported some of the possible reasons for such low removal efficiencies. Most ECs are hydrophilic in nature, having a low octanol/water partition coefficient (k_{ow} <10), making them difficult to degrade in the STPs. Hence, they stay in the environment for an extended period. Sequential batch reactors (SBRs) are also not effective in terms of removing ECs. Wei et al. (2018) reported around 30–85% removal for different ECs using an aerobic SBR. Researchers suggested higher efficiency can be achieved by hybrid SBR. Many pilot/field-scale ASP treatment units, when combined with other AOPs or membrane units, provided more than 80% removal of the ECs (Majumder et al., 2021a). Thus, integrating ASP with advanced oxidation processes (AOPs) like ozonation, UV treatment, membrane

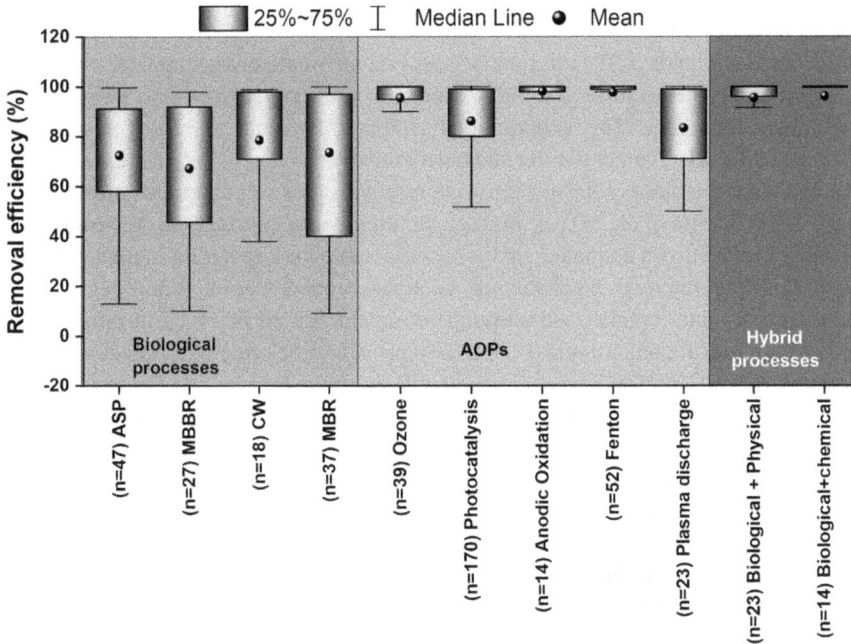

FIGURE 10.4 Box-plot showing the performance of different treatment systems for the removal of emerging contaminants. *n* is the number of studies considered.

Source: Majumder et al., 2021b, 2019; Saidulu et al., 2021.

filtration, and photodegradation can further improve the overall removal efficiency of ECs.

10.2.2 Moving Bed Biofilm Reactor

Moving bed biofilm reactor (MBBR) based systems have shown good performance in ECs removal. MBBR exhibits some inherent benefits over other treatment techniques like compact volume requirement, uniform biomass distribution, ability to handle high OLR, low HRT, and less sludge production with minimal recirculation requirement (Leyva-Díaz et al., 2020; Saidulu et al., 2021). The performance of MBBR-based processes in terms of removal of ECs have been provided in Figure 10.4. The average removal efficiency of ECs using MBBR was observed to be around 66%. However, a significant number of ECs showed higher removal efficiencies. Zupanc et al. (2013) investigated the removal of analgesic pharmaceuticals using MBBR and found high removal efficiencies for ibuprofen (94%) followed by diclofenac (85%) and naproxen (80%). However, it was observed that carbamazepine and clofibric acid showed poor removal (Zupanc et al., 2013). More than 85% removal for ECs like salicylic acid, ibuprofen, naproxen, and bisphenol-A was observed in other studies (Saidulu et al., 2021). MBBR thus serves as an effective treatment unit for removing most of the ECs. However, the efficiency may be improved by combining MBBR with other treatment units.

10.2.3 CONSTRUCTED WETLANDS

Constructed wetlands (CW) offer a low-cost organic treatment technology consisting of an integrated vegetation, naturally occurring microorganisms, and substrate for wastewater treatment. The general idea is that the plants, microorganisms, and substrates all work together to filter and purify the wastewater, maintain the wastewater flow such that the plant roots and substrate remove larger solid particles present (Jain et al., 2020; Varma et al., 2020). Further, pollutants and nutrients in wastewater are naturally broken down and taken up by bacteria and plants, resulting in their removal. Since multiple removal mechanisms, such as biodegradation, photodegradation, volatilization, plant uptake, and sorption on media, are all involved in removing or degrading harmful contaminants, CWs have proven to be effective (Jain et al., 2020; Varma et al., 2020). The performance of CW-based processes in terms of removal of ECs has been provided in Figure 10.4. Vymazal et al. (2017) observed varying removal efficiencies for different ECs (41–91%). CWs were found to be very effective in treating endocrine-disrupting chemicals (EDCs), with average removal efficiencies ranging from 75–100%, for bisphenol-A, nonylphenol, and estrone-1 (Saidulu et al., 2021). The average removal efficiency of ECs using CWs was found to be around 80%. Hence, CWs can be effectively used as a secondary treatment for wastewater comprising of ECs. Due to their robustness, CWs may be especially useful where the wastewater varies in quality and quantity.

10.2.4 MEMBRANE BIOREACTOR

Membrane bioreactors (MBRs) are wastewater treatment systems that combine a membrane process like microfiltration or ultrafiltration with ASP. Due to superior removal efficiencies of various ECs, it is now widely used in the treatment of municipal and industrial wastewater. The retention of sludge on the membrane aids in microbial degradation as well as the physical retention of molecules larger than the pore size of the membrane. As a result, when compared with traditional ASP and CWs, MBRs are more effective at removing ECs. The performance of MBR-based processes in terms of removal of ECs have been provided in Figure 10.4. Generally, the antibiotics and analgesics have shown a good removal by MBR due to relatively long SRT compared to conventional treatments techniques. Many pilot and field-scale MBR treatment units have achieved more than 90% removal. However, these units were combined with other AOPs or membrane processes (Majumder et al., 2021a). Thus, integrating MBR technology with AOPs like ozonation, UV treatment, photodegradation, and other tertiary treatments may further improve the overall removal efficiency of ECs.

10.3 ADVANCED OXIDATION PROCESS

Advanced oxidation processes (AOPs) are dedicated to removing the recalcitrant organic fraction of the wastewater. AOPs are known to generate different oxidizing radicals, such as hydroxyl ($\bullet OH$), superoxide radicals ($\bullet O_2^-$), and others, which react with the organics molecules and break down the complex organic structure. Since conventional treatment methods are often not able to completely remove the ECs, the AOPs have gained immense popularity over the past few decades. The different AOPs and their performance in removing different ECs have been provided in the following sections. The performance of different AOPs in terms of removal of ECs are discussed in Figure 10.4.

10.3.1 Ozone-Based AOPs

Ozone-based AOPs are capable of oxidizing the organic matter with high stability, especially compounds having more than a single bond and higher electron density (Miklos et al., 2018; Rein Munter, 2001). Ozonation can occur directly through molecular ozone oxidation or indirectly via ozone decomposition and the generation of •OH. Ozone-based AOPs have shown more than 90% removal for different ECs (Boczkaj and Fernandes, 2017; Majumder et al., 2019). The average removal efficiency of ECs using ozone-based treatment units was found to be around 95%. However, these processes have some limitations like incomplete degradation and high ozone consumption, which makes the process expensive. Often ozonation is carried out in the presence of H_2O_2 (hydrogen peroxide), which is another strong oxidizing agent. The presence of H_2O_2 leads to the formation of peroxide anions that produces •OH upon reaction with ozone (Merényi et al., 2010; Miklos et al., 2018). This reduces the overall cost of the process since a lesser amount of ozone is consumed. The reactions involved in the generation of •OH radicals from different ozone-based processes have been shown in Equations 10.1 to 10.10.

Ozone only

$$O_3 + H_2O \rightarrow O_2 + H_2O_2 \qquad \text{10.1}$$

$$H_2O_2 \rightarrow 2OH^• \qquad \text{10.2}$$

Ozone and UV

$$O_3 + uv \rightarrow O_2 + 2O^• \qquad \text{10.3}$$

$$O^• + H_2O \rightarrow 2OH^•1 \qquad \text{10.4}$$

$$2O^• + H_2 \rightarrow OH^• + OH^• \rightarrow H_2O_2 \qquad \text{10.5}$$

Ozone and H_2O_2

$$H_2O_2 \rightarrow HO_2^- + H^+ \qquad \text{10.6}$$

$$HO_2^- + O_3 \rightarrow HO_2^• + O_3^{•-} \qquad \text{10.7}$$

$$O_3^{•-} + H^+ \rightarrow HO_3^{•-} \qquad \text{10.8}$$

$$HO_3^{•-} \rightarrow O_2 + OH^• \qquad \text{10.9}$$

$$OH^• + H_2O_2 \rightarrow HO_2^• + H_2O \qquad \text{10.10}$$

In order to improve the performance of this process, different metals and metal oxides are added along with the ozone. The degradation involving the use of added catalysts is known as catalytic ozonation (Boczkaj and Fernandes, 2017; Majumder et al., 2019).

10.3.2 Photocatalysis

Photocatalysis is a process where the electrons from the valence band of semiconductor materials are excited to the conduction band in the presence of external light energy. This leads to the formation of electron and hole pairs. The holes react with water to form •OH, while the electrons react with oxygen present in the water to form $•O_2^-$. These active radicals take part in the degradation of ECs (Majumder et al., 2021b, 2019). The generation of the active radicals largely depends upon the different properties of the photocatalyst and the light irradiation. Photocatalysts with a small bandgap and high charge-carrying capacity are more likely to generate more active radicals. On the other hand, lights with smaller wavelengths have more energy since the wavelength of light is indirectly proportional to the energy. However, in order to make the process more sustainable, researchers have developed different photocatalytic materials that are active in visible lights, which have lower energy. In terms of the removal of ECs, photocatalysts have shown tremendous potential in terms of degrading ECs. The average removal efficiency of ECs using photocatalysis was found to be 84% (Figure 10.4). The major disadvantage of this process is the cost of preparation of the photocatalysts, operating cost, and the generation of sludge. In this context, research on the development of cost-effective materials and prolonging the life of the photocatalysts are being carried out.

10.3.3 Anodic Oxidation

Water is oxidized to form •OH in anodic oxidation using high O_2 evolution overvoltage anodes (Majumder et al., 2019; Moreira et al., 2017). The average removal efficiency of ECs using anodic oxidation was found to be greater than 95% (Figure 10.4). The most commonly used materials for the anodes are platinum, lead oxide, tin oxide, and boron-doped diamond. Boron-doped diamond anodes and platinum anodes are highly efficient in generating oxidizing radicals, but the major drawback is the cost of the materials. Furthermore, when metal oxide-based anodes are used, they tend to leach and cause secondary pollution (Majumder et al., 2019; Moreira et al., 2017). Hence, research on the development of efficient anode is still being carried out.

10.3.4 Fenton Process

In Fenton process, •OH are generated from the reaction between H_2O_2 and Fe^{2+} ions. A schematic of the Fenton process has been provided in Figure 10.5. Although this process has shown a good removal of ECs, excessive consumption of H_2O_2 and

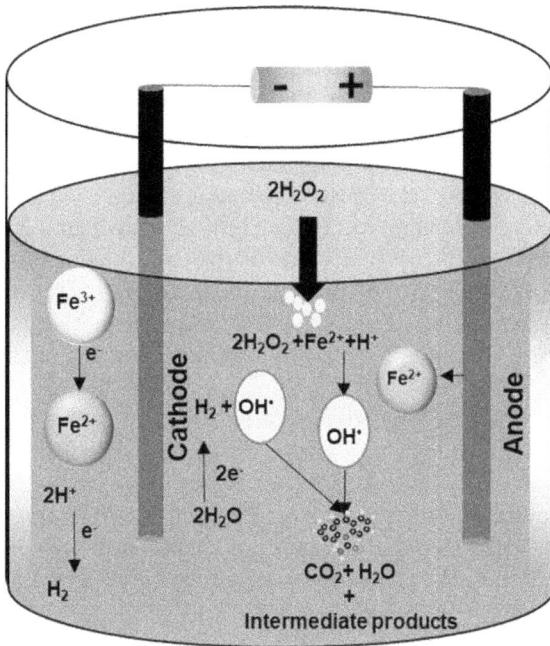

FIGURE 10.5 Schematic of Fenton process.

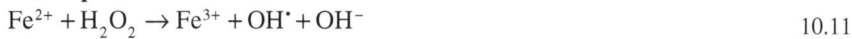

conversion of Fe^{2+} to Fe^{3+} are significant disadvantages. In order to overcome this issue, electro-Fenton processes are used, where H_2O_2 is generated "in-situ" at the cathode (Majumder et al., 2019; Wang et al., 2012). In photo-Fenton processes, the Fe^{3+} ions absorb light and get converted to Fe^{2+} ions, and further •OH radicals are generated. Recently, photo-electron Fenton processes are used to overcome the drawbacks of conventional Fenton processes (Majumder et al., 2019; Wang et al., 2012). The equations driving the different Fenton processes are provided in Equations 10.11 to 10.15.

Fenton process

$$Fe^{2+} + H_2O_2 \rightarrow Fe^{3+} + OH^{\bullet} + OH^{-}$$ 10.11

Photo-Fenton process

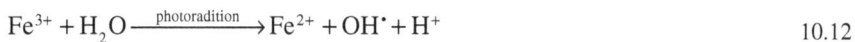

$$Fe^{3+} + H_2O \xrightarrow{\text{photoradition}} Fe^{2+} + OH^{\bullet} + H^{+}$$ 10.12

Electro-Fenton process

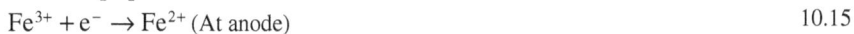

$$O_{2(g)} + 2\,H^{+} + 2e^{-} \rightarrow H_2O_2 \text{ (At cathode)}$$ 10.13

$$Fe^{2+} + H_2O_2 \rightarrow Fe^{3+} + OH^{\bullet} + OH^{-}$$ 10.14

$$Fe^{3+} + e^{-} \rightarrow Fe^{2+} \text{ (At anode)}$$ 10.15

Although modified Fenton processes can provide high removal of ECs, they require large quantities of chemicals and therefore can be expensive.

10.3.5 Other AOPs

Among other AOPs, sonolyis or ultrasound is the most widely used technique for the degradation of ECs. In sonolysis, free radicals are formed through cavitation, especially when ultrasound is irradiated at a low frequency. Cavitation is a physical phenomenon that describes the formation, growth, and collapse of small bubbles in liquids (Arefi-Oskoui et al., 2019; Mahamuni and Adewuyi, 2010; Majumder et al., 2019). During cavitation, the extremely high temperature and high pressure of imploding cavitation bubbles lead to the thermal dissociation of water molecules, resulting in the generation of •OH. Plasma discharge and microwave discharge are also used (Ghernaout and Elboughdiri, 2020; Majumder et al., 2019). However, they are seldom used because of the associated high costs and inconsistent removal efficiencies.

It was observed that the performance of AOPs in terms of removal of ECs was better than that of the biological processes. However, it should be noted that most of the studies conducted using AOPs do not involve high-strength wastewater. As a matter of fact, they are only used as tertiary treatment. The AOPs might not show such good removal efficiencies if different interfering agents are present in raw wastewater. Hence, these processes are better suited for targeting only the recalcitrant fraction of the contaminants present in the wastewater.

10.4 ADSORPTION AND MEMBRANE-BASED TECHNOLOGIES

Adsorption has certain advantages over other wastewater treatment processes, such as simple design, economical, easy to use, and not sensitive to pollutant toxicity. Currently, a number of adsorbents are being used for the adsorption of ECs. Activated carbons developed from household and agricultural wastes, sawdust, and sludge have proved to be very efficient in removing ECs (Srivastava et al., 2021). Metal oxides, metal–organic framework, graphene-based adsorbents are also in use for the adsorption of ECs due to their unique physicochemical properties (Lu and Astruc, 2020; Patel et al., 2019). However, adsorbents have a limited capacity and eventually get exhausted. If the adsorbents are used to treat raw wastewater, they get exhausted early. Hence, they are usually used as a polishing unit and implemented as a tertiary treatment process. Due to the exhaustive nature of the adsorbents, research has focused on their regeneration, which may increase the life of these adsorbents. However, another disadvantage associated with adsorption is their disposal since all the contaminants are being transferred from the wastewater to the adsorbent, instead of being broken down into simpler compounds (Majumder et al., 2019; Srivastava et al., 2021).

Different membrane-based technologies including reverse osmosis (RO), nanofiltration (NF), and ultrafiltration (UF) are being used to remove ECs. Usually, RO and NF have shown better performance in terms of removal of ECs (Majumder et al., 2021a; Yang et al., 2017). However, these membrane-based technologies are also being used as a post-treatment process. In the field scale, they are integrated with some biological process or AOPs to remove ECs (Majumder et al., 2021a; Yang et al.,

2017). The problems associated with membrane is the bio-fouling of the membranes and the membrane-based technologies have high operating cost. However, membrane-based technologies are being widely used as a tertiary treatment process in various pilot-scale and full-scale wastewater treatment units because they are relatively easier to maintain and operate, especially in comparison to other chemical-intensive processes.

10.5 HYBRID PROCESSES

Since ECs cannot be effectively removed using conventional wastewater treatment processes and AOPs require a certain degree of pre-treatment, different integrated treatment systems are being evolved to increase the overall removal efficiency of the ECs. Recently, a variety of hybrid treatment technologies are being used to treat ECs. Most hybrid systems are composed of biological treatment systems followed by physical or chemical treatment systems. Few of the biological hybrid treatment methods have been discussed in the following sections and their performance in terms of removal of ECs have been provided in Figure 10.4. It was observed from Figure 10.4 that the average removal efficiency of ECs using biological treatment systems was around 72% compared to AOPs, which was around 88%. On the other hand, the average removal efficiency of ECs using hybrid processes was found to be greater than 95% (Figure 10.4).

Biological–physical hybrid system combines biological systems like ASP, MBR, MBBR, and others with physical treatment systems, such as RO, NF, UF, and adsorption. In a study, Luo et al. (2017) used osmotic MBR combined with RO to remove 31 ECs and reported more than 95% removal efficiency for most of the ECs. Kovalova et al. (2013) discussed the performance of MBR integrated with ozone, powdered activated carbon, and UV for treating wastewater. The system showed more than 95% effective removal of ECs. Most of the MBR-based hybrid systems showed very good removal efficiencies for different ECs. The average removal of MBR-based integrated systems was estimated to be 90–95%. Among them, the hybrid of MBR and RO system has demonstrated encouraging results for all of the ECs studied, including carbamazepine, codeine, atenolol, octocrylene, benzophenone-3, and others. Nguyen et al. (2013) observed an increased removal of ECs (i.e., from 85% to 100%) when MBR was coupled with NF and RO. The MBR-RO hybrid system was found to remove analgesics, such as diclofenac, naproxen, ibuprofen, and ketoprofen almost completely from the influent. Sahar et al. (2011) evaluated the performance of the combined system comprising ASP and UF and found that ibuprofen, bisphenol-A, and salicylic acid were all removed with greater than 92% efficiency. Several chemical treatment processes, such as AOPs, chemical precipitation, and others, have been combined with biological treatment methods, such as ASP, MBR, CWs, and others, in order to achieve higher removal of ECs (Saidulu et al., 2021). These cases can be collectively called hybrid biological-chemical treatment processes and have been effective in removing many ECs (Figure 10.4).

It can be observed from Figure 10.4 that most of the biological units could remove around 70–80% of ECs, while AOPs could remove more than 90% of the ECs. However, the cost of treatment using AOPs is substantially higher and these processes are not stand-alone processes. Hybrid treatment systems have shown great performance in terms of both EC removal and removal of other treatment methods. Although the cost of treatment using hybrid methods may be higher than the conventional processes, they pave the way for wastewater reuse and energy recovery, which may compensate the cost of their treatment. Over the years, different full-scale STPs have adapted advanced treatment systems to remove the ECs. Case studies of few of the full-scale STPs and their performance in terms of removal of ECs have been discussed in the following sections.

10.6   CASE STUDIES ON THE TREATMENT OF EMERGING CONTAMINANTS

Removal of ECs from small laboratory settings can be more easily achieved than removing them in full-scale wastewater treatment plants. In small laboratory settings, the removal is carried out in a controlled environment with synthetic wastewater, which do not host the numerous impurities present in real wastewater. As discussed earlier, this is largely because the physicochemical properties of the ECs prevent them from being biodegraded. Since sewage hosts a considerable fraction of organic loading apart from the ECs, so directly implementing AOPs or other advanced treatment processes is also not an option. However, many countries have necessitated the removal of the ECs prior to their reuse or discharge, therefore necessitating changes to wastewater treatment plant design and operations.

Wetland-based treatment systems have been used to remove ECs only when the treated water is to be discharged into natural water streams (Conkle et al., 2008; Matamoros et al., 2009). On the other hand, membrane-based treatment systems have been used when there is water shortage, and the wastewater needs to be reused (Egea-Corbacho et al., 2019). However, the overall cost of wastewater treatment is substantially increased if AOPs or membrane-based processes are incorporated. It is necessary to consider the characteristics of the influent and the desired effluent quality before adopting a treatment. Hence, the selection of treatment techniques is a critical factor in the removal of ECs. In this context, few case studies of full-scale treatment plants targeting the removal of ECs have been discussed in the subsequent sections.

10.6.1   CASE STUDY 1 – USE OF CONSTRUCTED WETLAND NEAR NEW ORLEANS, LA

The wetland-based treatment plant is located at Mandeville, Louisiana, USA, which is to the north of Lake Pontchartrain. The raw wastewater flows through three aerated lagoons ($61 \times 183 \times 3$ m). Each of the aerated lagoons had a hydraulic retention time (HRT) of nine days. Hence a total of 27 days retention time was provided for the

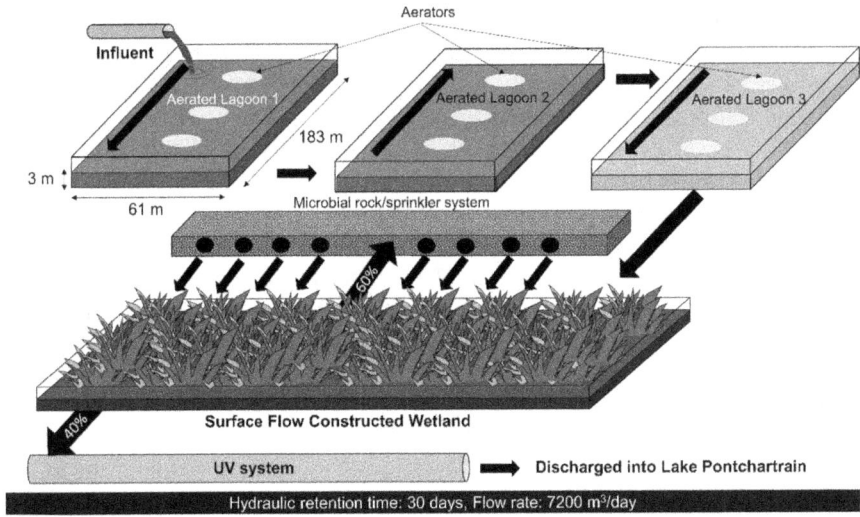

FIGURE 10.6 Schematic of the constructed wetland-based wastewater treatment plant at Mandeville, Louisiana. (Modified and reprinted with permission from Conkle et al., 2008. Copyright 2023 Elsevier).

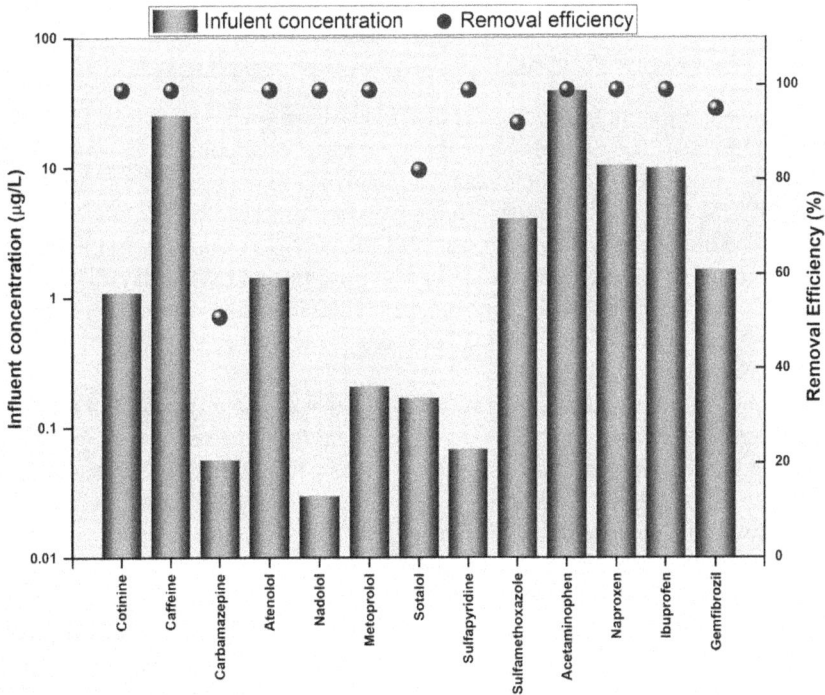

FIGURE 10.7 Influent concentration and removal efficiency of the emerging contaminants in the wastewater treatment plant at Mandeville, Louisiana.

Source: Conkle et al., 2008.

raw wastewater. The effluent of the aerated lagoons flowed through a surface flow CW with crushed gravel bed. Due to the presence of the crushed gravel bed, the wastewater could be evenly distributed. The two main plant species in the wetland were *Hydrocotyle tripartita and Phragmites australis*. After passing through the wetland, the wastewater is collected in rock basins (Conkle et al., 2008). Around 60% of the wastewater is recycled back into the wetland by a series of sprinklers to further aerate the wastewater. On the other hand, the remaining 40% of the wastewater is pumped through a UV-irradiation channel comprising 176 UV bulbs for disinfection purposes (Conkle et al., 2008). The disinfected water is discharged to Lake Pontchartrain via the stream, Bayou Chinchuba. The average retention time of the wetland-based treatment plant is 30 days, and it discharges around 7,200 m³/day (Conkle et al., 2008). The schematic of the Southeast Louisiana treatment plant is shown in Figure 10.6.

The system was efficient in ECs removal. Except for carbamazepine and sotalol, most of the ECs showed more than 90% removal. The concentration of the ECs in the raw wastewater and the removal efficiency of the treatment plant are shown in Figure 10.7. ECs, such as caffeine, acetaminophen, nadolol, propranolol, metoprolol, and sulfapyridine, were not detected in the effluent. This system proved to be efficient in terms of the removal of ECs (Conkle et al., 2008). Further, the system is driven by aeration. Hence generation of foul smells or unwanted gases is not a problem.

10.6.2   CASE STUDY 2 – CONSTRUCTED WETLANDS NEAR LEON, SPAIN

At Leon, three different full-scale treatment plants have been used to treat wastewater containing ECs. At Fresno de la Vega, the wastewater was subjected to a metallic bar screen, followed by two parallel anaerobic ponds (surface area: 335 m², depth: 3.75 m, and HRT: 0.4 days). The effluent from the two parallel anaerobic ponds was passed through a facultative pond (surface area: 8481 m², depth: 2 m, and HRT: 4.1 days) and a maturation pond (surface area: 3169 m², depth: 1.5 m, and HRT: 1 day). The specific surface area of the unit or the amount of area required to treat wastewater generated by a single person was 38 m²/person equivalent. The average flow rate was around 3,200 m³/day (Hijosa-Valsero et al., 2010).

At Cubillas de los Oteros, the raw wastewater was initially passed through bar screens and treated with a septic tank. The effluent from the septic tank was passed through a facultative pond rich with Lemna minor colony (surface area: 1,073 m², depth: 1.6 m, and HRT: 75.9 days). The effluent from the facultative pond was passed through a surface flow CW (surface area: 44 m², depth: 0.3 m gravel, and HRT: 1.2 days) planted with *Typha latifolia*. The depth of the water was around 0.4 m. The effluent of the surface flow CW passed through a subsurface flow CW (surface area: 585 m², depth: 0.55 m gravel, and HRT: 5.7 days) planted with *Salix atrocinerea*. The specific surface area of the system was 17 m²/person equivalent. The average flow rate was around 20 m³/day (Hijosa-Valsero et al., 2010).

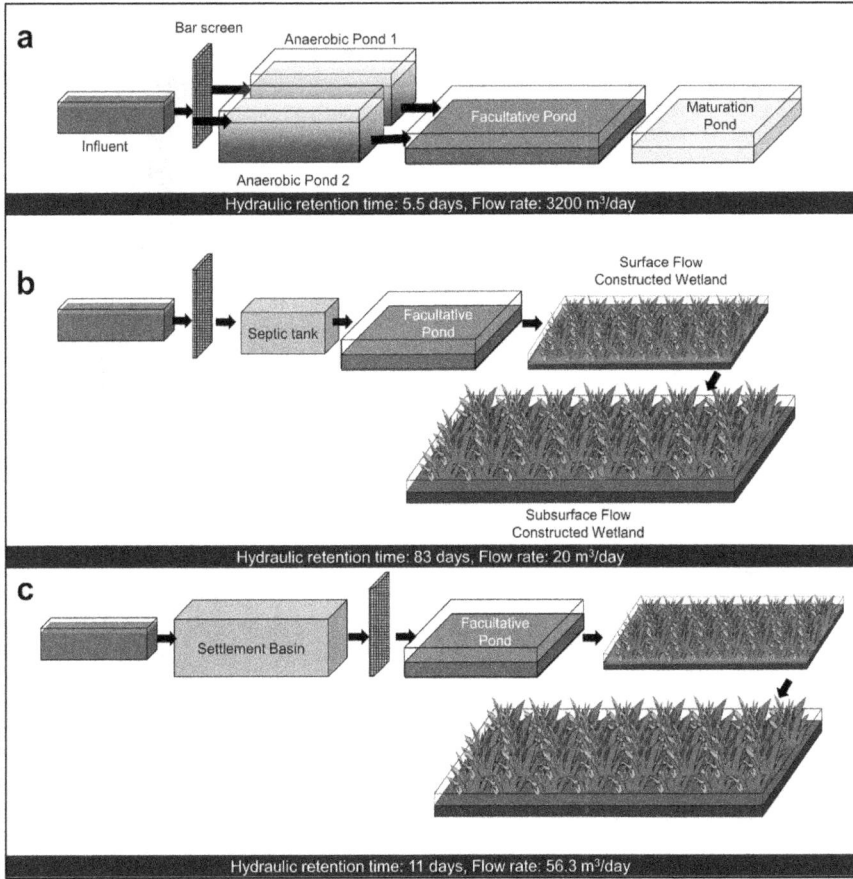

FIGURE 10.8 Schematic of the wastewater treatment plant at (a) Fresno de la Vega, Leon; (b) Cubillas de los Oteros, Leon; and (c) Bustillo de Cea, Leon. (Modified and reprinted with permission from Hijosa-Valsero et al., 2010. Copyright 2023 Elsevier)

At Bustillo de Cea, the raw wastewater was initially passed through a settling basin and bar screens. The effluent was then retained at a facultative pond (surface area: 230 m², depth: 1.5–2.0 m, and HRT: 4.2 days). The effluent from the facultative pond was passed through a surface flow CW (surface area: 210 m² and HRT: 3.5 days) planted with *Typha latifolia*. The effluent of the surface flow CW passed through a subsurface flow CW (surface area: 362.5 m², and HRT: 3.2 days) planted with *Salix atrocinerea*. The specific surface area of the system was 22 m²/person equivalent. The average flow rate was around 56.3 m³/day. The schematic of the three systems is given in Figure 10.8a–c (Hijosa-Valsero et al., 2010). The influent concentration of ECs in three of the treatment plants of Leon and their removal efficiency has been shown in Figure 10.9.

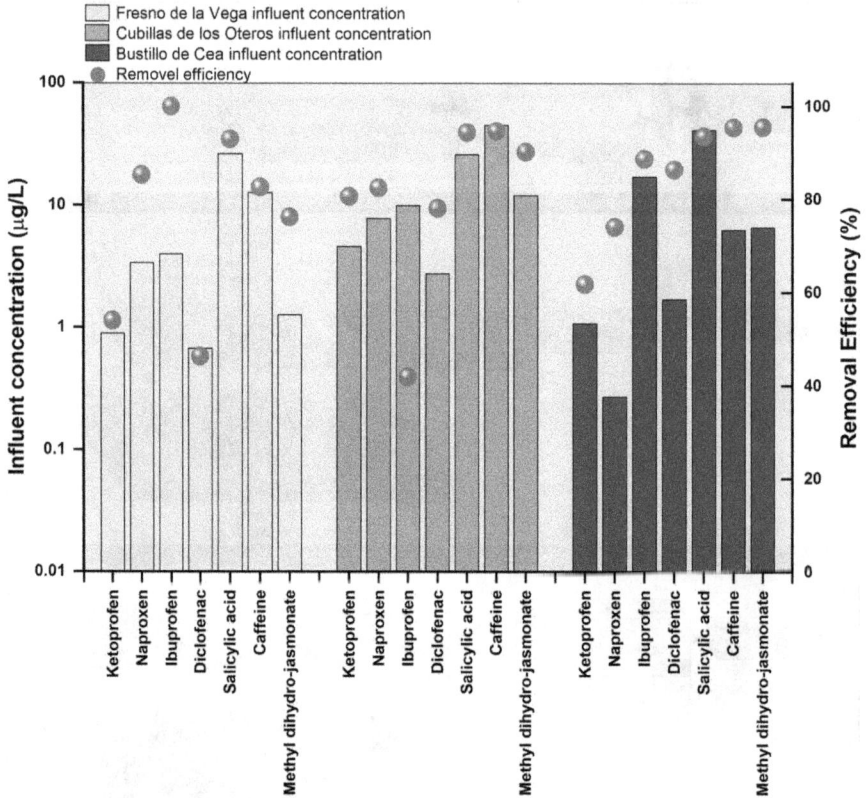

FIGURE 10.9   Influent concentration and removal efficiency of the emerging contaminants in the wastewater treatment plant at Leon, Spain.

Source: Hijosa-Valsero et al., 2010.

10.6.3   CASE STUDY 3 – MICROFILTRATION-REVERSE OSMOSIS, EAST ANGLIA, UK

The microfiltration-reverse osmosis (MF-RO) based treatment plant located in the East Anglian region of the United Kingdom receives wastewater from the secondary-treated sewage from neighboring STPs (Garcia et al., 2013a). The flow rate in the full-scale MF-RO treatment plant was around 1,200 m^3/day. The effluent undergoes a 150 µm screen prior to the MF unit. The hollow fiber MF undergoes regular backwashing and cleaning with hypochlorite, alkali, and acid. The filtrate from the MF plant is stored in a holding tank before being treated by the RO. The MF unit rejects 14% of the influent wastewater, while the RO unit further rejects 27% of the MF filtrate. As a result, the 1,200 m^3/day of treated water is generated from an initial feed of 1,910 m^3/day. The treated wastewater is used for reuse in different industries. The schematic of the MF-RO treatment plant at East Anglia, United Kingdom has been shown in Figure 10.10. The removal of most of the compounds by the treatment plant was more than 75%. ECs, such as 22'47'-tetrabromodiphenyl ether (BDPE47),

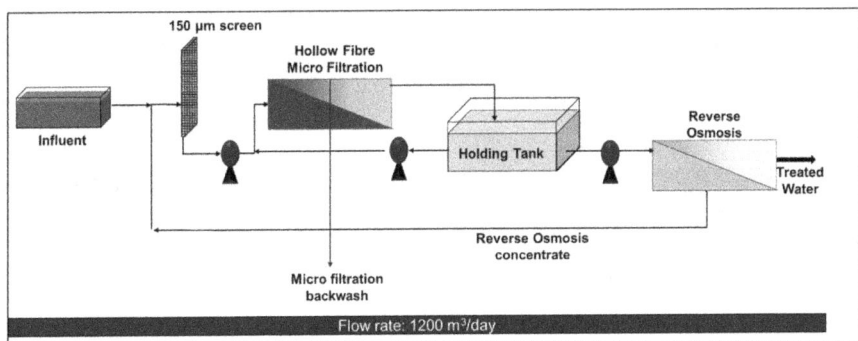

FIGURE 10.10   Schematic of the wastewater treatment plant at East Anglia, United Kingdom. (Modified and reprinted with permission from Garcia et al., 2013b. Copyright 2023 Taylor & Francis).

22'44'5-pentabromodiphenyl ether (BDPE99), 22'44'6-pentabromodiphenyl ether (BDPE100), 22'44'55'-hexabromodiphenyl ether (BDPE153), bis-(2-ethylhexyl) phthalate (DEHP) were removed substantially in the MF. On the other hand, ECs, such as ibuprofen, diclofenac, estradiol, and others showed negligible removal in the MF. While most of these compounds were significantly removed, ibuprofen and endocrine-disrupting compounds were only slightly removed in the RO process (Garcia et al., 2013b).

10.6.4 CASE STUDY 4 – ADVANCED OXIDATION TREATMENT, MELBOURNE, AUSTRALIA

The treatment plant located in Melbourne, Australia treats 350,000 m³/day of wastewater and uses a combination of ozonation and biological media filtration for advanced wastewater treatment. The secondary-treated water is first passed through two parallel contactors providing an average ozone dosing of 9.7± 0.5 mg/L. This ozone dosing is called the pre-ozonation. The effluent from the pre-ozonation treatment unit is passed through the biological media filter and taken to the post-ozonation unit, where two parallel rectangular contactors provide an ozone dosing of 0.7 and 4.7± 0.5 mg/L, respectively. The water from the post-ozonation unit was subjected to 23 mJ/cm² of UV irradiation. This was followed by a chlorine dose of 3.29 (±0.07) mg Cl₂/L. The treated wastewater is reused for irrigation, dual circulation, and firefighting (Blackbeard et al., 2016). The schematic of the ozone-based treatment plant at Melbourne, Australia has been provided in Figure 10.11.

The final concentration of the different ECs in the influent and after each treatment has been shown in Figure 10.12. Most of the ECs were found to be removed after UV treatment. Acesulfame K, Diatrizoate Sod., Oxazepam, Primidone, Temazepam, Atrazine, and Metolachlor were found to be retained even after chlorination. However, the concentrations of these compounds in the effluent were mostly below 0.1 µg/L (Figure 10.12).

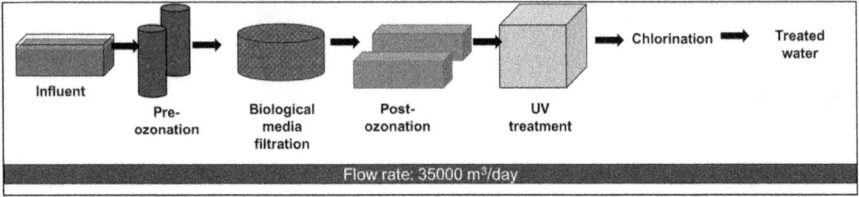

FIGURE 10.11 Schematic of the ozone-based wastewater treatment plant at Melbourne, Australia. (Modified and reprinted with permission from Blackbeard et al., 2016. Copyright 2023 RSC).

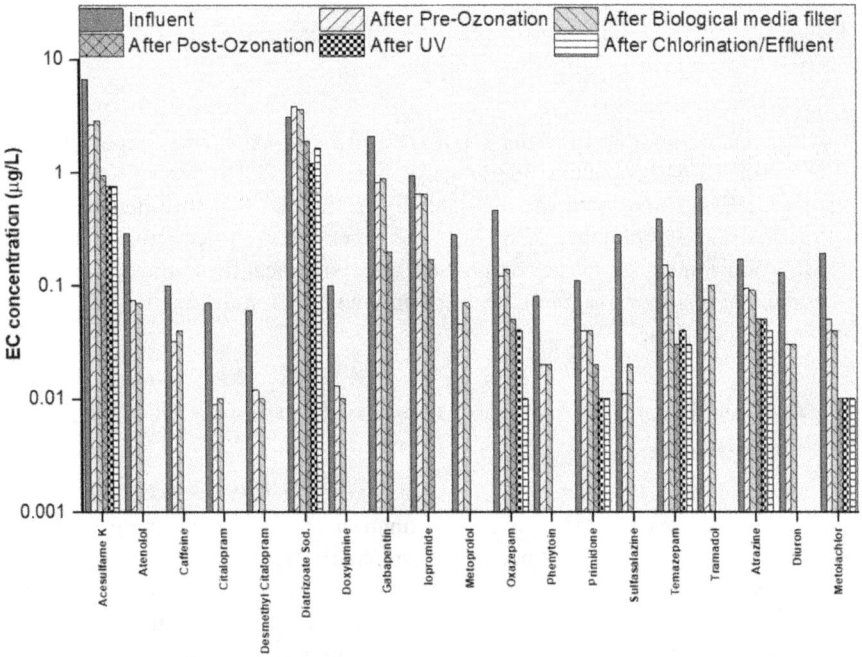

FIGURE 10.12 The concentration of different emerging contaminants after various stages in the treatment plant at Melbourne, Australia.

Source: Blackbeard et al., 2016.

10.6.5 Case Study 5 – Wastewater Treatment in Sicily Region, Italy

Sgroi et al. (2017) studied the removal efficiencies of various ECs at 10 different treatment plants at Sicily, Italy. The treatment train of the 10 treatment plants have been provided in Table 10.2.

The performance of the different treatment plants in terms of the removal of various ECs have been depicted in Figure 10.13. It can be observed that removal

TABLE 10.2
Details about 10 Different Wastewater Treatment Plants at Sicily, Italy

Sl.No	Population Equivalent	Primary Treatment	Secondary Treatment	Tertiary Treatment
			Treatment Train	
Treatment plant-1	5300	Preliminary or primary treamtent	Activated sludge (extended aeration)	
Treatment plant-2	34150		Activated sludge (nitrifying)	
Treatment plant-3	30000		Rotating biological contactors	
Treatment plant-4	6500		Rotating biological contactors	
Treatment plant-5	55000		Activated sludge (nitrifying)	
Treatment plant-6	55000		Activated sludge (denitrifying)	
Treatment plant-7	25000		Activated sludge (nitrification, denitrification, combined phosphorous removal by ferric chloride)	Sand filtration, UV disinfection
Treatment plant-8	432500		Activated sludge (nitrification, denitrification)	Sand filtration, UV disinfection
Treatment plant-9	75000		Activated sludge (nitrification, denitrification)	Chlorination
Treatment plant-10	40000		Activated sludge (nitrification, denitrification)	Chlorination

Source: Modified with permission from Sgroi et al., 2017a. Copyright 2023 Elsevier.

of more than 90% was observed for caffeine, triclosan, ibuprofen, and atenolol in some of the treatment plants. Low removal for various ECs was also observed. The different removal rates for various ECs at the same treatment plant is due to the variation in physicochemical properties of the compounds (Sgroi et al., 2017a). Negative removal was observed for sucralose, carbamazepine, and sulfamethoxazole (Sgroi et al., 2017a).

Negative removal of ECs in STPs is a common phenomenon (Sgroi et al., 2017b, 2017a; Tran et al., 2018a). Negative removal means the concentration of the contaminant was higher in the effluent than in the influent of the treatment plant. This may be because the ECs get encapsulated in the sludge during secondary or primary treatment. However, after undergoing subsequent treatment processes, the ECs may escape from the sludge and find their way into the effluent, thereby increasing its concentration in the effluent. There is a possibility of the ECs getting partially broken down into their metabolites. However, ECs recombine with other metabolites in the effluent to form the parent compound. This will increase the concentration of the contaminant in the effluent (Sgroi et al., 2017a; Tran et al., 2018b).

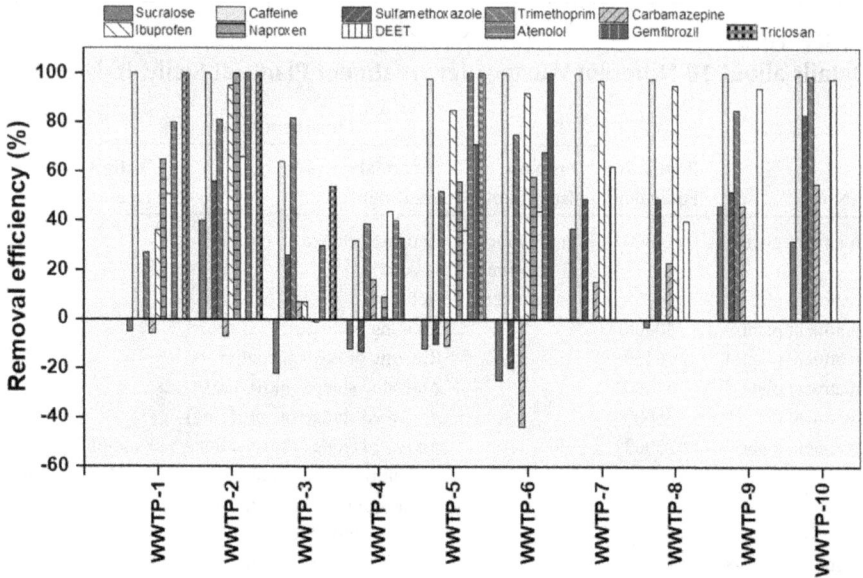

FIGURE 10.13 The removal efficiencies of different emerging contaminants in 10 different treatment plants at Sicily, Italy.

Source: Sgroi et al., 2017a.

10.6.6 CASE STUDY 6 – WASTEWATER TREATMENT AT VALLE DE BRAVO, MEXICO

Estrada-Arriaga et al. (2016) observed the removal efficiency of a treatment plant located at Valle de Bravo, Mexico treating wastewater of around 21,000 inhabitants. The treatment plant has a treatment capacity of 12,960 m^3/day. The influent is passed through screens and a vortex grit chamber to remove the larger particles present in the wastewater. The wastewater from the grit chamber is passed through six micro-screens. The wastewater is then treated via an anaerobic, anoxic, and aerobic ASP. The HRT for the anaerobic process is 2 h, the anoxic process is 3 h, and the aerobic process is 14 h (Estrada-Arriaga et al., 2016). The overall HRT is 19 h and solid retention time is 40 days. The wastewater out of the biological treatment units is passed through a chlorine disinfection tank (ClO$_2$ dosing is 1.4 g/h) and a chamber of UV irradiation (7.68 mJ/cm^2 of 254 nm). The treated water was discharged into the Amanalco river. The schematic of the treatment plant at Valle de Bravo, Mexico, has been depicted in Figure 10.14.

The performance of the treatment plant in terms of EC removal during a dry season has been shown in Figure 10.15. Most of the compounds showed a very high removal and were not detected in the effluent. Carbamazepine showed the least removal, which was around 33%. Among other ECs which were not completely removed were Diazepam, *N,N*-diethyl-meta-toluamide, Benzoylecgonine, and Metoprolol. This may be due the toxic nature of these compounds, which prevented them from being biodegraded. The complex structure of these

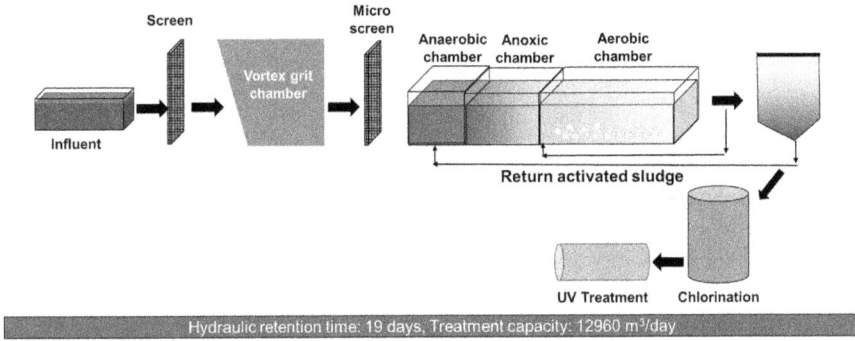

FIGURE 10.14 The schematic of the treatment plant at Valle de Bravo, Mexico. (Modified and reprinted with permission from Estrada-Arriaga et al., 2016 Copyright 2023 Elsevier).

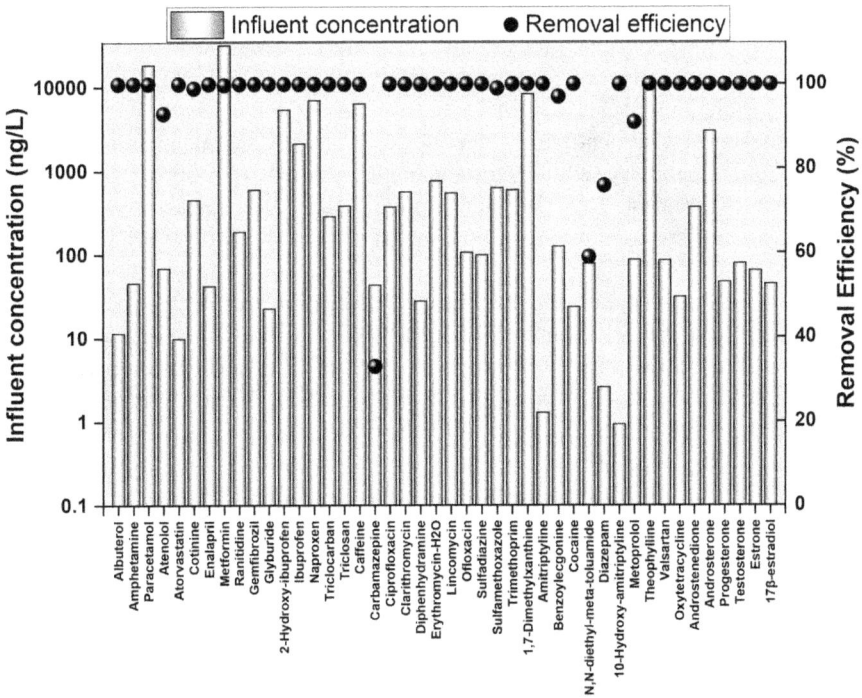

FIGURE 10.15 Influent concentration and removal efficiency of the emerging contaminants in the wastewater treatment plant at Valle de Bravo, Mexico.

Source: Estrada-Arriaga et al., 2016.

compounds prevent them from being oxidized when exposed to UV irradiation and chloride radicals (Estrada-Arriaga et al., 2016; Saidulu et al., 2021). The study showed that most of the ECs can be completely removed in this process except for a few compounds.

10.6.7   CASE STUDY 7 – WASTEWATER TREATMENT AT DASPOORT, ZEEKOEGAT, AND PHOLA, SOUTH AFRICA

The removal of different ECs, such as pharmaceutically active compounds, endocrine disruptors, personal care products, and pesticides from wastewater, has been studied more compared to PFAS. The PFAS are highly resistant to degradation because of their C–F bond (Kibambe et al., 2020a; Xiao, 2017). As a result, they are not easily removed in the treatment plants. Although there are many studies on the removal of PFAS in water treatment plants, their removal in wastewater treatment plants is limited. Kibambe et al. (2020a) studied the performance of three different treatment plants in South Africa and assessed their performance in terms of the removal of PFAS.

The sewage at Daspoort treatment plant passes through screens, grit chamber and then to two primary sedimentation tanks. The effluent from the primary sedimentation tank splits into three parallel three-stage Phoredox process. Phoredox process is consecutively treating the wastewater in an anaerobic, anoxic, and aerobic reactor. The effluent is then collected in a secondary clarifier, and the clarifier output is given UV treatment or chlorination. The treated effluent is discharged into the Apies River (Kibambe et al., 2020b).

The Zeekoegat treatment plant comprises four primary sedimentation tanks. The effluent from the primary sedimentation tank goes through a pre-anoxic zone, anaerobic, anoxic, and aerobic zone. The sludge is recirculated from the aerobic zone to the anoxic zone, and the solid retention time in this treatment plant was around 45 days (Kibambe et al., 2020b).

The Phola treatment plant comprises a set of screens and two vortex grit chambers. The effluent from the grit chamber flows through a series of anaerobic ponds, and then through a trickling filter (Kibambe et al., 2020b). The mechanism of removal in this system might be the biological filters (Kibambe et al., 2020a). The performance of the three treatment plants in terms of removal of total PFAS in the treatment plant is depicted in Figure 10.16.

It can be seen in Figure 10.16 that the removal of total PFAS was the most in Zeekoegat treatment plant, followed by Daspoort treatment plant and Phola treatment plant. The primary mechanism for PFAS removal in these systems is adsorption onto the activated sludge (Kibambe et al., 2020a). Negative removal was observed for a few of the PFAS. This may be because the ECs get encapsulated in the sludge during secondary or primary treatment. However, after undergoing subsequent processes, the ECs may escape from the sludge and find their way into the effluent, increasing its concentration (Tran et al., 2018b). The trickling filter showed lower efficiency in the removal of PFAS, with the total removal of PFAS only reaching around 41%. As a result, it can be suggested that in treatment plants, implementing three-stage anaerobic, anoxic, and aerobic treatment followed by secondary clarifier may reduce a significant load of PFAS and other contaminants. The effluent may be further polished by treating with RO or NF systems, which have shown a high degree of removal in water treatment plants (Appleman et al., 2014).

The different case studies indicated that for efficient removal of ECs, a post-treatment involving the use of membrane filtration or oxidation process is required.

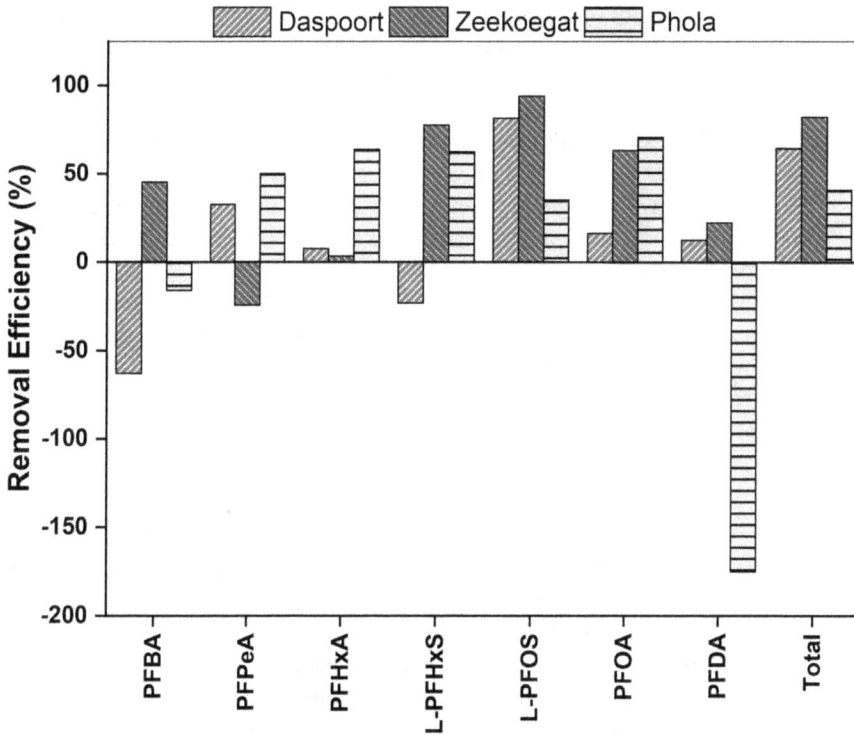

FIGURE 10.16 Performance of Daspoort treatment plant, Zeekoegat treatment plant, and Phola treatment plant for the removal of PFAS.

Source: Kibambe et al., 2020a, 2020b.

Primary and secondary treatment units helped in bringing down the concentration of solids and biodegradable organics. In most of the field-scale studies, UV or chlorination was provided, which not only removed the microorganisms but also helped to break down the ECs. Constructed wetlands were primarily used when the treated water was discharged into lakes or rivers, while membrane filtration was favored when the treated water needed to be reused.

10.7   CHAPTER SUMMARY

- Physicochemical properties, such as complex molecular structure, hydrophilicity, and toxic nature, are the main reasons behind the inability of conventional treatment units to remove ECs.
- The two primary mechanisms involved in removing ECs in biological processes are biodegradation and adsorption onto the sludge particles.
- Many pilot/field-scale ASP treatment units, when combined with other AOPs or membrane units, provided more than 80% removal of the ECs.

- The average removal efficiency of ECs using CWs was found to be around 80%. Hence, CWs can be effectively used as a secondary treatment for wastewater comprising of ECs. Due to its robustness, CWs may be preferred where wastewater quality and quantity are fluctuating.
- The effectiveness of AOPs in removing ECs was far superior to that of biological processes. It should be noted, however, that the majority of studies involving AOPs do not involve high-strength wastewater. As a result, these processes are better suited for tertiary treatment in STPs that are designed to target only the most recalcitrant contaminants in the wastewater.
- The case studies revealed that for the removal of ECs from the system, a post-treatment with UV, chlorination, or membrane filtration is required.
- The water treated with membrane-based processes is usually reused. This technology is also adopted to comply with very stringent discharge regulations.

10.8   CONCLUDING REMARKS

The different processes and case studies described in this chapter highlight the potential of the treatment technologies in removing ECs. Addition of advanced treatment technologies for the removal of ECs incur excessive cost to the overall treatment process. Hence, in order to make up for the cost due to the incorporation of the advanced processes, it is necessary to recover resource from the wastewater. Wastewater has the potential to be the source of resource recovery in the form of nutrient, energy, water, and other products. In this context, the different processes involved in resource recovery from wastewater has been described in the next chapter.

REFERENCES

Ahmed, M.B., Zhou, J.L., Ngo, H.H., Guo, W., Thomaidis, N.S., Xu, J., 2017. Progress in the biological and chemical treatment technologies for emerging contaminant removal from wastewater: A critical review. *J. Hazard. Mater.* 323, 274–298. https://doi.org/10.1016/J.JHAZMAT.2016.04.045

Appleman, T.D., Higgins, C.P., Quiñones, O., Vanderford, B.J., Kolstad, C., Zeigler-Holady, J.C., Dickenson, E.R.V., 2014. Treatment of poly- and perfluoroalkyl substances in U.S. full-scale water treatment systems. *Water Res.* 51, 246–255. https://doi.org/10.1016/j.watres.2013.10.067

Arefi-Oskoui, S., Khataee, A., Safarpour, M., Orooji, Y., Vatanpour, V., 2019. A review on the applications of ultrasonic technology in membrane bioreactors. *Ultrason. Sonochemistry* 58, 104633. https://doi.org/10.1016/j.ultsonch.2019.104633

Blackbeard, J., Lloyd, J., Magyar, M., Mieog, J., Linden, K.G., Lester, Y., 2016. Demonstrating organic contaminant removal in an ozone-based water reuse process at full scale. *Environ. Sci. Water Res. Technol.* 2(1), 213–222. https://doi.org/10.1039/c5ew00186b

Boczkaj, G., Fernandes, A., 2017. Wastewater treatment by means of advanced oxidation processes at basic pH conditions: A review. *Chem. Eng. J.* 320, 608–633. https://doi.org/10.1016/j.cej.2017.03.084

Conkle, J.L., White, J.R., Metcalfe, C.D., 2008. Reduction of pharmaceutically active compounds by a lagoon wetland wastewater treatment system in Southeast Louisiana. *Chemosphere* 73, 1741–1748. https://doi.org/10.1016/j.chemosphere.2008.09.020

Debnath, D., Gupta, A.K., Ghosal, P.S., 2019. Recent advances in the development of tailored functional materials for the treatment of pesticides in aqueous media: A review. *J. Ind. Eng. Chem.* 70, 51–69. https://doi.org/10.1016/J.JIEC.2018.10.014

Egea-Corbacho, A., Gutiérrez Ruiz, S., Quiroga Alonso, J.M., 2019. Removal of emerging contaminants from wastewater using nanofiltration for its subsequent reuse: Full–scale pilot plant. *J. Clean. Prod.* 214, 514–523. https://doi.org/10.1016/j.jclepro.2018.12.297

Estrada-Arriaga, E.B., Cortés-Muñoz, J.E., González-Herrera, A., Calderón-Mólgora, C.G., de Lourdes Rivera-Huerta, M., Ramírez-Camperos, E., Montellano-Palacios, L., Gelover-Santiago, S.L., Pérez-Castrejón, S., Cardoso-Vigueros, L., Martín-Domínguez, A., García-Sánchez, L., 2016. Assessment of full-scale biological nutrient removal systems upgraded with physico-chemical processes for the removal of emerging pollutants present in wastewaters from Mexico. *Sci. Total Environ.* 571, 1172–1182. https://doi.org/10.1016/j.scitotenv.2016.07.118

Garcia, N., Moreno, J., Cartmell, E., Rodriguez-Roda, I., Judd, S., 2013a. The cost and performance of an MF-RO/NF plant for trace metal removal. *Desalination* 309, 181–186. https://doi.org/10.1016/j.desal.2012.10.017

Garcia, N., Moreno, J., Cartmell, E., Rodriguez-Roda, I., Judd, S., 2013b. The application of microfiltration-reverse osmosis/nanofiltration to trace organics removal for municipal wastewater reuse. *Environ. Technol. (United Kingdom)* 4(24), 3183–3189. https://doi.org/10.1080/09593330.2013.808244

Ghernaout, D., Elboughdiri, N., 2020. Disinfecting water: Plasma discharge for removing coronaviruses. *OALib* 07, 1–29. https://doi.org/10.4236/oalib.1106314

Hijosa-Valsero, M., Matamoros, V., Martín-Villacorta, J., Bécares, E., Bayona, J.M., 2010. Assessment of full-scale natural systems for the removal of PPCPs from wastewater in small communities. *Water Res.* 44(5), 1429–1439. https://doi.org/10.1016/j.watres.2009.10.032

Jain, M., Majumder, A., Ghosal, P., Gupta, A.K., 2020. A review on treatment of petroleum refinery and petrochemical plant wastewater: A special emphasis on constructed wetlands. *J. Environ. Manage.* 272, 111057. https://doi.org/https://doi.org/10.1016/j.jenvman.2020.111057

Kasprzyk-Hordern, B., Dinsdale, R.M., Guwy, A.J., 2009. The removal of pharmaceuticals, personal care products, endocrine disruptors and illicit drugs during wastewater treatment and its impact on the quality of receiving waters. *Water Res.* 43, 363–380. https://doi.org/10.1016/J.WATRES.2008.10.047

Kibambe, M.G., Momba, M.N.B., Daso, A.P., Coetzee, M.A.A., 2020a. Evaluation of the efficiency of selected wastewater treatment processes in removing selected perfluoroalkyl substances (PFASs). *J. Environ. Manage.* 255, 109945. https://doi.org/10.1016/j.jenvman.2019.109945

Kibambe, M.G., Momba, M.N.B., Daso, A.P., Van Zijl, M.C., Coetzee, M.A.A., 2020b. Efficiency of selected wastewater treatment processes in removing estrogen compounds and reducing estrogenic activity using the T47D-KBLUC reporter gene assay. *J. Environ. Manage.* 260, 110135. https://doi.org/10.1016/j.jenvman.2020.110135

Kovalova, L., Siegrist, H., Von Gunten, U., Eugster, J., Hagenbuch, M., Wittmer, A., Moser, R., McArdell, C.S., 2013. Elimination of micropollutants during post-treatment of hospital wastewater with powdered activated carbon, ozone, and UV. *Environ. Sci. Technol.* 47, 7899–7908. https://doi.org/10.1021/es400708w

Leyva-Díaz, J.C., Monteoliva-García, A., Martín-Pascual, J., Munio, M.M., García-Mesa, J.J., Poyatos, J.M., 2020. Moving bed biofilm reactor as an alternative wastewater treatment process for nutrient removal and recovery in the circular economy model. *Bioresour. Technol.* 299, 122631. https://doi.org/10.1016/j.biortech.2019.122631

Lu, F., Astruc, D., 2020. Nanocatalysts and other nanomaterials for water remediation from organic pollutants. *Coord. Chem. Rev.* 408, 213180. https://doi.org/10.1016/j.ccr.2020.213180

Luo, W., Phan, H.V., Xie, M., Hai, F.I., Price, W.E., Elimelech, M., Nghiem, L.D., 2017. Osmotic versus conventional membrane bioreactors integrated with reverse osmosis for water reuse: Biological stability, membrane fouling, and contaminant removal. *Water Res.* 109, 122–134. https://doi.org/10.1016/j.watres.2016.11.036

Mahamuni, N.N., Adewuyi, Y.G., 2010. Advanced oxidation processes (AOPs) involving ultrasound for waste water treatment: A review with emphasis on cost estimation. *Ultrason. Sonochem.* 17, 990–1003. https://doi.org/10.1016/j.ultsonch.2009.09.005

Majumder, A., Gupta, A.K., Ghosal, P.S., Varma, M., 2021a. A review on hospital wastewater treatment: A special emphasis on occurrence and removal of pharmaceutically active compounds, resistant microorganisms, and SARS-CoV-2. *J. Environ. Chem. Eng.* 9, 104812. https://doi.org/10.1016/j.jece.2020.104812

Majumder, A., Gupta, B., Gupta, A.K., 2019. Pharmaceutically active compounds in aqueous environment: A status, toxicity and insights of remediation. *Environ. Res.* 176, 108542. https://doi.org/10.1016/j.envres.2019.108542

Majumder, A., Saidulu, D., Gupta, A.K., Ghosal, P.S., 2021b. Predicting the trend and utility of different photocatalysts for degradation of pharmaceutically active compounds: A special emphasis on photocatalytic materials, modifications, and performance comparison. *J. Environ. Manage.* 293, 112858. https://doi.org/10.1016/j.jenvman.2021.112858

Matamoros, V., Arias, C., Brix, H., Bayona, J.M., 2009. Preliminary screening of small-scale domestic wastewater treatment systems for removal of pharmaceutical and personal care products. *Water Res.* 43(1), 55–62. https://doi.org/10.1016/j.watres.2008.10.005

Merényi, G., Lind, J., Naumov, S., Sonntag, C. von, 2010. Reaction of ozone with hydrogen peroxide (peroxone process): A revision of current mechanistic concepts based on thermokinetic and quantum-chemical considerations. *Environ. Sci. Technol.* 44, 3505–3507. https://doi.org/10.1021/es100277d

Miklos, D.B., Remy, C., Jekel, M., Linden, K.G., Drewes, J.E., Hübner, U., 2018. Evaluation of advanced oxidation processes for water and wastewater treatment – A critical review. *Water Res.* 139, 118–131. https://doi.org/10.1016/J.WATRES.2018.03.042

Moreira, F.C., Boaventura, R.A.R., Brillas, E., Vilar, V.J.P., 2017. Electrochemical advanced oxidation processes: A review on their application to synthetic and real wastewaters. *Appl. Catal. B Environ.* 202, 217–261. https://doi.org/10.1016/j.apcatb.2016.08.037

Nguyen, L.N., Hai, F.I., Kang, J., Price, W.E., Nghiem, L.D., 2013. Removal of emerging trace organic contaminants by MBR-based hybrid treatment processes. *Int. Biodeterior. Biodegrad.* 85, 474–482. https://doi.org/10.1016/j.ibiod.2013.03.014

Parida, V.K., Saidulu, D., Majumder, A., Srivastava, A., Gupta, B., Gupta, A.K., 2021. Emerging contaminants in wastewater: A critical review on occurrence, existing legislations, risk assessment, and sustainable treatment alternatives. *J. Environ. Chem. Eng.* 9, 105966. https://doi.org/10.1016/J.JECE.2021.105966

Patel, M., Kumar, R., Kishor, K., Mlsna, T., Pittman, C.U., Mohan, D., 2019. Pharmaceuticals of emerging concern in aquatic systems: Chemistry, occurrence, effects, and removal methods. *Chem. Rev.* 119, 3510–3673. https://doi.org/10.1021/acs.chemrev.8b00299

Rein Munter, 2001. Proceedings of the Estonian Academy of Sciences, Chemistry – Google Books.

Sahar, E., David, I., Gelman, Y., Chikurel, H., Aharoni, A., Messalem, R., Brenner, A., 2011. The use of RO to remove emerging micropollutants following CAS/UF or MBR treatment of municipal wastewater. *Desalination* 273, 142–147. https://doi.org/10.1016/J.DESAL.2010.11.004

Saidulu, D., Gupta, B., Gupta, A.K., Ghosal, P.S., 2021. A review on occurrences, eco-toxic effects, and remediation of emerging contaminants from wastewater: Special emphasis on biological treatment based hybrid systems. *J. Environ. Chem. Eng.* 9, 105282. https://doi.org/10.1016/j.jece.2021.105282

Sgroi, M., Roccaro, P., Korshin, G. V., Greco, V., Sciuto, S., Anumol, T., Snyder, S.A., Vagliasindi, F.G.A., 2017a. Use of fluorescence EEM to monitor the removal of emerging contaminants in full scale wastewater treatment plants. *J. Hazard. Mater.* 323, 367–376. https://doi.org/10.1016/j.jhazmat.2016.05.035

Sgroi, M., Roccaro, P., Korshin, G.V., Vagliasindi, F.G.A., 2017b. Monitoring the behavior of emerging contaminants in wastewater-impacted rivers based on the use of fluorescence excitation emission matrixes (EEM). *Environ. Sci. Technol.* 51, 4306–4316. https://doi.org/10.1021/acs.est.6b05785

Srivastava, A., Gupta, B., Majumder, A., Gupta, A.K., Nimbhorkar, S.K., 2021. A comprehensive review on the synthesis, performance, modifications, and regeneration of activated carbon for the adsorptive removal of various water pollutants. *J. Environ. Chem. Eng.* 9(5), 106177. https://doi.org/10.1016/J.JECE.2021.106177

Subedi, B., Balakrishna, K., Sinha, R.K., Yamashita, N., Balasubramanian, V.G., Kannan, K., 2015. Mass loading and removal of pharmaceuticals and personal care products, including psychoactive and illicit drugs and artificial sweeteners, in five sewage treatment plants in India. *J. Environ. Chem. Eng.* 3, 2882–2891. https://doi.org/10.1016/J.JECE.2015.09.031

Tran, N.H., Reinhard, M., Gin, K.Y.H., 2018a. Occurrence and fate of emerging contaminants in municipal wastewater treatment plants from different geographical regions-a review. *Water Res.* 133, 182–207. https://doi.org/10.1016/j.watres.2017.12.029

Tran, N.H., Reinhard, M., Yew-Hoong Gin, K., 2018b. Occurrence and fate of emerging contaminants in municipal wastewater treatment plants from different geographical regions-a review. *Water Res.* 133, 182–207. https://doi.org/10.1016/j.watres.2017.12.029

Varma, M., Gupta, A.K., Ghosal, P.S., Majumder, A., 2020. A review on performance of constructed wetlands in tropical and cold climate: Insights of mechanism, role of influencing factors, and system modification in low temperature. *Sci. Total Environ.* 755, 142540. https://doi.org/10.1016/j.scitotenv.2020.142540

Vymazal, J., Dvořáková Březinová, T., Koželuh, M., Kule, L., 2017. Occurrence and removal of pharmaceuticals in four full-scale constructed wetlands in the Czech Republic – the first year of monitoring. *Ecol. Eng.* 98, 354–364. https://doi.org/10.1016/j.ecoleng.2016.08.010

Wang, J.L., Xu, L.J., 2012. Advanced oxidation processes for wastewater treatment: Formation of hydroxyl radical and application. *Crit. Rev. Environ. Sci. Technol.* 42, 251–325. https://doi.org/10.1080/10643389.2010.507698

Wei, C.H., Wang, N., HoppeJones, C., Leiknes, T.O., Amy, G., Fang, Q., Hu, X., Rong, H., 2018. Organic micropollutants removal in sequential batch reactor followed by nanofiltration from municipal wastewater treatment. *Bioresour. Technol.* 268, 648–657. https://doi.org/10.1016/j.biortech.2018.08.073

Xiao, F., 2017. Emerging poly- and perfluoroalkyl substances in the aquatic environment: A review of current literature. *Water Res.* 124, 482–495. https://doi.org/10.1016/j.watres.2017.07.024

Yang, Y., Ok, Y.S., Kim, K.-H., Kwon, E.E., Tsang, Y.F., 2017. Occurrences and removal of pharmaceuticals and personal care products (PPCPs) in drinking water and water/sewage treatment plants: A review. *Sci. Total Environ.* 596–597, 303–320. https://doi.org/10.1016/j.scitotenv.2017.04.102

Zupanc, M., Kosjek, T., Petkovšek, M., Dular, M., Kompare, B., Širok, B., Blažeka, Ž., Heath, E., 2013. Removal of pharmaceuticals from wastewater by biological processes, hydrodynamic cavitation and UV treatment. *Ultrason. Sonochem.* 20, 1104–1112. https://doi.org/10.1016/j.ultsonch.2012.12.003

11 Sustainable Treatment Technologies for Achieving Circular Economy

CHAPTER OBJECTIVES

This chapter focuses on how to achieve sustainable wastewater treatment. The life cycle assessment of wastewater treatment plants has been discussed. Since wastewater treatment is a costly affair, resource recovery during wastewater treatment has become essential. In this context, different technologies for resource recovery have also been discussed in this chapter.

11.1 INTRODUCTION

The advances in technology for only removing pollutants from wastewater and discharging it are a thing of the past. Recently, apart from treating municipal wastewater, a significant focus has been placed on developing technologies to recover resources in the form of nutrients, energy, metals, and reclaim or reuse waste. One of the most essential components of sustainable development and the circular economy is the recuperation of resources (Schroeder et al., 2019). A circular economy is a management process that reduces the burden on natural resources by recovering resources from waste, thereby generating social, economic, and environmental benefits (Lieder and Rashid, 2016). A circular economy is finding a new way to recycle waste material into another valuable product. In this way, the waste generated today can become the resource for tomorrow (Kakwani and Kalbar, 2020). In circular economy, wastewater is a valuable resource as it contains nutrients, energy, and heavy metals (Vaneeckhaute et al., 2018; Guerra-Rodríguez et al., 2020). Although the recovery of resources may seem very lucrative, it is essential to assess the market potential of these products. Furthermore, it is essential to carry out the treatment and recovery in a sustainable manner. In order to assess the sustainability of a process, life cycle assessment (LCA) is a necessary technique. LCA is a method for assessing the environmental impacts of a product, service, or process from start to finish. In the case of wastewater treatment, LCA entails examining the environmental impacts of wastewater treatment technologies from the beginning to the end of their service life (Corominas et al., 2013; Gallego-Schmid and Tarpani, 2019). In this context, the various steps involved in carrying out an LCA of a wastewater treatment plant have been discussed in the following section.

DOI: 10.1201/9781003364450-11

11.2 LIFE CYCLE ASSESSMENT

The development of a complete LCA requires four main phases as per ISO standards (ISO 14040:2006, 2006; ISO 14044:2006, 2006) (Gallego-Schmid and Tarpani, 2019). The four phases are goal and scope definition, inventory analysis, impact assessment, and interpretation.

The LCA of a wastewater treatment plant can be performed on the system's performance and the product, that is, the quality of the treated water, the amount of wastewater treated, and other factors. On that basis, functional units, system boundaries, and inventories can be defined. Following the definition of the goal and scope, the data pertaining to the system is gathered. Following that, the system's environmental impact is evaluated. The assessment of impact is followed by the interpretation of the findings. For better interpretation, the goal and scope, or inventory analysis, may need to be re-adjusted. The detailed description of the different steps involved in LCA is as follows. A pictorial representation of the different processes involved in a typical LCA of a sewage treatment plant (STP) has been provided in Figure 1.5 in Chapter 1.

I. *Goal and scope definition:* This is the first and most important step in an LCA. In this step, the functional unit and the system boundaries are defined.

The functional unit tells us what is going to be assessed in the LCA. In the case of wastewater treatment, the most commonly used functional unit is the volume of wastewater treated or the treatment efficiency (Corominas et al., 2013; Gallego-Schmid and Tarpani, 2019; Tabesh et al., 2019).

The system boundaries are decided in the initial stage. System boundaries tell us the limitations or define the start and endpoints of an LCA. In the case of an STP, an LCA is usually conducted during the operational phase of the STP. However, LCA implies that the assessment should be from the cradle to the grave of the product or service. Hence, for STPs, LCAs can also include the construction phase of the plant and the decommissioning phase of the plant (Corominas et al., 2013; Gallego-Schmid and Tarpani, 2019; Tabesh et al., 2019).

II. *Inventory analysis:* Data collection involving the system's environmental inputs and outputs is carried out in the inventory analysis. This may involve the characteristics of influent and effluent of the STP, quantity and characteristics of produced sludge, the amount of energy required for the operation, amount of emissions during the treatment process, the chemicals required for treatment, and others (Corominas et al., 2013; Gallego-Schmid and Tarpani, 2019; Tabesh et al., 2019). The total environmental inputs and outputs of the system defined in phase I are quantified in phase II.

III. *Impact assessment:* In this phase, based on the data quantified in phase II, the significant environmental impact of the different inventories is assessed. This phase involves defining or classifying impact categories. Common impact categories associated with STPs are global warming potential, eutrophication potential of marine water and freshwater, acidification potential, toxicity toward humans, toxicity toward the ecosystem, to name a few (Corominas et al., 2013; Gallego-Schmid and Tarpani, 2019; Tabesh et al., 2019). After classification, the environmental impact is measured. Normalization and weighting are usually applied during impact assessment to facilitate better comparability with other LCAs. Normalization facilitates comparing all of the environmental impacts on the same scale, while weighting is used to convert and aggregate indicator results across impact categories into one single indicator (Corominas et al., 2013).

IV. *Interpretation:* In this phase, the results arriving from the previous phases are discussed and checked for inconsistencies. The recommendations are provided based on the limitations which were identified. The comparison with other LCAs is carried out in this phase.

LCA is vital to assess which treatment or recovery process is best suited for a particular type of wastewater. Based on the findings of LCA, different treatment and recovery technologies can be implemented. In this context, discussions on the different resources that can be recovered from wastewater and the reclamation of water are provided in the following sections.

11.3 RESOURCE RECOVERY

It has been assessed that an STP can generate a significant portion of the total energy required for its operation (Schopf et al., 2018). It was found that the amount of chemical energy stored in an STP is five times higher than the energy required for ASP. Researchers have previously recovered 94 petajoules of energy per year from a Dutch STP (Kehrein et al., 2020). Such results show that the STPs are major storehouses of energy. Similarly, wastewater reuse is an area that has attracted significant attention as well. Different countries have established guidelines and standards for their reuse in different applications. Although reclaimed wastewater is largely used for non-potable purposes, such as agriculture, landscaping, washing, cooling, and other purposes, many countries have recently started to use reclaimed wastewater for drinking purposes after it has gone through sufficient treatment and met the desired standards (Alcalde Sanza and Gawlik, 2014; CPHEEO, 2013; USEPA, 2012; Yang et al., 2006). Municipal wastewater also hosts a significant fraction of heavy metals, which is also a valuable resource and can be recovered (Kehrein et al., 2020). A brief overview of the different resources that can be recovered has been provided in Figure 11.1.

11.3.1 Nutrient Recovery

The disposal of wastewater frequently results in high nutrient concentrations in aquatic ecosystems, which can promote unwanted phytoplankton growth. Recovering nutrients from wastewater is a long-term solution to wastewater management that contributes to social sustainability (Kehrein et al., 2020; Rout et al., 2021). High-strength municipal wastewater is known to contain around 25 mg/L of organic nitrogen and 12 mg/L of phosphorus. Considering the amount of wastewater a

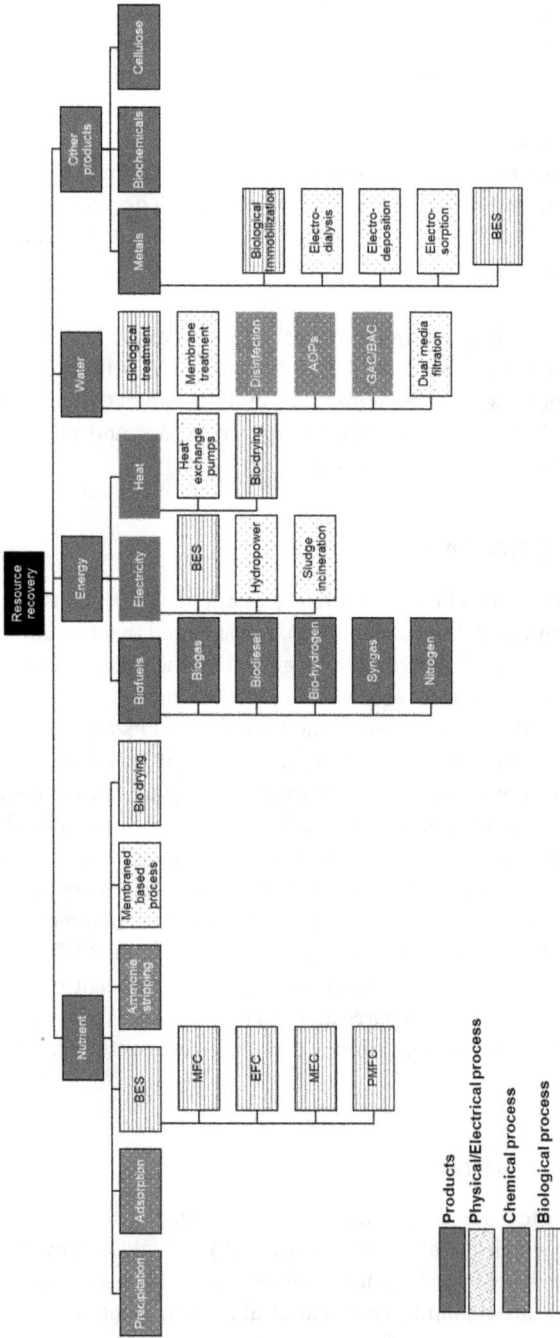

FIGURE 11.1 An overview of different resources that can be recovered from municipal wastewater and the process associated with the recovery.

full-scale STP treats, the amount of nitrogen and phosphorus recovery may be significant (Tchobanoglous et al., 2003).

Phosphorous has varying uses in agriculture since it is an essential component required for the growth of plants. It is also used in various industries. As a result, more attempts are being made to recover phosphorus from wastewater (Cornel and Schaum, 2009). Phosphorus can be recovered from wastewater, sewage sludge, and the ash of incinerated sewage sludge. The phosphorus recovery rate can reach up to 40–50% from the liquid phase, whereas recovery rates from sewage sludge and sewage sludge ash can be as high as 90% (Cornel and Schaum, 2009). On the other hand, nitrogen, which primarily exists in ammonium in wastewater (NH_4^+), is regarded as one of the best sources for fertilizer production. It is generally wasted during wastewater treatment and can be recovered for global environmental sustainability. Ammonia recovery from organic waste streams has been studied using a variety of techniques (Ata et al., 2017; Menkveld and Broeders, 2018; Winkler et al., 2013).

11.3.1.1 Chemical Recovery

Adsorption and precipitation are the most common methods used in recovering different nutrients from wastewater. In chemical precipitation techniques, magnesium and calcium-based materials are used to precipitate nutrients. The calcium- and magnesium-based materials react with the nutrients to form hydroxyapatite $Ca_5(OH)$ $(PO_4)_3$ and struvite ($MgNH_4PO_4 \cdot 6H_2O$), respectively. The general equations driving the formation of hydroxyapatite and struvite are provided in Equations 11.1 and 11.2, respectively.

$$5Ca^{2+} + 3PO_4^{2-} + OH^- \rightarrow Ca_5(OH)(PO_4)_3 \downarrow \qquad (11.1)$$

$$Mg^{2+} + PO_4^{3-} + NH_4^+ + 6H_2O \rightarrow MgNH_4PO_4 \cdot 6H_2O \downarrow \qquad (11.2)$$

The different factors that affect precipitation are pH, temperature, and concentration of the nutrients present in the wastewater. The most favorable pH for precipitation to occur is in the range of 7–11 (Tansel et al., 2018). This is mainly because the precipitation rate is lower at acidic pH (Ye et al., 2020). Although the lower temperature is beneficial for struvite formation, precipitation at lower pH is favored when the temperature is high. The phosphorus concentration in municipal wastewater limits the effectiveness of its recovery by struvite precipitation. The phosphorus concentration has a major impact on the driving force and kinetics of struvite precipitation. Extensive experimental results revealed that adequate struvite precipitation could only be achieved when the phosphorus concentration was higher than 100 mg/L (Kehrein et al., 2020; Xie et al., 2016). Research also suggested that the ratio of Mg:P and Ca:P for efficient precipitate formation should be more than 1.0 and 1.67, respectively (Ye et al., 2020).

Selectrodialysis (SED) may be used to increase the efficiency of phosphate recovery from a struvite reactor. SED was used to desalt and transfer the phosphate-containing wastewater to the struvite reactor. A cost analysis was also performed in the above study, which showed that 1 kWh electricity could produce 60 g of phosphate using a full-scale unit, with a rate of desalination of 95% on the feed wastewater. The struvite precipitation experiment demonstrated that 93% of the phosphate could be recovered (Zhang et al., 2013). As a result, an integrated SED-struvite reactor technique can be used to improve phosphate recovery. The formed struvite obtained may be used as a fertilizer, while hydroxyapatite is a key ingredient in the phosphate industry.

Adsorption is another process that is quite efficient in terms of nutrient recovery. Metal-based adsorbents with large specific surface areas are usually employed for the adsorption of nutrients. The process is economical and associated with simple design and operation. The three primary mechanisms involved in phosphate adsorption are surface precipitation, electrostatic attraction, and ion exchange. As a result, the pH of the wastewater plays an integral role in the adsorption of phosphate. Since phosphate ions are negatively charged, adsorbents having a positive surface charge are more likely to perform better. Hence, adsorbents having an isoelectric point higher than that of the pH of the wastewater are preferred. Unlike phosphate adsorption, ammonium adsorption is a physical phenomenon. Also, ammonium ions may be converted to volatile ammonia through high reaction temperature or alkaline pH after which they may be adsorbed onto acidic solutions like sulfuric acid. The formed ammonium sulfate is often used in fertilizers (Equation 11.3) (Ye et al., 2020; Zhang et al., 2019).

$$2NH_3 + H_2SO_4 \rightarrow (NH_4)_2SO_4 \downarrow \qquad (11.3)$$

11.3.1.2 Ammonia Air Stripping for Nitrogen Recovery

Ammonia air stripping is a commonly used method for recovering nitrogen as ammonia from industrial and municipal wastewater (Figure 11.2). A conventional ammonia stripping tower is generally loaded with packing media. After going through the preheating process, the ammonia-saturated water then enters the stripper tower from the very top, while the counter-flow air stream comes in from the very bottom. The water that has been stripped is then sent through the tower's bottom to be treated some more, while the treated ammonia-laden gas is collected from the top of the tower and then delivered to the ammonia acid absorbers. Here, the ammonia gas reacts with sulfuric acid, which absorbs the ammonia and leads to the formation of ammonium sulfate (commonly used for fertilizer production). However, stripping ammonia from a conventional stripping tower is replete with several challenges, like controlling the pH. A pH of more than 9 is favorable for stripping ammonia to maximize ammonia stripping and reduce energy waste. In order to address these issues, researchers are focusing on reactor modification to improve ammonia stripping performance and economic effectiveness.

Ata et al. (2017) discussed the performance of microwave radiation integrated with the ammonia stripping process for ammonia removal from an aqueous solution with an initial ammonia concentration of 1,800 mg/L. The system was operated with a minimum airflow rate of 7.5 L/min, microwave power of 200 W with an optimum radiation time of 180 min, and at a stripping temperature of 60°C. The results showed that a maximum of 94.2% ammonia was removed via an acidic medium at an optimum pH of 11. This absorbed ammonia can be converted to ammonium minerals for ammonia

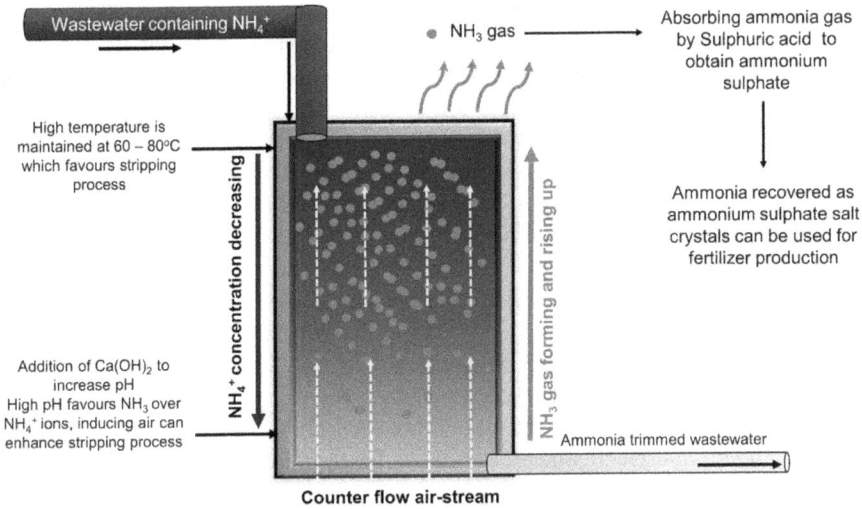

FIGURE 11.2 The working mechanism of the ammonia air stripping method for nutrient recovery.

nitrogen recovery. Microwave radiation accelerates ammonia stripping by heat and improves the efficiency of liquid-gas stripping. Hence, the process can be quickened, and a large amount of ammonia can be removed or recovered from wastewater (Ata et al., 2017). Menkveld and Broeders (2018) discussed the performance of the Nijhuis ammonium recovery system (NAR) for removing and recovering ammonia nitrogen from wastewater. The NAR system comprises two specially designed towers – one for carrying out the ammonia stripping process and the other with a scrubber at the bottom for absorbing the ammonia through a sulfuric acid medium. The main advantage of the NAR system is that it is very compact in size and has an availability of a pH controlling option just prior to the first tower, as a high pH (greater than 10) is required to strip ammonia from ammonium. The ammonia-rich gas is absorbed by a specially designed scrubber containing sulfuric acid, which converts it to ammonium sulfate. The test results showed that the NAR system could achieve an average ammonia removal efficiency of 80–90% for anaerobically digested manure and municipal wastewater (Menkveld and Broeders, 2018). A cost analysis was also performed to determine the cost-effectiveness of the NAR system, which is primarily determined by the amount of NH_3-N in the influent. Moreover, it was found that the NAR system was much more cost-effective when compared to other improved air stripping methods. However, compared to conventional air stripping methods, it was slightly more expensive but had more advantages (Menkveld and Broeders, 2018).

11.3.1.3   Bioelectrochemical Systems

Bioelectrochemical systems (BES) are such systems that convert the chemical energy of organic waste, such as wastewater and lignocellulose biomass, into electrical energy or hydrogen/chemical products. In this process, the organic compounds are oxidized by the microorganisms present in the wastewater. During the oxidation process, electrons are generated, which can produce energy for valuable products. Several research studies have found that BES systems can effectively remove and recover nutrients (nitrogen and phosphorus) from wastewater (Kelly and He, 2014; Pant et al., 2012; Wang and Ren, 2014). BES systems can use a wide range of organic substrates and not rely on expensive metals as catalysts. The most common types of BES available depending upon the biocatalysts used or based on their mode of application are microbial fuel cells (MFC), enzymatic fuel cell (EFC), microbial electrolysis cell (MEC), plant microbial fuel cell (PMFC), and microbial solar cell (MSC) (Kehrein et al., 2020). A typical BES and the mechanism behind nutrient recovery have been shown in Figure 11.3.

Zhang et al. (2011) performed an experiment using a BES to remove and recover nitrogen and phosphorus from wastewater. In order to produce algal biomass, the process involved a synergistic interaction between microalgae (*Chlorella vulgaris*) (at the cathode) and electrochemically active bacteria (at the anode). The growing of microalgae in conjunction with the BES technology aided the electrochemically active bacteria that were located at the anode of the cell in their process of oxidizing organic materials in order to liberate electrons and protons. In the meantime, as a result of the solar irradiation, the microalgae that were located in the cathode were able to take nitrogen and phosphorus from the anode in order to begin the process of photosynthesis. Additionally, the oxygen that was produced as a byproduct of algae photosynthesis was put to use as an electron acceptor in the cathode's process of nitrification, which led to the production of energy. Through this synergetic interaction, nitrogen and phosphorus could be effectively recovered from the algal biomass without extra cost. The study also reported a maximum removal efficiency of 87.6% and 69.8% for nitrogen and phosphorus, respectively. The mass balance analysis suggested the main mechanism behind the removal/recovery of nitrogen and phosphorus was the algal biomass assimilation, with a maximum uptake of 75% and 93%, respectively (Zhang et al., 2011).

In another study involving the use of BES, Kelly and He (2014) tried to recover nitrogen from municipal wastewater. During the conventional denitrification process, nitrate accepts electrons from organic compounds to reduce to nitrogen gas. In this process, a similar electron transfer concept was implemented to use nitrate as a terminal electron acceptor in a BES. In BES, bioelectrochemical denitrification uses autotrophic denitrifying bacteria that can take electrons from a solid electron donor (e.g., a cathode electrode), unlike that in conventional denitrification processes, which uses heterotrophic denitrifying bacteria. The denitrification process led to the generation of nitrogen gas. Further, this nitrogen gas can be converted to ammonia gas by nitrifying bacteria. A sulfuric acid medium scrubber can be used to absorb the ammonia gas and convert it to ammonium sulfate (Kelly and He, 2014).

11.3.1.4   Combination of BES and Precipitation

BES and precipitation techniques may be combined to enhance the recovery of nutrients. Ichihashi and Hirooka (2012) studied the effectiveness of phosphorus removal and recovery from wastewater using air-cathode single MFC. The MFC

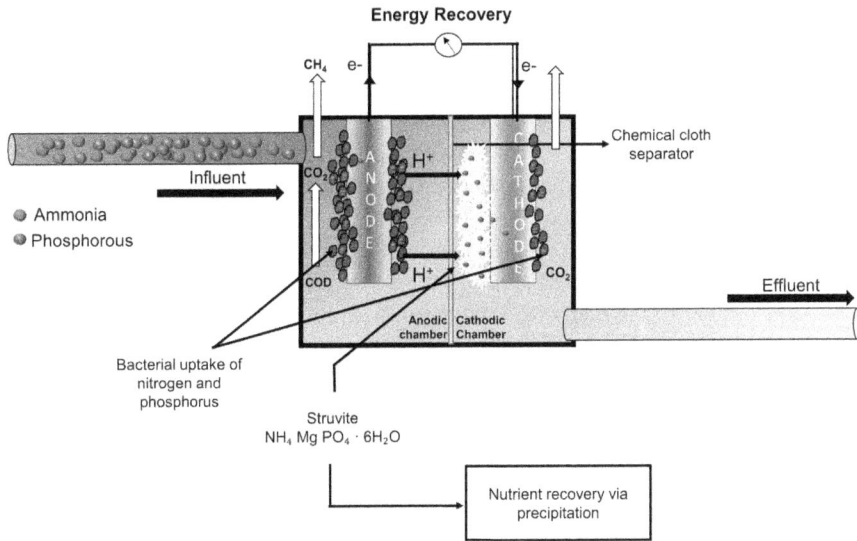

FIGURE 11.3 Nutrient recovery, methane, and electricity production using in bioelectrochemical system.

was composed of an anode made of carbon felt (7.7 cm diameter and 1 cm thick), a cathode prepared by coating a wet-proof porous carbon paper with 0.5 mg/cm^2 platinum/carbon catalyst, and a polyester nonwoven cloth separated both the cathode and anode. The mechanism behind such a combination of BES and precipitation has been depicted in Figure 11.3. It was discovered that for effective phosphorus recovery from struvite precipitation, the struvite crystal must first be formed, which occurs when the pH is between 8 and 9. The formation of struvite precipitate depends on pH. Hence, keeping an alkaline environment near the cathode can stimulate its formation. During the MFC operation, struvite precipitates are generated due to the oxygen reduction near the cathode. The amount of phosphorus in these precipitates, which can be recovered effectively, was calculated to be 4.6–27 % of the influent phosphorus. The main purpose of using single-chambered MFC with swine wastewater in the above study is to enhance the concentration of orthophosphate in the struvite, which is formed at the cathode due to the cathodic reaction, and thus more phosphorus can be recovered from the influent wastewater (Ichihashi and Hirooka, 2012).

Fischer et al. (2011) performed an experiment for effective orthophosphate recovery by mobilizing iron phosphate ($FePO_4$) from digested sewage sludge using an MFC system consisting of an anode made up of six carbon felt electrodes, and a cathode made up of the same material. In this process, MFC served as a power source for the generation of stoichiometric amounts of electrons and protons at the cathode, mainly due to the metabolic activity of *Escherichia coli*. The electrons reduced iron cations, and protons replaced the charges. This led to the mobilization of iron phosphates to orthophosphates (H_3PO_4, $H_2PO_4^-$, HPO_4^{2-} and PO_4^{3-}). Finally, phosphate could be recovered as fertilizer from struvite precipitates by reacting the orthophosphate-containing supernatant solution with stoichiometric amounts of $MgCl_2$ and NH_4OH (Fischer et al., 2011)

11.3.1.5 Membrane-Based Processes

Membrane separation methods have proved to be a potential technology for nutrient (phosphorus and nitrogen) removal and recovery because they are often designed to be contaminant-specific and can be targeted to separate nutrients from other toxic substances in the wastewater. Compared with other nutrient recovery techniques such as BES, ion exchange and precipitation, the membrane separation process is more stable and simpler to operate with less waste generation and high nutrient recovery. Various studies reported that microfiltration (MF), ultrafiltration (UF), nanofiltration (NF), forward osmosis (FO), and reverse osmosis (RO), which exhibit high rejection of phosphates, show great potential for nutrient recovery. The fundamental process involves recovering the nutrients from the concentrate by precipitation or ammonia stripping (Figure 11.4).

Qiu et al. (2015) operated an FO membrane and an MF membrane in parallel inside the bioreactor. The FO membrane was mainly employed for rejecting the phosphates (PO_4^{3-}) from the incoming wastewater influent, which resulted in its enrichment in the bioreactor. On the other hand, the primary function of MF was to extract phosphate. Subsequently, phosphorus is then recovered from phosphate-enriched MF permeate through precipitation. The FO membrane could reject almost 97.9% of phosphate phosphorus $(PO_4^{3-}-P)$ inside the reactor. Also, this process could recover more than 90% of phosphorus at an optimum pH of 9 as amorphous calcium phosphate through precipitation with phosphorus content ranging from 11.1 to 13.3% (Qiu et al., 2015).

Kekre et al. (2021) used reactive electrochemical membranes based on the electrochemical filtration approach for the recovery of nutrients from wastewater. These are special membranes in which the surface characteristics and mass transfer to and from membranes can be altered by applying an external electric field or potential to the membrane. The study showed that due to the increase in magnesium concentration in the feed solution and adjusting the pH near the membrane surface, the reactive

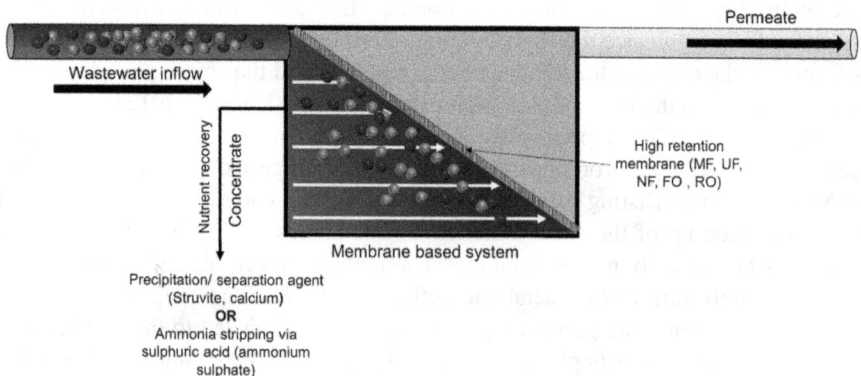

FIGURE 11.4 The working mechanism of a membrane-based system for nutrient recovery.

electrochemical membranes could successfully precipitate and separate struvite from the wastewater. In this process, about 65% of nitrogen and phosphorus were successfully removed within the first 30 min of electrochemical filtration (Kekre et al., 2021).

Pradhan et al. (2019) used a nitrogen–phosphorus harvest technique for recovering both nitrogen and phosphorus from human urine using a gas-permeable hydrophobic membrane (GPHM) via calcium sediment precipitation (Equation 11.4) and nitrogen stripping techniques (Equation 11.3). The process involved a combination of both membrane and precipitation techniques. In this process, the urine pH was increased with $Ca(OH)_2$, which resulted in the precipitation of phosphorus as calcium sediment, and the conversion of NH_3-N into ammonia gas simultaneously. This ammonia gas was then passed through GPHM, where it reacted with sulfuric acid to form ammonium sulfate (Equation 11.3). Finally, both phosphorus and nitrogen were recovered as calcium phosphate and ammonium sulfate, respectively, in the same run. The test results showed more than 98% (w/w) phosphorus, and nitrogen can be recovered from urine with a time run of 8 h at 30°C. The recovered ammonium sulfate contained almost 19% (w/w) nitrogen, whereas calcium sediment contained 1–2% (w/w) phosphorus (Pradhan et al., 2019).

$$5Ca(OH)_2 + 3PO_4^{3-} \rightarrow Ca_5(PO_4)OH \qquad (11.4)$$

11.3.1.6 Phosphorus Recovery from Incinerated Sewage Sludge Ash

Phosphorus recovery from incinerated sewage sludge ash (ISSA) has recently received a lot of attention from researchers because it offers higher rates of influent phosphorus recovery than other methods. However, specific types of incinerators are needed to achieve high recovery efficiency, which can reach high temperatures and completely burn waste sludge. The process requires highly skilled staff for proper handling and operation. There are different methods for extracting phosphorus from ISSA, such as chemical extraction techniques, electro-dialytic techniques, and thermochemical extraction techniques. ISSA can retain 75–98% phosphorus during sewage sludge incineration, and the approximate phosphorus content in ISSA is in the range of 10–25% as P_2O_5. Chemical extraction is the most widely used method because it achieves high efficiency at a low cost while simultaneously leaching metals/metalloids present in the ISSA. The primary mechanism for recovering phosphorus from ISSA is precipitation. Phosphorus is precipitated from sludge leachate by lowering the pH to 4 and then transforming the phosphorus metal precipitates into plant-available fertilizers by adding Ca^{2+}, Mg^{2+}, or other cations (Figure 11.5).

Fang et al. (2018) discussed the chemical extraction technique for phosphorus recovery from ISSA and compared the performance of different chemical agents, including inorganic acids, organic acids, and chelating agents for leaching phosphorus from ISSA. The primary aim of the study was to optimize a leaching process that could recover phosphorus leachate, which could be transformed into high-purity phosphorus fertilizer. Characterization studies on leached ISSA confirmed the presence of various crystalline minerals, including whitlockite ($Ca_9(MgFe)$ $(PO_4)6PO_3OH)$, which is commonly used as a phosphorus-containing fertilizer. The test results showed that inorganic acids were the best agents for extracting phosphorus

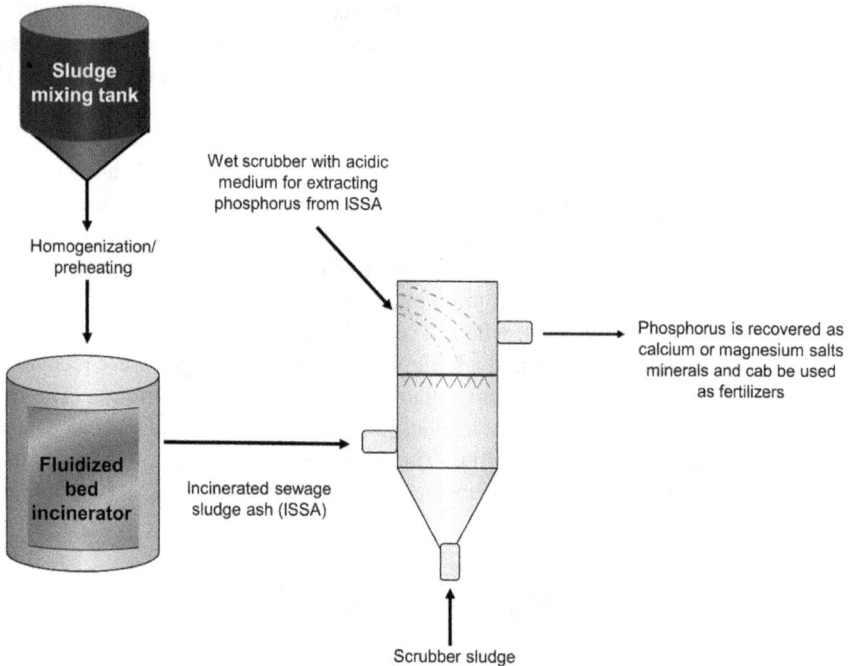

FIGURE 11.5   Schematic showing phosphorus recovery from incinerated sewage sludge ash.

from sludge leachate, out of which sulfuric acid was the most efficient (Fang et al., 2018). Similarly, Adam et al. (2009) conducted a two-step thermochemical treatment process operating at a temperature of 1,000°C in a laboratory-scale rotary furnace by treating seven different sewage sludge ashes exhibiting high phosphorus contents (approximately 20% of P_2O_5) for phosphorus recovery. The first stage of treatment included mono-incineration, which completely removed all organic compounds from the sludge, after which phosphorus and heavy metals were the primary leftovers present in the resulting sewage ashes. A second step treatment was used to effectively recover phosphorus from ashes using chemical agents, such as calcium chloride and magnesium chloride, so that the majority of the phosphorus was transformed into mineral phases (calcium and magnesium phosphate compounds) that can be used as fertilizers, and this step also removed heavy metals (Adam et al., 2009). Guedes et al. (2014) experimented using the electrodialytic separation process (ED) for phosphorus recovery from ISSA. The ED setup consisted of three compartments, having an anode and cathode (platinum-coated electrode) separated by an electrolyte solution with a magnetic stirrer. The process began with suspending ISSA in the sulfuric acid, which was agitated constantly for 14 days. After 14 days, the negatively charged orthophosphates were transported toward the anode from ash suspension, whereas the heavy metals in the ash were mobilized toward the cathode. The test results showed that after 14 days, 59–69% of phosphorus was transported toward the anode, which could be effectively recovered. It has also been reported that at pH levels ranging from

1.3 to 3.5, most of the phosphorus in the ash existed in the form of $H_2PO_4^-$ and H_3PO_4, which could be successfully used in fertilizer production (Guedes et al., 2014).

11.3.1.7 Bio-drying Concept of Sewage Sludge for Nitrogen Recovery

Almost 40% of the organic matter is converted to biomass (sludge) during the biological wastewater treatment process. This excess sludge is generated as a byproduct and poses serious challenges because treating this biomass (via incineration, oxidation, or digestion) is a costly process. A common and effective method for dealing with such large amounts of organic waste is biological composting or bio-drying, which converts the biomass to odorless and pathogen-free humus that can be successfully applied to the land. One significant advantage of this method is that bacteria actinomycetes, molds, fungi, and yeast oxidize the short- and long-chain fatty acids, paper products, and other pollutants to produce heat during the composting process, resulting in waste reduction through microbial conversion and water evaporation. Furthermore, due to the biological conversion of sludge, ammonium products get converted to ammonia gas (Choi et al., 2001; Navaee-Ardeh et al., 2006; Winkler et al., 2013).

Winkler et al. (2013) used the bio-drying concept to recover nitrogen from sewage sludge. The procedure of bio-drying begins with the collection of dewatered sewage sludge with an average solid content of 25% and an organic fraction of 65% from STPs (Netherlands). The sludge was combined with a coarse proportion of previously dried sludge. The sludge mixing is carried out to inoculate the microbial population, which is well adapted for aerobic breakdown of organic matter under thermophilic temperatures (65–75°C). The sludge was transported to large drying tunnels, where air blowers aerated the sludge. The primary function of aeration was to promote microbial activity while also evaporating water from the sludge. Microbial activity also aided in odor control, heat production, and nitrogen recovery. During the biological conversions of the biomass, ammonium was generated (sludge contains a high amount of nitrogen as ammonia), which was converted to ammonia gas by increasing the pH. The ammonia-rich gas was then reacted with sulfuric acid to produce ammonium sulfate. It was also reported that this method could achieve almost 60–80% nitrogen recovery (Winkler et al., 2013). The mechanism involved in bio-drying has been depicted in Figure 11.6.

11.3.1.8 Ion Exchange Process for Ammonia Nitrogen and Phosphorus Recovery

Ion exchange refers to purifying aqueous solutions using a solid polymeric ion-exchange resin such as zeolite, montmorillonite, clay, and soil humus. The term more precisely refers to a wide range of processes involving the exchange of ions between two electrolytes. Ion exchange is associated with high nitrogen and phosphorus recovery efficiency (Williams, 2013).

- *Ammonia recovery:* Ammonia can be removed or recovered from municipal wastewater by ion exchange technique using a specially designed media known as clinoptilolite ($(Na,K,Ca)_{2-3}A_{13}(Al,Si)_2Si_{13}O_{36}.12H_2O$), which is a

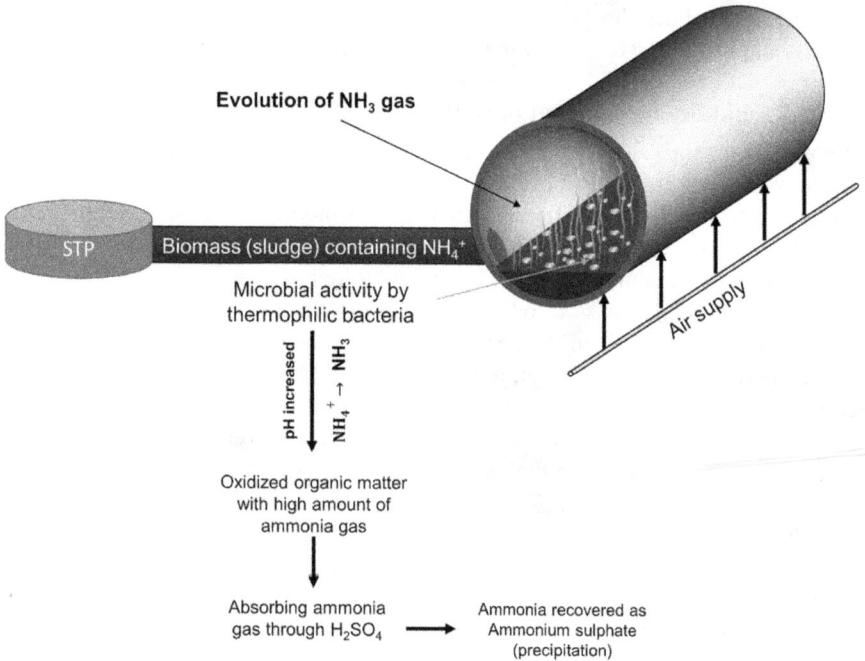

Evolution of NH₃ gas

STP

Biomass (sludge) containing NH₄⁺

Air supply

Microbial activity by
thermophilic bacteria

pH increased

$NH_4^+ \rightarrow NH_3$

Oxidized organic matter
with high amount of
ammonia gas

Absorbing ammonia
gas through H_2SO_4 → Ammonia recovered as
Ammonium sulphate
(precipitation)

FIGURE 11.6 The working mechanism of a bio-drying method for nutrient recovery.

natural mineral and acts as a zeolite that can absorb ammonium efficiently. The exchange of ions takes place either between the two layers of silica oxide or aluminum oxide tetrahedral. The cations like K^+, Ca^{2+}, and Mg^{2+} are very weakly bonded to the surface and thus take part in the ion exchange process in exchanging ammonium from wastewater. The main purpose of using clinoptilolite is that it can select ammonium ions over other ions. At optimum conditions, with pH more than 10, it is reported that clinoptilolite could effectively remove 94–98% ammonia from the wastewater (Williams, 2013).

- *Phosphate recovery:* Phosphorus, as PO_4-P, can also be removed or recovered using the ion exchange process. However, this process has a few downsides, including poor selectivity for phosphate over other ions. Despite a number of challenges, the process can be optimized for effective phosphate removal by employing PO_4-P selective resins (containing Cu or Fe) that form strong coordination bonds with HPO_4^{2-} and H_2PO_4. It has also been reported that polymeric ligand exchange resins impregnated with metal nanoparticles or their combination are very efficient in selecting phosphate compounds because these resins attract Lewis bases like PO_4-P over other compounds in the wastewater (Williams, 2013).

11.3.2 ENERGY RECOVERY

According to the International Energy Outlook 2019 (IEO2019) report by the U.S. Energy Information Administration (EIA), worldwide energy demand is predicted to climb by approximately 50% between 2018 and 2050. As a result, fossil-related emissions are expected to rise, and fossil fuels are expected to meet about 80% of the aforementioned demand. The EIA also predicted that between 2018 and 2050, global energy-related CO_2 emissions will increase at a rate of 0.6% per year. These forecasts underscore the need to minimize the energy intensity of STPs by focusing on energy efficiency and recovery when designing treatment procedures. This will help reduce the amount of energy that is required to treat wastewater. Both the United States and the United Kingdom's municipal wastewater treatment facilities currently account for approximately 4% of their respective countries' total national electricity usage. Hence, it is imperative to look for an alternative source of energy to run STPs. Furthermore, wastewater is a storehouse of a vast amount of energy. If this energy can be harnessed, many problems pertaining to the energy crisis may be addressed. In this context, the following sections discuss the different energies that can be harnessed from wastewater.

11.3.2.1 Biofuels

In the last few decades, biofuels have been used worldwide, and the biofuel industry is booming in Europe, Asia, and North and South America. They have many advantages over fossil-based fuels since they are sulfur-free and emit low levels of carbon monoxide and hazardous pollutants. Biofuels can reduce greenhouse gas emissions and provide an alternative to fossil fuels, hence increasing energy security. Some of the common forms of biofuels that are commercially produced are biogas, biohydrogen, biodiesel, syngas, and nitrogenous fuels.

Biogas: The most common way to get energy back is to turn sludge into biogas through anaerobic digestion. This method is used on a wide range of scales all over the world. In reactors with full mixing, it is possible to turn about 80% of the biochemical oxygen demand (BOD) in the sludge into biogas, which can then be collected. There is a chance that more advanced reactor configurations could make biodegradation work better and make it easier to get the methane out of the broth. When the temperature is kept at a normal level, up to 40% of the methane that is produced can be dissolved in the broth. Hence maintaining temperature is a key factor cost in digesters, which also affects the cost of the process. This dissolved methane could, in the long run, end up playing a role in climate change. One way to increase the amount of biogas that is recovered could be to increase the amount of primary sludge that is digested after capturing the maximum amount of chemical oxygen demand (COD) at the entry of the treatment unit. When this kind of energy recovery is used in treatment units, the net amount of energy used drops by about 40%. (Kehrein et al., 2020). The mechanism behind biogas or methane generation has been depicted in Figure 11.3.

Mata-Alvarez et al. (2014) compared traditional anaerobic digestion to anaerobic co-digestion, which is the simultaneous anaerobic digestion of two or more substrates. When two different digestion processes are used, they produce less methane than when they are used together. As a direct result of this, anaerobic co-digestion has become a common practice in agriculture. Most of the time, when people talk about manure-based digesters, they talk about two types of co-substrates: agro-industrial waste and the organic part of municipal solid waste. Co-digestion with anaerobic bacteria is closely linked to the process of making systems that are financially independent. In this situation, research has been done over the past few years to find ways to increase biogas production by creating the best conditions, using fresh residues by using pre-treatments, and increasing plant profits by using digested sludge as fertilizer (Mata-Alvarez et al., 2014).

Bio-hydrogen: Hydrogen gas is a good source of energy that can be used to clean up a wide range of pollutants in water. As a source of energy, it is especially appealing because chemical fuel cells can run on it. Using a two-step anaerobic sludge treatment process that includes hydrolysis and acidogenic fermentation by phototrophic and/or lithotrophic microorganisms, hydrogen can be recovered biologically from wastewater. Since the dark fermentation method only turns about one-third of the COD into hydrogen and the rest into volatile fatty acids, photofermentation is usually used with it. Lee et al. (2010) have discussed the different methods of producing biological hydrogen (Bio-H_2), which is renewable and carbon neutral. It was reported that Bio-H_2 could be generated in three different ways: fermentation, photosynthesis, and microbial electrolysis cells. These processes work on the principle that microorganisms can use protons (H^+) as an electron sink for two electron equivalents ($2H^+ + 2e^- \rightarrow H_2$). Hydrogen has low solubility in water; hence it can be harvested as gas in large quantities from water.

Biodiesel: Most of the organic fraction in municipal wastewater comprises lipids that can be assimilated and accumulated with the help of specialized microorganisms anaerobically. By skimming the surface of wastewater treatment reactors, this lipid-rich biomass could be collected and used as a feedstock for making high-yield biodiesel. The use of phototrophic microalgae in high-rate ponds to treat wastewater is one of the well-studied biodiesel production routes.

Muller et al. (2014) discussed the different alternatives for producing bio-oils. In Fischer–Tropsch process, short-chain alcoholic liquid biofuels are produced from wastewater biomass. The production of bio-oils from waste sludge by pyrolysis can be considered a viable option for renewable biofuels from sludge. Another alternative method for producing liquid biofuels, such as biodiesel and short-chain alcohols, is carried out using wastewater-derived gasified biomass. The process involves microbial syngas fermentation and the conversion of CO and H_2 into hydrocarbons. Pastore et al. (2013) investigated different alternatives for the production of biodiesel from domestic sludge. Using dewatered sludge at the initial stage could cut down the cost of production of biodiesel as it will be exempted from processes like sludge drying.

Syngas: Syngas is mostly made up of 35–40% hydrogen gas and 30–60% carbon monoxide. It is burned in gas engine boilers to make steam that is used to make electricity. It is made from sewage sludge by using supercritical water treatment processes. During supercritical water gasification or partial oxidation, the temperature and pressure are raised above the critical point of water (374°C and 221 bar). This releases syngas from biomass. This technology is better than other ways of dealing with sludge because it only takes a few minutes to turn the sludge into an energy carrier. Also, sludge from wastewater treatment plants that has too much water does not have to be dried out before it is put into supercritical water reactors.

Tyagi and Lo (2013) discussed the widespread popularity of the microwave irradiation process as an effective thermal method for sludge treatment primarily due to its rapid and selective heating, energy efficiency, ability to increase product yield and quality, and reduced hazardous product formation and emissions. It was reported that microwave pyrolysis of sludge for short reaction time and high heating rates yields oils (biofuels) with high aliphatic and oxygenated properties and no environmentally harmful (carcinogenic and mutagenic) compounds. Therefore, pyrolysis and gasification of sewage sludge using microwave irradiation could significantly increase the production of energy-rich syngas with a high calorific value.

Nitrogenous fuels: Wastewater can also be used to recover nitrogenous fuels as well. Scherson et al. (2013) discussed one such method for the recovery of nitrogenous fuels using three steps: (i) nitration of NH_4^+ to NO_2^-; (ii) partial anoxic reduction of NO^{2-} to N_2O; and (iii) lastly, chemical conversion of N_2O to N_2 with energy recovery. Gao et al. (2014) suggested NH_3 can be ignited to generate power or used as a transport fuel. It can even be converted into N_2O by nitridation. BES systems can also be used to generate nitrogen from municipal wastewater. The autotrophic denitrifying bacteria take up electrons from the cathode to reduce nitrate to nitrogen gas (Kelly and He, 2014).

11.3.2.2 Electricity

Bioelectrochemical systems: The BES are very effective methods for producing energy and other valuable products, which works on the principle that microorganisms oxidize COD, and the electrons generated during the process are efficiently used for producing energy. Rabaey and Rozendal (2010) discussed microbial electrosynthetic process, where electricity-driven reduction of CO_2 and redox reaction of wastewater occurs. The bioelectrochemical systems comprise an anode and a cathode compartment separated by a membrane. A simultaneous redox reaction occurs in the system, with an oxidation process occurring on the anode side and reduction at the cathode. Electrons can be transferred directly between cells and electrodes or via soluble molecules that can be reduced and oxidized while receiving electrons from cells and transporting them to the electrode, leading to the effective generation of electricity (Puyol and Batstone, 2017). The mechanism of electricity generation in BES has been provided in Figure 11.3.

Hydropower: It is a well-known method of recovering electricity by making use of the constant discharge from wastewater treatment plants that can be accomplished through the application of hydropower technologies to effluents. The amount of power that can be generated by hydropower technology is primarily determined by the flow rate and the hydraulic head. Power et al. (2014) discussed the advancement in the sector's sustainability via energy recovery at the outlets of wastewater treatment plants using hydropower turbines. Flow and head data were collected from outlet pipes from around 100 plants in Ireland and the UK. A sensitivity analysis revealed that the flow rate, turbine selection, electricity pricing, and financial incentives significantly affected power generation.

The different types of equipment involved in electricity generation must be made of stainless steel if wastewater is used (Kehrein et al., 2020).

Sludge incineration: When the sewage sludge is burned in the presence of oxygen, the organic content in the sludge is completely oxidized, resulting in the formation of CO_2, water, and inert material (ash) as end products. The heat that is generated when something burns can be used to produce electricity. Raw sewage sludge has a heating value that is 30–40% higher than digested sewage sludge. This means that it could be used as a fuel to produce electricity. To get energy from organic matter, plants are set up in different ways to burn large amounts of biomass, such as dried sewage sludge. Biomass combustion plants could make electricity with an average efficiency of 25–30%. Fluidized bed technology in combustion plants can boost electricity efficiency by up to 40% while lowering costs and increasing fuel flexibility. Another extensively used approach in Europe is sludge co-combustion in coal-fired power stations, which yields similar efficiencies (Faaij, 2006).

11.3.2.3   Heat

The thermal energy in wastewater treatment effluents comes mainly from household water heating and slightly from heat released during the treatment process from microbial reactions. Some effective methods that are commonly used for the recovery of heat energy from municipal wastewater are as follows.

Heat exchange pumps: The use of heat exchange pumps can be a stable source for heat recovery since the effluent temperature is relatively stable compared to atmospheric temperatures. The effluent can also be used as the source of water for heat pumps to take in, since the influent still has a lot of impurities that can clog up the equipment. Heat pumps use electricity to get heat from wastewater. They usually get three to four units of heat energy for every unit of electricity they use (Kehrein et al., 2020; Mo and Zhang, 2013). In many regions of the world, large-scale heating systems based on wastewater thermal energy have been developed (Mo and Zhang, 2013). Heating and cooling systems that use wastewater have significantly reduced energy use, particularly in Japan. The city administration of Osaka, for example, obtained energy savings of 20–30% by implementing thermal energy recovery from effluents. Every winter, effluents are utilized directly to melt vast amounts of snow in Sapporo, Japan (Zarook Shareefdeen, 2015).

Bio-drying concept: The bio-drying concept can be considered as an innovative technology for producing energy from sewage sludge (Kehrein et al., 2020). Winkler et al. (2013) experimented using a full-scale bio-drying system that could treat 150 kilotons of waste-activated sludge per year. The waste was treated in a two-step forced aeration process at thermophilic conditions (65–75°C), reducing the wet sludge's total weight by 73%. The final product could allow combustion for energy generation in external facilities having a high caloric value. It was observed that the system consumed less than 0.5 MW of energy and recovered almost 9.3 MW of biologically produced heat used to heat office buildings (Winkler et al., 2013).

11.3.3   Water Reclamation

Due to the severe water shortage in recent times, wastewater reuse is increasingly practiced. The reclaimed water has been used for different purposes, including agriculture, horticulture, landscaping, aquifer recharge, impoundments, washing, and even drinking purposes. Different organizations around the world have set up different guidelines for their reuse. These have been previously discussed in our earlier chapters. In the following section, the different technologies involved in reclaiming wastewater have been discussed.

11.3.3.1   Scenario in the United States

The treatment processes involved in water reclamation in the United States involve preliminary treatment (screening and grit removal) and primary treatment (sedimentation), followed by secondary treatment and disinfection. Secondary treatment includes either low-rate processes (stabilization ponds or aerated lagoons) or high-rate processes (ASP, RBC, TF, and MBRs) (National Research Council, 2012). The wastewater from the secondary treatment may be disinfected and released as reclaimed water. Almost 50% of the STPs undergo this treatment.

Most of the treatment plants in the United States go beyond secondary treatment to advanced treatment. The wastewater from secondary treatment undergoes suspended solid removal using chemical coagulation techniques or filtration. After that, they undergo phosphorous removal by chemical (precipitation) or biological means (algae uptake). The phosphorous-free wastewater is then subjected to nitrification/denitrification, gas stripping, and chlorination before being released as reclaimed water.

In certain instances, if the wastewater has a lower loading of phosphorus, nitrogen, and suspended solids, the secondary treated effluent is subjected to dissolved solids removal and organics removal. The typical steps for dissolved solids removal are electrodialysis, reverse osmosis, and nanofiltration. In the case of organic removal, activated carbon adsorption, oxidation using ozone and other advanced oxidation processes, and nanofiltration are the most commonly used technologies.

Natural processes are often used in combination with the above-mentioned processes to reclaim water. Subsurface-managed natural systems include vadose zone wells, riverbank filtration wells, and surface spreading basins to provide natural storage to the reclaimed water or enhance the quality of the reclaimed water. They can also be used to recharge aquifers. Reclaimed water is often directly injected into unsaturated aquifers by constructing pumping wells (National Research Council, 2012).

Treatment wetlands have also been used to treat reclaimed water. They are either sub-surface flow or surface flow systems with plants growing. The sub-surface wetlands have plants growing from the substrate, while in surface flow systems, the plants are growing at a depth of 0.15–0.6 m from the surface of the flowing water. These systems are efficient in removing a large range of contaminants because of the different removal mechanisms, such as adsorption, filtration, precipitation, microbial degradation, plant uptake, and others, all taking place simultaneously. If the wetlands are well aerated, it can help remove ammonia, while denitrification is favored when anaerobic conditions prevail in wetlands with more depth (National Research Council, 2012; Varma et al., 2020).

The details of few projects in the United States that practice the non-potable reuse of reclaimed water are provided in Figure 11.7. It can be seen that most of the reclaimed water was used in groundwater recharge and surface augmentation (US EPA, 2017). Various projects tried to reuse the reclaimed water for direct potable uses as well. The details of such projects are provided in Figure 11.8.

11.3.3.2 Asian Scenario

Over the years, countries from Asia, such as India, Singapore, Japan, Korea, and others, have emphasized reusing reclaimed water. In this context, numerous guidelines have been issued over the years (CPHEEO, 2013; Gaulke, 2006). In Chennai, India, the wastewater from the urban areas was used for cooling water at GMR Vasavi Power Plant. The treatment unit comprised an equalization tank, screens, aeration tank, recarbonation chamber, lime addition, $FeCl_3$ dosing, reverse osmosis, and ion exchange. Lime addition and $FeCl_3$ were provided to remove the phosphate since phosphate should be absent in water used for cooling purposes.

The reuse plant at Indira Gandhi International Airport, Delhi, India, treated the sewage using dissolved air floatation, extended aeration with biological nitrification and denitrification, $FeCl_3$ treatment for precipitation phosphorous, dual media filter and ultrafiltration to remove suspended solids, and spiral would reverse osmosis. The reclaimed water was used for maintaining greenery and for cooling tower makeup water (CPHEEO, 2013). At Mumbai International Airport Limited, India, the reuse plant treated 4,000 m^3/day of wastewater using circular SBR comprising floor-mounted bubble diffusers, floating floor level anoxic mixers, and floating decanter. The effluent from the SBR was subjected to hypo chlorination, pressure sand filtration, and ultrafiltration. The treated effluent was primarily used for toilet flushing (CPHEEO, 2013). Since the 1980s, Japan has been promoting the reuse of treated wastewater for toilet flushing, irrigation, stream flow augmentation, landscape irrigation, direct heating and cooling, and recreational activities (Takeuchi and Tanaka, 2020). The secondary treated effluent from the STPs was further treated with advanced processes like rapid sand filtration or membrane filtration. Reclaimed wastewater in Japan is being used to fill up artificial streams or ponds, and supplied to railways, industries, and firefighting. In order to remove odor and color from the reclaimed water, the secondary treated water was subjected to pre-ozonation, bio-filtration, ozonation, and microfiltration (CPHEEO, 2013; Takeuchi and Tanaka, 2020). In Tokyo, sand filtration was combined with ozonation for streamflow augmentation. At Osaka, 34,000 m^3/day of wastewater was treated using fiber filtration and ozonation. The reclaimed water was used for toilet flushing, landscape irrigation, recreational purposes, and as cooling water. In Korea, out of the 16.07 billion tons of treated water, 48.7 million tons of the reclaimed water (0.3 %) is reused. A significant amount of the reclaimed water generated is used for cleaning roads, cleaning inside and outside of treatment plants, and cleaning coaches of subway trains (CPHEEO, 2013; Maksimović et al., 2015).

FIGURE 11.7   Treatment processes involved and type of non-potable reuse of reclaimed water in the United States.

Source: Adapted from US EPA, 2017.

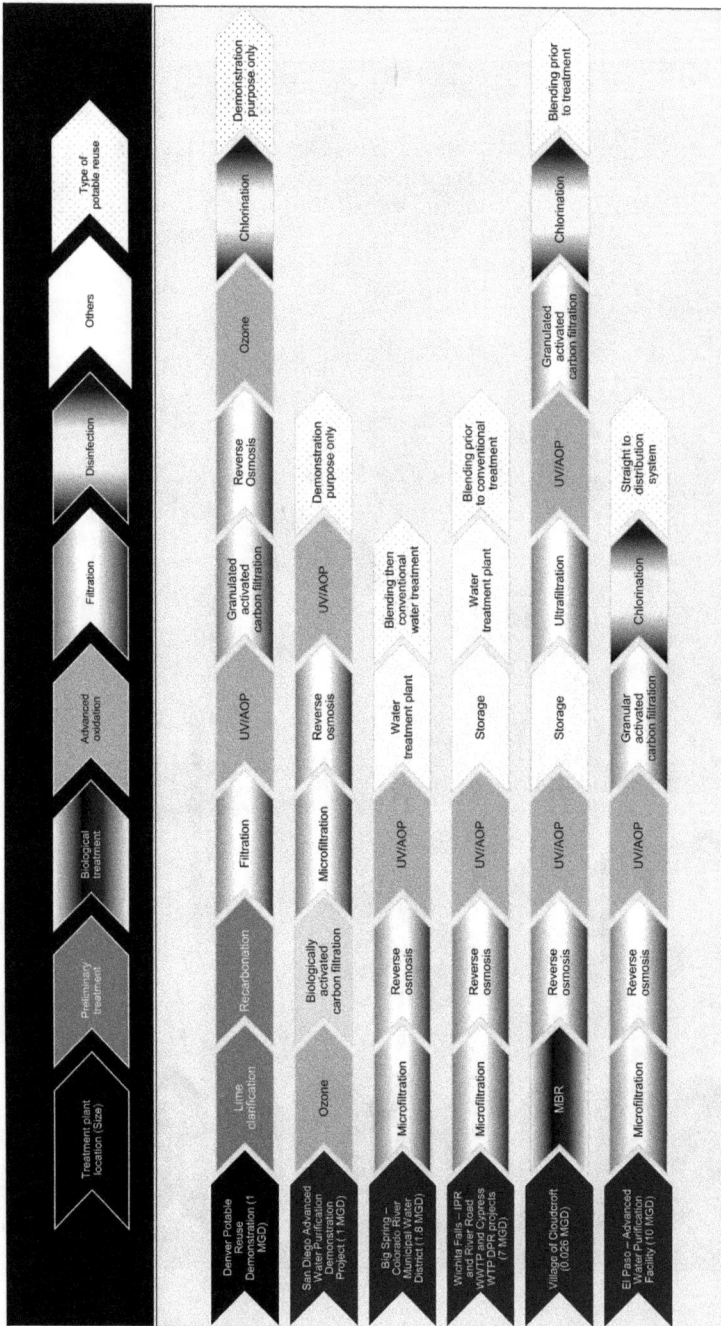

FIGURE 11.8 Treatment processes involved and type of potable reuse of reclaimed water in the United States.

Source: Adapted from US EPA, 2017.

Singapore came up with the idea of the NEWater process, which recycles the municipal wastewater into high-quality water that can meet the country's water supply during the dry seasons. The first stage of the process involves microfiltration or ultrafiltration. This step aims at removing the microscopic particles and bacteria. The second stage involves reverse osmosis, where various organic and inorganic contaminants are removed. Finally, the water undergoes UV disinfection, where any remaining pathogenic substances are removed. The reclaimed water is used for non-potable purposes, such as cooling at different industries and commercial buildings. During dry seasons, the reclaimed water is mixed with raw water at reservoirs and supplied to consumers as tap water (Lee and Tan, 2016; Tan, 2018).

11.3.3.3   Australian Scenario

As per the Water Recycling Guidelines for the state of Victoria, reclaimed water may be classified into four different categories (EPA Victoria, 2012; Radcliffe and Page, 2020). Class A or the highest quality of reclaimed water may be used for urban (non-potable) purposes with uncontrolled public access, cultivation of human food crops that are eaten raw, and industrial purposes (open systems with worker exposure potential). The desired treatment is primary treatment, secondary treatment, and tertiary treatment with pathogen reduction.

The class B reclaimed water may be used for agricultural purposes (cattle grazing) and industrial purposes (washdown water). Primary treatment followed by secondary treatment with pathogen reduction is the desired treatment required. The class C reclaimed water may be used for non-potable urban activities with controlled public access, agricultural purposes (food crops that need to be cooked, grazing, processed food), and industrial purposes with no potential exposure to the workers. The required level of treatment is secondary treatment with pathogen reduction. Finally, the class D reclaimed water may only be used for non-food crops, gardening, landscaping, and other similar activities that do not involve ingestion of such water by humans (EPA Victoria, 2012; Radcliffe and Page, 2020). Chlorination, UV light, and filtration are the most commonly used technologies for removing pathogens.

11.3.3.4   European Scenario

In Spain, a minimum of 390 million m^3 or 35% of the treated water is reclaimed every year. Almost all the coastal areas are heavily dependent on reclaimed water to overcome the poor water supply in these areas. The different projects for water reuse, with sizes ranging from 0.1 to 5 million m^3/year are located across Spain (Alcalde Sanza and Gawlik, 2014; Bixio et al., 2006). The most common end uses of reclaimed water are non-potable uses in agriculture, industries, and residential areas. In Belgium, the water reclamation plants located at Brussels and Flanders provide water for industrial cooling and other purposes. In Wulpen, 2.5 million m^3/year of water is reclaimed by microfiltration and reverse osmosis. The reclaimed water is stored in aquifers for 1–2 months before using it for water supply augmentation. Other water reclamation projects in Belgium and Netherlands focus on groundwater recharge and enhancement of nature. France has regularly been using reclaimed water for irrigation purposes (Alcalde Sanza and Gawlik, 2014; Bixio et al., 2006). Cirelli et al. (2012) used constructed wetlands to treat municipal wastewater in Italy. The reclaimed water was used to cultivate eggplants and tomatoes. Around 10–20 million m³/year of water in Germany is reused by treating it with constructed wetlands. The water gets further purified in the constructed wetland before it is discharged to the river. This step also helps in achieving environmental enhancement (Alcalde Sanza and Gawlik, 2014; Bixio et al., 2006).

11.3.4 OTHER PRODUCTS

Apart from nutrients, energy, and water, municipal wastewater is also rich in different products, such as biochemicals, metals, salts, and cellulose. Different metals, such as Cd, Zn, Cr, Cu, Mn, and others, can be recovered using a wide range of technologies. Previously BES, such as single-chamber microbial fuel cells, microbial electrolysis cells, and bio-cathode microbial fuel cells, have been used to recover metals and simultaneously generate electricity from municipal wastewater. Direct metal recovery through the use of abiotic cathodes, metal recovery through the use of abiotic cathodes that have been supplemented by external power sources, metal conversion through the use of bio-cathodes, and metal conversion through the use of bio-cathodes that have been supplemented by external power sources are the various mechanisms that are involved in metal recovery through the use of BES (Abourached et al., 2014; Huang et al., 2011; Qin et al., 2012; Tandukar et al., 2009; Wang and Ren, 2014).

Electrodeposition is another technique that is widely used for metal recovery. Electrodeposition uses an electric current to decrease dissolved metal ion concentration in the wastewater and generate a coherent metal coating at the electrode (Gu et al., 2020; Jin and Zhang, 2020; Stando et al., 2021). Several treatment methods like electrowinning, electrorefining, and slurry electrolysis have been effectively used in various reactors to recover different metal ions and impure metal oxide/sulfide/alloy. A good process design for selective metal electrodeposition is extremely desirable to simplify recovery operations, and current lab-scale approaches must be further refined for these practical applications (Jin and Zhang, 2020; Stando et al., 2021).

Electrosorption, another water treatment technique based on electric double-layer capacitance and adsorption, has been used to recover metals. In the presence of a potential difference, the charged metal ions are pushed toward the electrodes, forming an electric double layer for further adsorption (Chen et al., 2020; Jin and Zhang, 2020). Electrosorption is gaining popularity because of its easy charge-discharge features, the development of porous nanomaterial electrodes, and past desalination experiences. The use of membranes can reduce "common-ion effects" and make adsorption/desorption easier, while the use of Faradic and capacitive electrodes and flow systems can improve things even more (Chen et al., 2020; Jin and Zhang, 2020; Ziati, 2017). Electrodialysis is another option for recovering metals. Electrodialysis is a type of membrane separation technology that uses electricity to selectively recover or concentrate target metal ions while purifying water without generating secondary pollutants. The synergistic effect of electricity and membrane is critical for this technology, especially in terms of separation rate and efficiency (Gurreri et al., 2020; Jin and Zhang, 2020).

Among other products, organic acids can be recovered by hydrolyzing and fermenting primary sludge with mixed microbial communities. Polyhydroxyalkanoates can be recovered from wastewater via three steps. First, the COD is fermented in an acidogenic reactor to form volatile fatty acids. Simultaneously, polyhydroxyalkanoates-producing biomass is to be maintained in a separate reactor. Finally, the biomass is put into the reactor of volatile fatty acids to form polyhydroxyalkanoates. Recently, the aerobic granular sludge process has been used to treat wastewater. The granules require extracellular polymeric substances to maintain their physical and chemical structure. The extracellular polymeric substances may be recovered using sodium carbonate (Na_2CO_3) and calcium ions to extract the extracellular polymeric substances from the sludge formed during the aerobic granular sludge process. Among other products, cellulose can be obtained by sieving the influent. Cellulose is not easily degradable and brings down the performance of the STP. Hence, recovering them significantly lowers the aeration requirements of the biological processes (Kehrein et al., 2020). Municipal wastewater is a host to a large variety of products and recovering them may help improve the STP's performance and bring down the overall cost of the treatment.

11.4   CHAPTER SUMMARY

- Wastewater is a vast storehouse of resources, such as nutrients, energy, reclaimed water, metals, biochemicals, and others.
- Bioelectrochemical systems are efficient for recovering resources in the form of electricity, nutrients, metals, and biogas.
- Struvite formation is the most common and efficient technique used for recovering phosphorous.
- Converting NH_4^+ to NH_3 gas and then converting it into ammonium sulfate is one of the best ways of recovering nitrogen since ammonium sulfate is an essential ingredient in fertilizers.
- Reclaimed water is largely used for toilet flushing, washing, groundwater recharge, cooling towers, irrigation, and streamflow augmentation across the world.
- In order to reuse water for potable purposes, the reclaimed water needs to be treated to a superior quality using reverse osmosis and providing adequate disinfection. Then it must be mixed with other raw water and sent to a water treatment plant before being supplied to customers.

11.5   CONCLUDING REMARKS

Wastewater treatment is an expensive process and often leads to secondary pollution. Hence, in order to achieve sustainable wastewater treatment, proper LCA should be carried out to identify the kind of treatment suitable for the wastewater. Also, resource recovery technologies should be implemented into the STP to recover resources and generate revenue compensating for the high treatment cost.

REFERENCES

Abourached, C., Catal, T., Liu, H., 2014. Efficacy of single-chamber microbial fuel cells for removal of cadmium and zinc with simultaneous electricity production. *Water Res.* 51, 228–233. https://doi.org/10.1016/J.WATRES.2013.10.062

Adam, C., Peplinski, B., Michaelis, M., Kley, G., Simon, F.-G., 2009. Thermochemical treatment of sewage sludge ashes for phosphorus recovery. *Waste Manag.* 29(3), 1122–1128. https://doi.org/10.1016/j.wasman.2008.09.011

Alcalde Sanza, L., Gawlik, B.M., 2014. Water Reuse in Europe: Relevant Guidelines, Needs for and Barriers to Innovation, *JRC Science and Policy Reports.* https://doi.org/10.2788/29234

Ata, O.N., Kanca, A., Demir, Z., Yigit, V., 2017. Optimization of ammonia removal from aqueous solution by microwave-assisted air stripping. *Water, Air, Soil Pollut.* 228, 1–10. https://doi.org/10.1007/s11270-017-3629-5

Bixio, D., Thoeye, C., Wintgens, T., Hochstrat, R., Melin, T., Chikurel, H., Aharoni, A., Durham, B., 2006. Wastewater reclamation and reuse in the European Union and Israel: Status quo and future prospects. *Int. Rev. Environ. Strateg.* 6(2), 251–268.

Chen, R., Sheehan, T., Ng, J.L., Brucks, M., Su, X., 2020. Capacitive deionization and electrosorption for heavy metal removal. *Environ. Sci. Water Res. Technol.* 6, 258–282. https://doi.org/10.1039/C9EW00945K

Choi, H.L., Richard, T.L., Ahn, H.K., 2001. Composting high moisture materials: Biodrying poultry manure in a sequentially fed reactor. *Compost Sci. Util.* 9(4), 303–311. https://doi.org/10.1080/1065657X.2001.10702049

Cirelli, G.L., Consoli, S., Licciardello, F., Aiello, R., Giuffrida, F., Leonardi, C., 2012. Treated municipal wastewater reuse in vegetable production. *Agric. Water Manag.* 104, 163–170. https://doi.org/10.1016/j.agwat.2011.12.011

Cornel, P., Schaum, C., 2009. Phosphorus recovery from wastewater: needs, technologies and costs. *Water Sci. Technol.* 59, 1069–1076.

Corominas, L., Foley, J., Guest, J.S., Hospido, A., Larsen, H.F., Morera, S., Shaw, A., 2013. Life cycle assessment applied to wastewater treatment: State of the art. *Water Res.* 47(15), 5480–5492. https://doi.org/10.1016/j.watres.2013.06.049

CPHEEO, 2013. Chapter 7: Recycling and reuse of sewage. *Man. Sewerage Sew. Treat. Syst.* 7.1–7.53.

EPA Victoria, 2012. Guidance for water recycling in Victoria – Use of reclaimed water. www.epa.vic.gov.au/for-community/environmental-information/water/alternative-water-supplies-and-their-use/review-on-the-use-of-recycled-water (accessed 9.27.21).

Faaij, A.P.C., 2006. Bio-energy in Europe: Changing technology choices. *Energy Policy* 34(3), 322–342.. https://doi.org/10.1016/j.enpol.2004.03.026

Fang, L., Li, J., Guo, M.Z., Cheeseman, C.R., Tsang, D.C.W., Donatello, S., Poon, C.S., 2018. Phosphorus recovery and leaching of trace elements from incinerated sewage sludge ash (ISSA). *Chemosphere* 193, 278–287. https://doi.org/10.1016/j.chemosphere.2017.11.023

Fischer, F., Bastian, C., Happe, M., Mabillard, E., Schmidt, N., 2011. Microbial fuel cell enables phosphate recovery from digested sewage sludge as struvite. *Bioresour. Technol.* 102(10), 5824–5830. https://doi.org/10.1016/j.biortech.2011.02.089

Gallego-Schmid, A., Tarpani, R.R.Z., 2019. Life cycle assessment of wastewater treatment in developing countries: A review. *Water Res.* 153, 63–79. https://doi.org/10.1016/j.watres.2019.01.010

Gao, H., Scherson, Y.D., Wells, G.F., 2014. Towards energy neutral wastewater treatment: methodology and state of the art. *Environ. Sci. Process. Impacts* 16(6), 1223–1246. https://doi.org/10.1039/C4EM00069B

Gaulke, L.S., 2006. On-site wastewater treatment and reuses in Japan. *Proc. Inst. Civ. Eng. Water Manag.* 159, 103–109. https://doi.org/10.1680/wama.2006.159.2.103

Gu, J. nan, Liang, J., Chen, C., Li, K., Zhou, W., Jia, J., Sun, T., 2020. Treatment of real deplating wastewater through an environmental friendly precipitation-electrodeposition-oxidation process: Recovery of silver and copper and reuse of wastewater. *Sep. Purif. Technol.* 248, 117082. https://doi.org/10.1016/J.SEPPUR.2020.117082

Guedes, P., Couto, N., Ottosen, L.M., Ribeiro, A.B., 2014. Phosphorus recovery from sewage sludge ash through an electrodialytic process. *Waste Manag.* 34(5), 886–892. https://doi.org/10.1016/j.wasman.2014.02.021

Guerra-Rodríguez, S., Oulego, P., Rodríguez, E., Singh, D.N., Rodríguez-Chueca, J., 2020. Towards the implementation of circular economy in the wastewater sector: Challenges and opportunities. *Water (Switzerland)* 12(5), 1431. https://doi.org/10.3390/w12051431

Gurreri, L., Tamburini, A., Cipollina, A., Micale, G., 2020. Electrodialysis applications in wastewater treatment for environmental protection and resources recovery: A systematic review on progress and perspectives. *Membr.* 10, 146. https://doi.org/10.3390/MEMBRANES10070146

Huang, L., Chai, X., Chen, G., Logan, B.E., 2011. Effect of set potential on hexavalent chromium reduction and electricity generation from biocathode microbial fuel cells. *Environ. Sci. Technol.* 45, 5025–5031. https://doi.org/10.1021/ES103875D

Ichihashi, O., Hirooka, K., 2012. Removal and recovery of phosphorus as struvite from swine wastewater using microbial fuel cell. *Bioresour. Technol.* 114, 303–307. https://doi.org/10.1016/j.biortech.2012.02.124

Jin, W., Zhang, Y., 2020. Sustainable electrochemical extraction of metal resources from waste streams: From removal to recovery. *ACS Sustain. Chem. Eng.* 8(12), 4693–4707. https://doi.org/10.1021/acssuschemeng.9b07007

Kakwani, N.S., Kalbar, P.P., 2020. Review of circular economy in urban water sector: Challenges and opportunities in India. *J. Environ. Manage.* 271, 111010. https://doi.org/https://doi.org/10.1016/j.jenvman.2020.111010

Kehrein, P., Van Loosdrecht, M., Osseweijer, P., Garfí, M., Dewulf, J., Posada, J., 2020. A critical review of resource recovery from municipal wastewater treatment plants-market supply potentials, technologies and bottlenecks. *Environ. Sci. Water Res. Technol.* 6(4), 877–910. https://doi.org/10.1039/c9ew00905a

Kekre, K.M., Anvari, A., Kahn, K., Yao, Y., Ronen, A., 2021. Reactive electrically conducting membranes for phosphorus recovery from livestock wastewater effluents. *J. Environ. Manage.* 282, 111432. https://doi.org/10.1016/j.jenvman.2020.111432

Kelly, P.T., He, Z., 2014. Nutrients removal and recovery in bioelectrochemical systems: A review. *Bioresour. Technol.* 153, 351–360. https://doi.org/10.1016/j.biortech.2013.12.046

Lee, H., Tan, T.P., 2016. Singapore's experience with reclaimed water: NEWater. *Int. J. Water Resour. Dev.* 32(4), 611–621. https://doi.org/10.1080/07900627.2015.1120188

Lee, H.S., Vermaas, W.F.J., Rittmann, B.E., 2010. Biological hydrogen production: Prospects and challenges. *Trends Biotechnol.* 28, 262–271. https://doi.org/10.1016/j.tibtech.2010.01.007

Lieder, M., Rashid, A., 2016. Towards circular economy implementation: A comprehensive review in context of manufacturing industry. *J. Clean. Prod.* 115, 36–51. https://doi.org/10.1016/j.jclepro.2015.12.042

Maksimović, Č., Kurian, M., Ardakanian, R., 2015. Case studies illustrating the multiple-use water services options, in: *Rethinking Infrastructure Design for Multi-Use Water Services*, pp. 69–92. https://doi.org/10.1007/978-3-319-06275-4_3

Mata-Alvarez, J., Dosta, J., Romero-Güiza, M.S., Fonoll, X., Peces, M., Astals, S., 2014. A critical review on anaerobic co-digestion achievements between 2010 and 2013. *Renew. Sustain. Energy Rev.* 36, 412–427. https://doi.org/10.1016/j.rser.2014.04.039

Menkveld, H.W.H., Broeders, E., 2018. Recovery of ammonia from digestate as fertilizer. *Water Pract. Technol.* 13, 382–387. https://doi.org/10.2166/wpt.2018.049

Mo, W., Zhang, Q., 2013. Energy–nutrients–water nexus: Integrated resource recovery in municipal wastewater treatment plants. *J. Environ. Manage.* 127, 255–267. https://doi.org/https://doi.org/10.1016/j.jenvman.2013.05.007

Muller, E. EL, Sheik, A.R., Wilmes, P., 2014. Lipid-based biofuel production from wastewater. *Curr. Opin. Biotechnol.* 30, 9–16. https://doi.org/10.1016/j.copbio.2014.03.007

National Research Council, 2012. *Water Reuse: Potential for Expanding the Nation's Water Supply Through Reuse of Municipal Wastewater.* National Academies Press. https://doi.org/10.17226/13303

Navaee-Ardeh, S., Bertrand, F., Stuart, P.R., 2006. Emerging biodrying technology for the drying of pulp and paper mixed sludges. *Dry. Technol.* 24(7), 863–878. https://doi.org/10.1080/07373930600734026

Pant, D., Singh, A., Van Bogaert, G., Irving Olsen, S., Singh Nigam, P., Diels, L., Vanbroekhoven, K., 2012. Bioelectrochemical systems (BES) for sustainable energy production and product recovery from organic wastes and industrial wastewaters. *RSC Adv.* 2, 1248–1263. https://doi.org/10.1039/c1ra00839k

Pastore, C., Lopez, A., Lotito, V., Mascolo, G., 2013. Biodiesel from dewatered wastewater sludge: A two-step process for a more advantageous production. *Chemosphere* 92(6), 667–673. https://doi.org/10.1016/j.chemosphere.2013.03.046

Power, C., McNabola, A., Coughlan, P., 2014. Development of an evaluation method for hydropower energy recovery in wastewater treatment plants: Case studies in Ireland and the UK. *Sustain. Energy Technol. Assessments* 7, 166–177. https://doi.org/10.1016/j.seta.2014.06.001

Pradhan, S.K., Mikola, A., Heinonen-Tanski, H., Vahala, R., 2019. Recovery of nitrogen and phosphorus from human urine using membrane and precipitation process. *J. Environ. Manage.* 247, 596–602. https://doi.org/10.1016/j.jenvman.2019.06.046

Puyol, D., Batstone, D.J., 2017. Resource recovery from wastewater by biological technologies. *Front. Microbiol.* 8, 998. https://doi.org/10.3389/fmicb.2017.00998

Qin, B., Luo, H., Liu, G., Zhang, R., Chen, S., Hou, Y., Luo, Y., 2012. Nickel ion removal from wastewater using the microbial electrolysis cell. *Bioresour. Technol.* 121, 458–461. https://doi.org/10.1016/J.BIORTECH.2012.06.068

Qiu, G., Law, Y.M., Das, S., Ting, Y.P., 2015. Direct and complete phosphorus recovery from municipal wastewater using a hybrid microfiltration-forward osmosis membrane bioreactor process with seawater brine as draw solution. *Environ. Sci. Technol.* 49, 6156–6163. https://doi.org/10.1021/es504554f

Rabaey, K., Rozendal, R.A., 2010. Microbial electrosynthesis – revisiting the electrical route for microbial production. *Nat. Rev. Microbiol.* 8(10), 706–716. https://doi.org/10.1038/nrmicro2422

Radcliffe, J.C., Page, D., 2020. Water reuse and recycling in Australia – history, current situation and future perspectives. *Water Cycle* 1, 19–40. https://doi.org/10.1016/J.WATCYC.2020.05.005

Rout, P.R., Shahid, M.K., Dash, R.R., Bhunia, P., Liu, D., Varjani, S., Zhang, T.C., Surampalli, R.Y., 2021. Nutrient removal from domestic wastewater: A comprehensive review on conventional and advanced technologies. *J. Environ. Manage.* 296, 113246. https://doi.org/10.1016/J.JENVMAN.2021.113246

Scherson, Y.D., Wells, G.F., Woo, S.-G., Lee, J., Park, J., Cantwell, B.J., Criddle, C.S., 2013. Nitrogen removal with energy recovery through N_2O decomposition. *Energy Environ. Sci.* 6, 241–248. https://doi.org/10.1039/C2EE22487A

Schopf, K., Judex, J., Schmid, B., Kienberger, T., 2018. Modeling the bioenergy potential of municipal wastewater treatment plants. *Water Sci. Technol.* 77(11), 2613–2623. https://doi.org/10.2166/wst.2018.222

Schroeder, P., Anggraeni, K., Weber, U., 2019. The relevance of circular economy practices to the sustainable development goals. *J. Ind. Ecol.* 23, 77–95. https://doi.org/10.1111/jiec.12732

Stando, G., Hannula, P.M., Kumanek, B., Lundström, M., Janas, D., 2021. Copper recovery from industrial wastewater – Synergistic electrodeposition onto nanocarbon materials. *Water Resour. Ind.* 26, 100156. https://doi.org/10.1016/J.WRI.2021.100156

Tabesh, M., Feizee Masooleh, M., Roghani, B., Motevallian, S.S., 2019. Life-cycle assessment (LCA) of wastewater treatment plants: A case study of Tehran, Iran. *Int. J. Civ. Eng.* 17, 1155–1169. https://doi.org/10.1007/s40999-018-0375-z

Takeuchi, H., Tanaka, H., 2020. Water reuse and recycling in Japan – History, current situation, and future perspectives. *Water Cycle* 1, 1–12. https://doi.org/10.1016/j.watcyc.2020.05.001

Tan, T.P., 2018. *NEWater in Singapore*, Global Water Forum.

Tandukar, M., Huber, S.J., Onodera, T., Pavlostathis, S.G., 2009. Biological chromium(VI) reduction in the cathode of a microbial fuel cell. *Environ. Sci. Technol.* 43, 8159–8165. https://doi.org/10.1021/ES9014184

Tansel, B., Lunn, G., Monje, O., 2018. Struvite formation and decomposition characteristics for ammonia and phosphorus recovery: A review of magnesium-ammonia-phosphate interactions. *Chemosphere.* 194, 504–514. https://doi.org/10.1016/j.chemosphere.2017.12.004

Tchobanoglous, G., Burton, F.L., Stensel, H.D., 2003. *Wastewater Engineering: Treatment, Disposal, and Reuse.* Metcalf and Eddy.

Tyagi, V.K., Lo, S.-L., 2013. Microwave irradiation: A sustainable way for sludge treatment and resource recovery. *Renew. Sustain. Energy Rev.* 18, 288–305. https://doi.org/10.1016/j.rser.2012.10.032

US EPA, 2017. *2017 Potable Reuse Compendium.* CDM Smith.

USEPA, 2012. *EPA Guidelines for Water Reuse.* CDM Smith.

Vaneeckhaute, C., Belia, E., Meers, E., Tack, F.M.G., Vanrolleghem, P.A., 2018. Nutrient recovery from digested waste: Towards a generic roadmap for setting up an optimal treatment train. *Waste Manag.* 78, 385–392. https://doi.org/10.1016/j.wasman.2018.05.047

Varma, M., Gupta, A.K., Ghosal, P.S., Majumder, A., 2020. A review on performance of constructed wetlands in tropical and cold climate: Insights of mechanism, role of influencing factors, and system modification in low temperature. *Sci. Total Environ.* 755, 142540. https://doi.org/10.1016/j.scitotenv.2020.142540

Wang, H., Ren, Z.J., 2014. Bioelectrochemical metal recovery from wastewater: A review. *Water Res.* 66, 219–232. https://doi.org/10.1016/J.WATRES.2014.08.013

Williams, A.T., 2013. *Ion exchange nutrient recovery from municipal wastewater.* ProQuest.

Winkler, M.-K.H., Bennenbroek, M.H., Horstink, F.H., van Loosdrecht, M.C.M., van de Pol, G.-J., 2013. The biodrying concept: An innovative technology creating energy from sewage sludge. *Bioresour. Technol.* 147, 124–129. https://doi.org/10.1016/j.biortech.2013.07.138

Xie, M., Shon, H.K., Gray, S.R., Elimelech, M., 2016. Membrane-based processes for wastewater nutrient recovery: Technology, challenges, and future direction. *Water Res.* 89, 210–221. https://doi.org/https://doi.org/10.1016/j.watres.2015.11.045

Yang, W., Cicek, N., Ilg, J., 2006. State-of-the-art of membrane bioreactors: Worldwide research and commercial applications in North America. *J. Memb. Sci.* 270, 201–211. https://doi.org/10.1016/j.memsci.2005.07.010

Ye, Y., Ngo, H.H., Guo, W., Chang, S.W., Nguyen, D.D., Zhang, X., Zhang, J., Liang, S., 2020. Nutrient recovery from wastewater: From technology to economy. *Bioresour. Technol. Reports.* 11, 100425. https://doi.org/10.1016/j.biteb.2020.100425

Zarook Shareefdeen, A.E.S.K., 2015. *Modern Water Reuse Technologies: Membrane Bioreactors*, 1st Edition. CRC Press.

Zhang, T., Wu, X., Fan, X., Tsang, D.C.W., Li, G., Shen, Y., 2019. Corn waste valorization to generate activated hydrochar to recover ammonium nitrogen from compost leachate by hydrothermal assisted pretreatment. *J. Environ. Manage.* 236, 108–117. https://doi.org/ https://doi.org/10.1016/j.jenvman.2019.01.018

Zhang, Y., Desmidt, E., Van Looveren, A., Pinoy, L., Meesschaert, B., Van Der Bruggen, B., 2013. Phosphate separation and recovery from wastewater by novel electrodialysis. *Environ. Sci. Technol.* 47, 5888–5895. https://doi.org/10.1021/es4004476

Zhang, Y., Noori, J.S., Angelidaki, I., 2011. Simultaneous organic carbon, nutrients removal and energy production in a photomicrobial fuel cell (PFC). *Energy Environ. Sci.* 4(10), 4340–4346. https://doi.org/10.1039/c1ee02089g

Ziati, M.F.M.S., 2017. Removal of chromium from tannery wastewater by electrosorption on carbon prepared from peach stones: effect of applied potential. *Carbon Lett.* 21, 81–85. https://doi.org/10.5714/CL.2017.21.081

Index

Note: Page numbers in **bold** refer to tables and those in *italic* refer to figures.

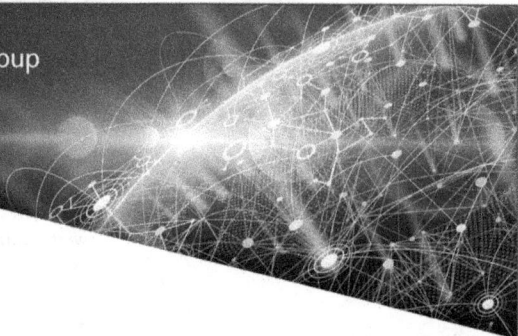

For Product Safety Concerns and Information please contact our EU
representative GPSR@taylorandfrancis.com
Taylor & Francis Verlag GmbH, Kaufingerstraße 24, 80331 München, Germany

9 781032 428208